Application and Integration of Omics-Powered Diagnostics in Clinical and Public Health Microbiology

Jacob Moran-Gilad • Yael Yagel
Editors

Application and Integration of Omics-Powered Diagnostics in Clinical and Public Health Microbiology

Springer

Editors
Jacob Moran-Gilad
Ben-Gurion University of the Negev
Beer Sheva, Israel

Yael Yagel
Ben-Gurion University of the Negev
Beer-Sheva, Israel

ISBN 978-3-030-62154-4 ISBN 978-3-030-62155-1 (eBook)
https://doi.org/10.1007/978-3-030-62155-1

This Springer imprint is published by the registered company Springer Nature Switzerland AG
The registered company address is: Gewerbestrasse 11, 6330 Cham, Switzerland

Contents

Chapter 1
Introduction to Advanced Diagnostics in Microbiology

Yael Yagel and Jacob Moran-Gilad

1.1 Introduction

Technological advancements involving new diagnostic platforms have revolutionised the microbiology field over recent years, allowing faster and more accurate diagnostics [1]. In this book, we aim to present the advanced technologies currently used in microbiology, their clinical applications and future prospectives. We will divide these methods into genomic-based, including whole-genome sequencing (Chaps. 2, 3, 4, and 5) and metagenomics (Chaps. 6, 7, and 8), and proteomic-based, focusing on MALDI-TOF (Chaps. 9 and 10) and FTIR (Chap. 11). Finally, we will discuss the utility of NGS in the field of forensic medicine. Together, this collection of chapters written by renowned experts, reflect the exciting and broad applications of these technological advancements with respect to the diagnostic workflow in clinical and public health microbiology.

The chapter ahead is an introduction to advanced diagnostics, focusing on genomic-based technologies. The following paragraphs will discuss the workflow of the various diagnostic methods currently used in microbiology laboratories either in the clinical or research settings and introduce basic definitions of terms used later in this book.

Y. Yagel (✉) · J. Moran-Gilad
Microbiology, Advanced Genomics and Infection Control Applications Laboratory (MAGICAL), Department of Health Systems Management, School of Public Health, Faculty of Health Sciences, Ben-Gurion University of the Negev, Beer-Sheva, Israel
e-mail: giladko@post.bgu.ac.il

J. Moran-Gilad, Y. Yagel (eds.), *Application and Integration of Omics-powered Diagnostics in Clinical and Public Health Microbiology*,
https://doi.org/10.1007/978-3-030-62155-1_1

1.2 Exploring Novel Diagnostic Techniques

Advanced diagnostics can be divided into several groups, according to their methodological approach as well as their practical applications. One such division differentiates between culture-dependent (culture-based) and culture-independent microbiology, followed by a subdivision according to the diagnostic methods used, either conventional (phenotypic) techniques, molecular assays targeting specific genes, proteomics (primarily using matrix-assisted laser desorption-ionisation time-of-flight mass spectrometry (MALDI-TOF-MS)) and genomics/metagenomics (Fig. 1.1).

With culture-based diagnostics, applicable mainly to bacterial and fungal pathogens, one or more culture phases are involved to yield growth of the suspected micro-organism from a clinical or non-clinical sample. Subsequently, growing isolates are characterised with respect to taxonomy, antimicrobial drug susceptibility and other traits (such as virulence and molecular subtypes) by a range of approaches. These mainly include characterisation by conventional (phenotypic) techniques and taxonomical identification using MALDI-TOF. Molecular assays performed on cultured isolates consist of polymerase chain reaction (PCR) amplification and detection of specific genes (for example, those inferring antibiotic resistance) and amplification of the 16S *rRNA* gene and subsequently using Sanger sequencing for identification. Single-cell whole-genome sequencing (WGS), powered by NGS, is performed by sequencing in parallel a very high number of bacterial DNA fragments, followed by bioinformatics analyses to reconstruct the fragmented DNA sequences back to a contiguous genome ("contig"). As opposed to 16S *rRNA* analysis which enables the identification of a bacteria at the species level at best, WGS, performed downstream to culture isolation, allows for an unprecedented accuracy and resolution in phylogenomic subtyping and has the potential to serve as a one-stop-shop for pathogen characterisation, especially inference of antibiotic resistance and virulence, by mapping the "resistome" and "virulome" [2].

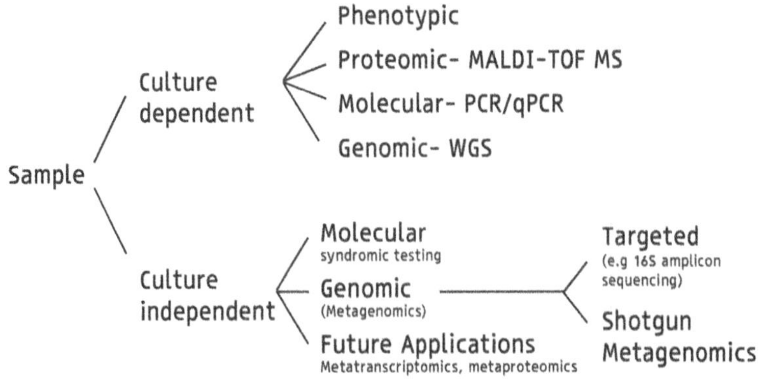

Fig. 1.1 Current and future diagnostic strategies in microbiology

On the other hand, culture-independent microbiology involves the application of diagnostic techniques directly on clinical or non-clinical samples, while obviating the need to recover an organism by culture. This approach has long been used in the field of virology, where virus isolation is rarely performed for routine diagnostic purposes. However, culture-independent detection methods are also applicable to bacterial, fungal and parasitic diseases. With culture-independent microbiology, several diagnostic strategies are now commonly used. The main one is the application of PCR assays targeting specific genes that relate to the presence of a pathogen and/or an important inferred phenotype, such as antimicrobial resistance to a key agent (such as the *mecA* gene of methicillin-resistant *Staphylococcus aureus*, MRSA). More recently, a massive increase in the availability of in-house and commercial multiplex PCR assays is evident, covering a wide range of diagnostic targets in a single run. These assays are increasingly designed for syndromic diagnosis, including the most common pathogens causing infection in the gastrointestinal, respiratory or genitourinary tracts [3]. When discussing "omics powered" diagnostic tools, however, the focus is on the application of NGS technology directly on samples, an approach also known as metagenomics. Metagenomics can further be split into targeted (or amplicon-based) and shotgun (or whole genome) metagenomics. In targeted metagenomics, the sample is subjected to an amplification step, usually of the 16S *rRNA* gene, and subsequent sequencing of the particular amplicon achieved through the targeted PCR step. This method is mostly used to describe the microbial population in a body site (i.e. the microbiota). It has the advantages of sequencing only microbial DNA (disregarding the human DNA in the sample) focusing on analysing the taxonomic data obtained through the 16S gene. As opposed to 16S sequencing from a single isolate grown in a culture, or even 16S amplification from a normally sterile site (CSF, blood) intending to isolate and sequence a single pathogen, in amplicon-based metagenomics, multiple DNA fragments are sequenced in parallel using NGS platforms, allowing the accurate mapping of the entire taxonomical composition of a sample rich in bacterial populations (e.g. gut, vagina).

Another approach for applying metagenomics is shotgun or whole-genome metagenomics, in which there is no pre-sequencing specific amplification phase, and the entire genomic content of a sample is being sequenced without introducing bias. Using shotgun metagenomics enables the identification of all the microorganisms present in a sample (including viruses, fungi and parasites) as well as the characterisation of other important elements such as antibiotic resistance determinants and virulence factors. The presence of human DNA is both an obstacle for the microbiological analysis, as it constitutes the majority of the genomic content, but also a potential for a complementary analysis of the host human genome or transcriptome for tailoring treatment and establishing a prognosis [4].

For the sake of consistency and coherence, throughout the book, NGS terminology will be divided into *WGS* – referring to culture-driven sequencing of growing isolates, and *metagenomics* – referring to culture-independent shotgun metagenomics. Targeted metagenomics usually using the 16S *rRNA* as the target gene, which is the most widely used method in microbiome studies will not be addressed (except with respect to forensic microbiology).

1.3 Introduction to NGS

The next few paragraphs will describe the significant milestones of the evolution of NGS technologies and clarify the basic terms used later in the book. Of note, it is not intended to include all the NGS platforms available in the market, nor is it meant to be a comprehensive manual to using these techniques in the lab. Rather we aim to describe the most widely used tools (and therefore the most commonly mentioned platforms throughout this book) and present the necessary information essential for understanding the role of NGS in the diagnostic scheme.

The first available sequencing technology was developed by Sanger in the 70's, using a chain termination method. This method produced one long sequence of DNA, allowing for the analysis of a single DNA molecule per reaction. This pioneering method, although used in the completion of the Human Genome Project is laborious, time-consuming and expensive. The need for a rapid, accurate "high throughput" (i.e. generating multiple results during a single machine run) gave rise to new sequencing methods (next-generation sequencing, NGS) such as the Roche 454's pyrosequencing system in 2005, and later on with various platforms produced by Illumina (e.g. MiSeq, HiSeq, NextSeq). These platforms, along with technologies provided by other companies, also referred to as short read methods, are based on the massively parallel sequencing of many short DNA fragments, generating millions of short sequencing outputs (reads). These can later be assembled into longer contiguous sequences (contigs) based on homology within the different reads, assuming the DNA fragmentation is random such that a single area is represented more than once within the total output. The integration of these platforms in the microbiology workflow commonly includes: (i) DNA extraction, (ii) library preparation – where extracted DNA is randomly fragmented into same-sized pieces, and then ligated to primers and adaptors, (iii) template preparation including amplification (iv) sequencing, which, in the case of sequencing by synthesis methods, involves the incorporation of a fluorescently labelled deoxyribonucleotide triphosphates (dNTPs) during each cycle of DNA synthesis, followed by the identification of fluorophore excitation [5]. Since sequencing errors are a critical issue when discussing NGS techniques, these platforms each established its limit for sequencing cycles, setting the numbers of bases sequenced per one machine operation (run). Even so, these technologies are still prone to sequencing errors, and thus when a single nucleotide position is represented more than once, the ability to establish a consensus call for a base improves the accuracy of the final sequence. The multiple representations of a single nucleotide position in a sequence establishes the "depth" or "coverage" of the sequencing run, which is an essential parameter for the quality assessment of each run.

Although the short-read technologies revolutionised the world of sequencing, the data produced by these platforms sometimes results in fragmented assemblies especially in cases of repetitive sequences within the genome. This has led to the advent of new sequencing technologies producing long-reads. The two main long-read platforms currently available are from Pacific Biosciences (PacBio RS) and Oxford Nanopore Technologies (MinION). In contrast to short-read platforms,, long-read technologies target single DNA molecules, resulting in real-time sequencing. The MinION by Oxford Nanopore Technologies, for example, identifies DNA bases by measuring the changes in electrical conductivity generated as DNA strands pass through a biological pore. Long-read sequencing has several advantages over short-read sequencing platforms, namely the ability to produce real-time results, and the production of tens of thousands of bases per read, immensely improving the ability to analyse large and complex genomes, as well as *de novo* sequencing of bacteria not well represented in public databases.

However, a significant drawback of certain long-read technologies is a higher error rate compared to short-read technologies [6]. This can be improved in some platforms by sequencing a single molecule multiple times resulting in a unique consensus. While *de novo* genome assemblies can be produced from short-read data, assembly continuity is often relatively poor, due to the limited ability of short reads to handle long repeats. Assembly quality can be significantly improved by using complementary long-read sequencing, making them an excellent adjunct to short-read produced data using a hybrid assembly approach [7].

To conclude, the transition of sequencing technologies from laborious, expensive, and of low throughput methods to the simultaneous sequencing of thousands to millions of DNA fragments in various sizes, has enabled enormous advancement in the applications and utilisation of these technologies. As the knowledge and experience of using these methods in real-life clinical scenarios expand, they will become a "must-have" technology for both research and clinical institutes.

1.4 Summary and Book Outline

As advanced diagnostic methods slowly enter clinical life, it is essential for researchers, microbiologists as well as clinicians to get familiarised with the general concepts and the current experience with those methods. This will gain importance as these technologies become cheaper, analytically robust and clinically validated and standardised, meeting clinical criteria for the diagnosis of various disease states. This book is intended to summarise the current experience with multiple methods that have either already established a niche in microbiological diagnostics or are of promise to do so in the near future.

In the chapter to follow, Couto and Rossen guide the readers through a comprehensive general overview of the possibilities of NGS, including whole-genome sequencing (WGS) and metagenomics, in the fields of clinical and public health microbiology, and speculate on its future importance. Next, we will deeply explore specific aspects of NGS applications in bacterial characterisation; In Chap. 3, Vourli, Kontos and Pournaras discuss WGS and metagenomics for identification and antimicrobial susceptibility testing with a particular emphasis on mycobacteria as an example of utilising NGS technologies on clinically relevant slow-growing bacteria. Tagini, Pillonel and Greub (Chap. 4) expand on the role of WGS in the identification of bacterial virulence factors while discussing the main technical approaches and limitations. Van Alphen, von Wintersdorff and Savelkoul (Chap. 5) complete the overview of NGS for bacterial characterisation with a thorough review of the general concepts of typing, as well as the various available typing techniques while discussing both the advantages and the challenges of using WGS for typing purposes.

Chapter 6 is dedicated to NGS in the field of clinical virology. Van Rijn-Klink, De Vries and Claas discuss viral pathogen detection and discovery by metagenomic sequencing and virome research. Next, the typing of viruses by NGS with a focus on resistance testing by deep sequencing is addressed.

In Chap. 7, Laura Filkins and Robert Schlaberg will review the advantages of clinical metagenomics, while in Chap. 8, Etienne Ruppé, Yannick Charretier, Vladimir Lazarevic and Jacques Schrenzel describe current common hurdles, such as sample preparation, wet lab issues and bioinformatics challenges, and discuss ways to overcome them. They also review the current experience with using metagenomics in clinical samples from different anatomical sites.

The next chapters review the use of proteomics for microbiological diagnosis. Cuénod and Egli (Chap. 9) and Rodríguez-Sánchez and Oviano (Chap. 10) report on the utilisation of MALDI-TOF MS for bacterial typing beyond species identification and advanced applications of MALDI-TOF-MS such as direct application on samples and identifying antibiotic resistance, respectively. Novais and Peixe introduce the role of Fourier Transform Infrared (FT-IR) spectroscopy in the armamentarium of high-throughput microbiological tools for bacterial diagnostics and review the established uses of FT-IR in food and water microbiology and its potential as an accurate and cost-effective method for diverse types of applications.

Lastly, Fernández-Rodríguez, González-Candelas and Arora present the exciting new role of NGS technologies in forensic microbiology, a relatively new scientific field, attempting to utilise our knowledge of the evolution and habitats of different microorganisms. They thoroughly review what is known on forensic microbiology in determining the cause of death, the study of pathogen transmission between donor-recipient pairs, the source(s) of outbreaks, the identification of body fluids or the identification of individuals through the microbial composition of their remains.

To conclude, the readers of this book, whether coming from the bench or the bedside, will be presented with a comprehensive view of advanced approaches to microbiological diagnosis, reviewing both their numerous advantages and the hurdles to their implementation. These technologies are currently at various stages of incorporation into the clinical workflow; from research-only to routine application in public health or hospital laboratories. However, all have a high potential of becoming an important tool for rapid and accurate diagnosis, making their recognition important for various professionals in the fields of microbiology, infectious disease and public health.

It is of note that writing and production of the book took place before the emergence of COVID-19 and thus the pandemic is not specifically addressed.

References

1. Motro Y, Carriço JA, Friedrich AW, Rossen JWA, Moran-Gilad J (2018) ESCMID postgraduate education course: regional capacity building for integration of next-generation sequencing in the clinical microlab. Microbes Infect 20(5):275–280
2. Bathoorn E, Chlebowicz MA, Couto N, Ferdous M, García-Cobos S, Kooistra-Smid AMD et al (2017) Application of next generation sequencing in clinical microbiology and infection prevention. J Biotechnol 243:16–24
3. Ramanan P, Bryson AL, Binnicker MJ, Pritt BS, Patel R (2018) Syndromic panel-based testing in clinical microbiology. Clin Microbiol Rev [Internet]. [cited 2020 Feb 15] 31(1). Available from: https://cmr.asm.org/content/31/1/e00024-17
4. Simner PJ, Miller S, Carroll KC (2018) Understanding the promises and hurdles of metagenomic next-generation sequencing as a diagnostic tool for infectious diseases. Clin Infect Dis 66(5):778–788
5. Besser J, Carleton HA, Gerner-Smidt P, Lindsey RL, Trees E (2018) Next-generation sequencing technologies and their application to the study and control of bacterial infections. Clin Microbiol Infect 24(4):335–341
6. Lu H, Giordano F, Ning Z (2016) Oxford nanopore MinION sequencing and genome assembly. Genomics Proteomics Bioinformatics 14(5):265–279
7. Sohn J, Nam J-W (2018) The present and future of de novo whole-genome assembly. Brief Bioinform 19(1):23–40

Chapter 2
Overview of Microbial NGS for Clinical and Public Health Microbiology

Natacha Couto and John W. Rossen

2.1 Introduction

For several decades, we have been confined in the clinical microbiology laboratory to techniques that are limited in the amount of information they provide, e.g. limited to species identification or antimicrobial susceptibility; limited with respect to the turnaround time, e.g. culture of slow-growing or obligate intracellular pathogens, and/or limited in the sensitivity of the tests due to, for example, previous antimicrobial therapy administered to the patient before sample collection. These limitations lead to significant consequences for both the patient and the health care system in general, like higher morbidity and mortality due to inappropriate antimicrobial therapy and increased medical costs due to the long turnaround time and limited sensitivity of the diagnostic assays and consequently a longer stay of the patient in the hospital. Next Generation Sequencing (NGS) has the potential to revolutionise the way we perform microbiology as it can become a 'one test fits all' [1]. With NGS, pathogen identification, therapeutic resistance, pathogenicity, outbreak transmission, and within-host evolution (in case of chronic infections) can be studied at the same time [1, 2]. NGS is already applied in several medical microbiology

N. Couto (✉)
Department of Medical Microbiology and Infection Prevention, University of Groningen, University Medical Center Groningen, Groningen, Netherlands

The Milner Centre for Evolution, Department of Biology and Biochemistry, University of Bath, Bath, UK
e-mail: nmgdc20@bath.ac.uk

J. W. Rossen
Department of Medical Microbiology and Infection Prevention, University of Groningen, University Medical Center Groningen, Groningen, Netherlands

IDbyDNA inc., Salt Lake City, UT, USA

J. Moran-Gilad, Y. Yagel (eds.), *Application and Integration of Omics-powered Diagnostics in Clinical and Public Health Microbiology*,
https://doi.org/10.1007/978-3-030-62155-1_2

laboratories, including our laboratory at the University Medical Center Groningen (UMCG), where it is used for outbreak management and infection prevention within the hospital and within the region, identification of bacteria using the 16S-23S rRNA encoding region, and metagenomics approaches for identification and typing of pathogens. However, numerous limitations need to be supplanted in order to make it feasible and affordable in any Medical Microbiology laboratory, independent of it being a local, regional or academic hospital, or a national reference centre.

Several decisions have to be made when applying NGS to clinical and public health microbiology. There is no flawless workflow and every step in the process needs constant optimisation. Usually, an NGS workflow comprises the following steps: (1) sample collection; (2) DNA/RNA extraction; (3) library preparation; (4) sequencing; and (5) bioinformatics analysis, as it is shown in Fig. 2.1. The description/optimisation of these steps is not the focus of this book chapter, which will concentrate first on the practical issues of implementing NGS in diagnostic microbiology and second on a series of case studies that show the potential value of NGS for the surveillance and control of microorganisms.

2.2 Implementation of NGS in Clinical Microbiology Laboratories

In most countries, NGS was first introduced to microbiology in academic and/or reference laboratories, due to capital investment, operational costs, and requirements for expertise in the laboratory and bioinformatics processes [3]. The implementation of the NGS competencies at the reference laboratory depends on the type of national health system and may mirror a hierarchal structure that favours a more centralised microbiological surveillance and reference functions [4]. This hierarchal structure reduces the costs per sample at the reference laboratory, by collecting samples from different sources; however, this comes with the cost of prolonged turnaround times [5]. Nevertheless, the decrease in sequencing costs, the introduction of bench-top or portable and low-to-medium throughput devices [6, 7], the growing availability of free, user-friendly bioinformatics tools [8, 9] and the availability of specialised technicians resulted in a broad and rapid introduction of the NGS technology into non-academic laboratories, enabling a transition from a hierarchical to a network-like structure [1]. This significantly reduces the turnaround time, empowers hospital-based microbiology, and positively impacts local efforts such as infection control interventions [10].

To implement NGS in routine diagnostics, several adjustments in the laboratory workflow are required [3]. Both parts of the procedure, i.e. the wet laboratory part (nucleic acid extraction, library preparation, sequencing), and the bioinformatics part (analyses of the sequence data and translating them into easy to understand reports) should be performed by dedicated staff members specialised in NGS. The use of NGS fits best in a batch-wise approach; however, this is typical for

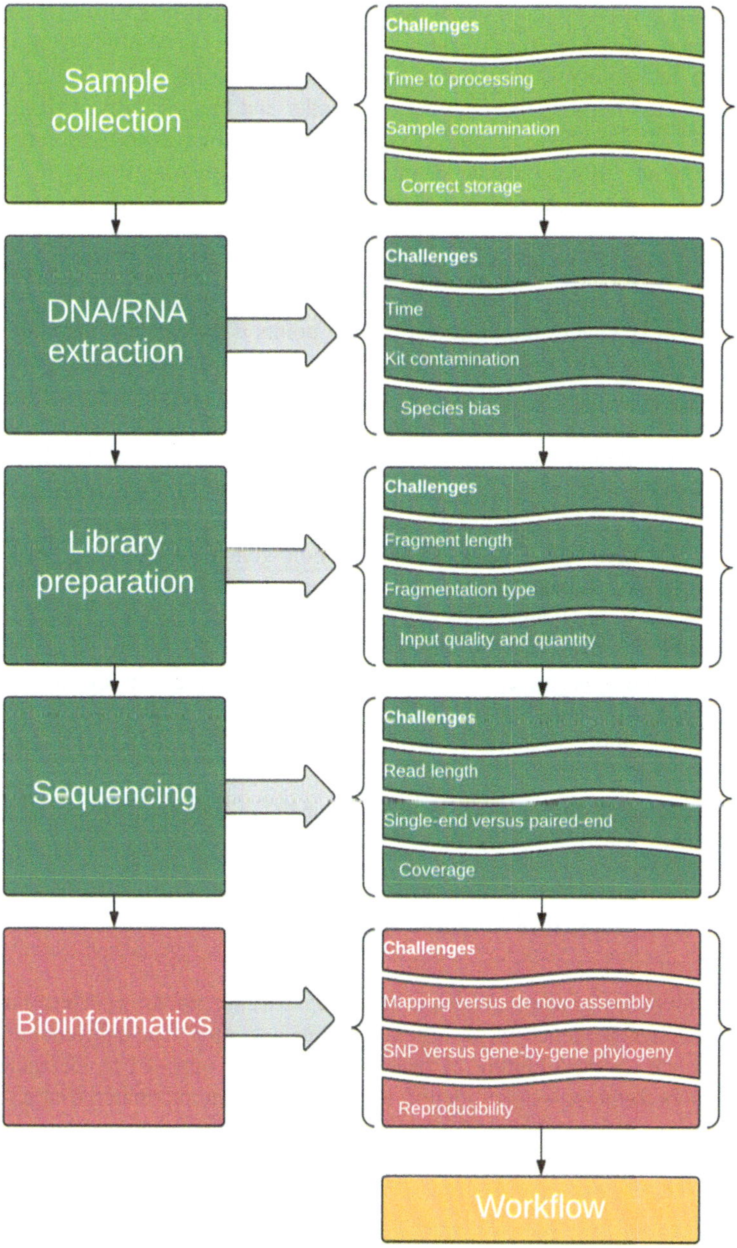

Fig. 2.1 Steps and challenges involved in an NGS workflow implemented in a diagnostic microbiology laboratory

high-throughput laboratories or surveillance projects, and it is far from ideal for routine diagnostics [3]. Recent equipment releases, like the MinION from Oxford Nanopore Technologies and the iSeq 100 from Illumina Inc., may overcome such limitations, as such sequencers either are smaller and less expensive (MinION) or provide a low-to-medium output (iSeq 100), which allows them to be more cost-effective. In any way, a balance should be kept between costs, quality (e.g. accuracy), turnaround time and complexity of the laboratory and bioinformatics processes.

Like for any new laboratory method, NGS requires validation. Yet, this process is far from being forthright and is required at both the laboratory and the bioinformatics level [3]. One of the most challenging points is the fact that there are many different kits, platforms and bioinformatics tools that can be used for NGS. The microbiologist should be aware of the stability, shelf life of the reagents and flow cells, and robustness of the bioinformatics tools used in the workflow, to ensure the repeatability and reproducibility of every step.

Additionally, NGS often is superior to other methods currently used within the laboratory; for example, it has higher discriminatory power compared to the current reference standard typing methods [3]. As whole-genome sequencing (WGS) can be used for all microbial species, it is almost impossible, with respect to time and costs, to perform an independent validation for all known species. Therefore, one may consider choosing several indicator species (e.g. one aerobe Gram-positive, one aerobe Gram-negative, one anaerobe Gram-positive, one anaerobe Gram-negative and one slow-growing microorganism) for the validation of the WGS workflow. The guidelines already developed for the validation of NGS in oncology and by the College of American Pathologists may serve as a model for worldwide guidelines of using NGS for pathogen detection [11, 12].

2.3 Whole-Genome Sequencing

Whole-genome sequencing (WGS) is here defined as the process of determining the complete DNA sequence of an organism from a sample that only contains that organism, e.g., a pure culture of bacterial isolate (as opposed to metagenomics which refers to the process of identifying the entire genomic material of a sample containing many organisms). This sample can contain the organism's genome and other genetic elements (i.e. plasmids or phages) that can be present within the organism's cell. Whole-genome sequencing has been applied in clinical and public health microbiology for several purposes, but we will focus on the four aspects we consider more important for this book chapter, i.e., outbreak management and infection prevention within the hospital, outbreak management and infection prevention within the region, transmission of zoonotic microorganisms between animals and humans and antimicrobial resistance characterisation.

2.3.1 Outbreak Management and Infection Prevention within the Hospital

Until recently, pathogen surveillance was performed by laborious techniques, that could only discriminate to a certain level (i.e. multi-locus sequence typing [MLST] for bacteria, or gene-specific sequencing for viruses), were limited to one specific pathogen (i.e. *spa* typing in *Staphylococcus aureus*) or could not be efficiently shared between laboratories (i.e. pulsed-field gel electrophoresis [PFGE] -based typing results) [13]. WGS, in principle, has the advantage of being applicable to any pathogen and, in fact, one of the most widely used applications of WGS today is for outbreak surveillance and infection prevention within healthcare institutes and foodborne related infections within a defined region. Several studies have proven the usefulness of WGS-based typing for disclosing and tracing the dissemination of microbial pathogens and, to a lesser extent, of mobile genetic elements (MGEs). In Fig. 2.2, we show a simple example of how this can be achieved. At the UMCG, it has been used to characterise both antimicrobial-resistant Gram-positive and Gram-negative bacterial outbreaks within the hospital and also for transmission of MGEs between different bacterial isolates obtained from the same or different patients. Additionally, since in the Netherlands there is a strict policy of "search and destroy", we have identified and characterised pathogens in the water and the environment that could have potentially resulted in transmission to patients, but that were under control through disinfection measures of the corresponding areas. A few examples are presented below.

In 2012, a newly emerging $bla_{\text{CTX-M-15}}$ producing *Klebsiella pneumoniae* clone with sequence type (ST) 1427 was detected in a patient previously hospitalised in Germany, South-Africa and Gambia, who was admitted to the UMCG university hospital [14]. After 2.5 months, regular surveillance screening (once per week) identified two $bla_{\text{CTX-M-15}}$ *K. pneumoniae* positive roommates of this patient. After whole-genome phylogenetic analysis and patient contact tracing, an epidemiological link between the affected patients was identified. In total, five patients were involved in the outbreak, of which three developed an infection. In addition, environmental contamination with the outbreak clone was found in the patients' rooms [14]. Interestingly, there was an in-host polymorphism detected among multiple isolates obtained from different body sites of the index patient, which were probably related to antibiotic treatment and/or host adaptation [14]. To prevent further spread, stringent infection control measures consisting of strict patient and staff cohorting were introduced. Contact screening up to 2 weeks after the discharge of all $bla_{\text{CTX-M-15}}$ *K. pneumoniae* positive patients revealed no further cases and the outbreak was declared to be under control after 3 months [14]. Unfortunately, due to the unavailability of a single room at the time of admission and because initial screening results for highly resistant microorganisms were negative, the index patient was placed in a room shared with multiple patients, which enabled the spread of the resistant

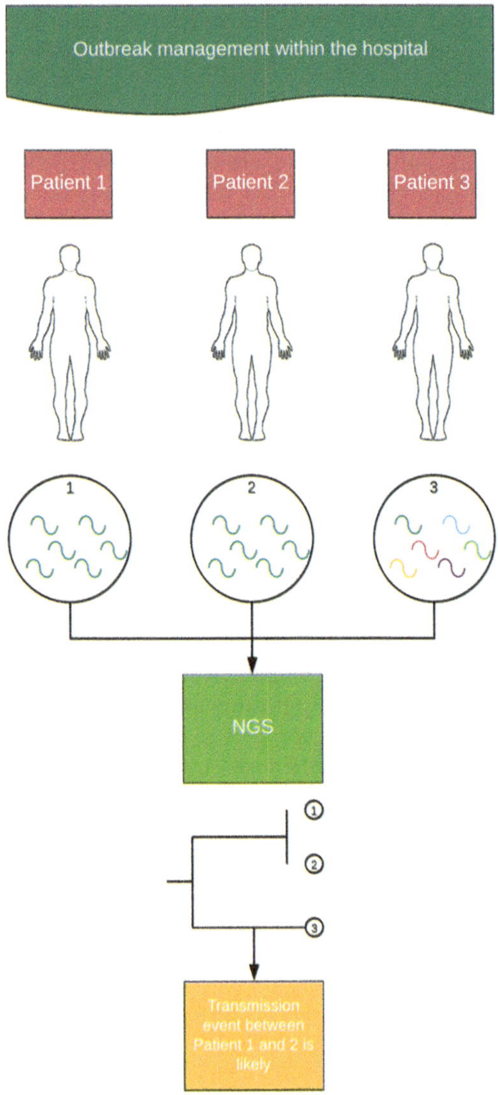

Fig. 2.2 Illustration of a possible outbreak episode within the hospital and the role of NGS to understand transmission events

K. pneumoniae and so, this study highlighted once more the importance of isolating patients previously hospitalised in countries with high rates of antimicrobial-resistant bacteria.

In 2014, a retrospective analysis of vancomycin-resistant *Enterococcus faecium* (VREfm) outbreaks that occurred in the UMCG was performed [15]. It included 75 patients, but only 36 VREfm isolates obtained from 34 patients from seven VREfm outbreak investigations were analysed. The core genome MLST (cgMLST) analysis further divided the ST into different cluster types (CTs), however, only four different *vanB* transposons were found among the isolates. Within VREfm isolates belonging to ST117 CT103, two different *vanB* transposons were found, while, VREfm isolates belonging to ST80 CT104 and CT106 harboured an identical *vanB* transposon [15]. The presence of the same *vanB* transposon in VREfm isolates belonging to distinct lineages combined with the epidemiological data suggested an exchange of genomic material between VREfm and vancomycin-susceptible *Enterococcus faecium* (VSEfm). Thus, transposon typing resolved this series of outbreaks and demonstrated that an outbreak can be caused by a mobile element rather than a specific strain. Transposons with low DNA sequence homology among them were also found indicating that they probably originated from other species [15]. The presence of insertion sequences originating from anaerobic bacteria suggested transposon acquisition from anaerobic gut bacteria by VSEfm [15]. The occurrence of these two events is an important factor in the emergence of (*vanB*) VREfm. This study highlighted the importance of analysing additional transposon structures to detect horizontal gene transfer between phylogenetically unrelated strains.

In 2017, we identified four isolates of *Legionella anisa* in water from dental chair units (DCUs) at the UMCG hospital dental ward [16]. Whole-genome sequencing combined with whole-genome MLST (wgMLST) analysis indicated that all four isolates (two isolates from the same chair) belonged to the same cluster with two to four allele differences. This suggested that a common contamination source was present in the dental unit waterlines, which was resolved by replacing the chairs and the main pipeline of the unit. *L. anisa*, the most common non-pneumophila *Legionella* species in the environment, has a role as the causative agent of Legionnaires' Disease (LD) and Pontiac fever [17] and it may be hospital-acquired [18]. Although a direct link between the dental unit and the patients is rarely shown, the water delivered by the dental unit waterlines has been shown to be one of many possible sources for *Legionella* infection [19]. This highlights the need to monitor water quality to protect patients and health-workers from acquiring legionella, or other potentially pathogenic bacteria.

2.3.2 Outbreak Management and Infection Prevention within the Region

In collaboration with other regional, national and international reference centres, the UMCG has been characterising the transmission of relevant pathogens between institutions within the region and at the national and international level.

Between May 2012 and September 2013, the transmission of a $bla_{CTX\text{-}M\text{-}15}$-producing *Klebsiella pneumoniae* ST15 occurred between patients treated in a single centre [20]. Additionally, one of these patients was treated in three different institutions located in two cities and was involved in further intra- and inter-institutional spread of this high-risk clone (local expansion, $bla_{CTX\text{-}M\text{-}15}$ producing, and containing hypervirulence factors). Environmental contamination and lack of consistent patient screening were identified as the responsible factors for the dissemination of this specific clone. The design of a tailor-made real-time -PCR specific for the outbreak clone based on the whole-genome sequences of the strains allowed the early detection of this *K. pneumoniae* high-risk-clone with prolonged circulation in the regional patient population [20] and helped prevent further spread. This study raised awareness to the necessity for inter-institutional/regional collaborations for infection/outbreak management of relevant pathogens [20].

In a large cohort study, WGS was used for molecular characterisation of Shiga toxin-producing *Escherichia coli* (STEC) isolated from faeces of patients obtained from two regions in the Netherlands to reveal the relation between molecular determinants and disease outcome [21]. STEC is a significant public health concern associated with both outbreaks and sporadic cases of human gastrointestinal illness worldwide [22]. A subpopulation of STEC, the enterohaemorrhagic *E. coli*, can cause bloody diarrhoea in humans, and some can cause haemolytic uremic syndrome (HUS) [23]. This study concluded that there was no clear correlation between serogenotype, *stx* subtype or ST and disease outcome and the latter was probably influenced by other host factors. Additionally, this study demonstrated that there was substantial genetic diversity and distinct phylogenetic groups observed in the two studied regions, showing that the STEC populations within these two geographically regions were not genetically linked [21].

More recently, a study was conducted to understand the epidemiology of resistant bacteria, including extended-spectrum β-lactamase (ESBL)- and plasmid AmpC (pAmpC)-, and carbapenemase (CP)-producing *Enterobacteriaceae* and vancomycin-resistant enterococci (VRE) across the Northern Dutch-German border region [24]. The Netherlands and Germany are bordering countries that created a cooperative network to prevent the spread of multidrug-resistant microorganisms (MDRO), such as ESBL and CP-producing *Enterobacteriaceae* and VRE, and to harmonise guidelines in healthcare settings as patients are regularly transferred between healthcare institutions within the two countries [25]. However, it was concluded that cross-border transmission of ESBL-producing *E. coli* and VRE was unlikely, based on the cgMLST analysis performed [24]. Yet, the authors reinforced that continuous monitoring is required to control the spread of these pathogens and to stay informed about their epidemiology, in order to implement effective infection prevention measures [24].

2.3.3 Transmission of Zoonotic Microorganisms Between Animals and Humans

Human health is influenced by several factors in the environment, including contact with animals, animal products or contaminated habitats. At the UMCG, we are working in collaboration with other non-hospital institutions to understand the dynamics of transmission of microbial pathogens between humans, animals and the

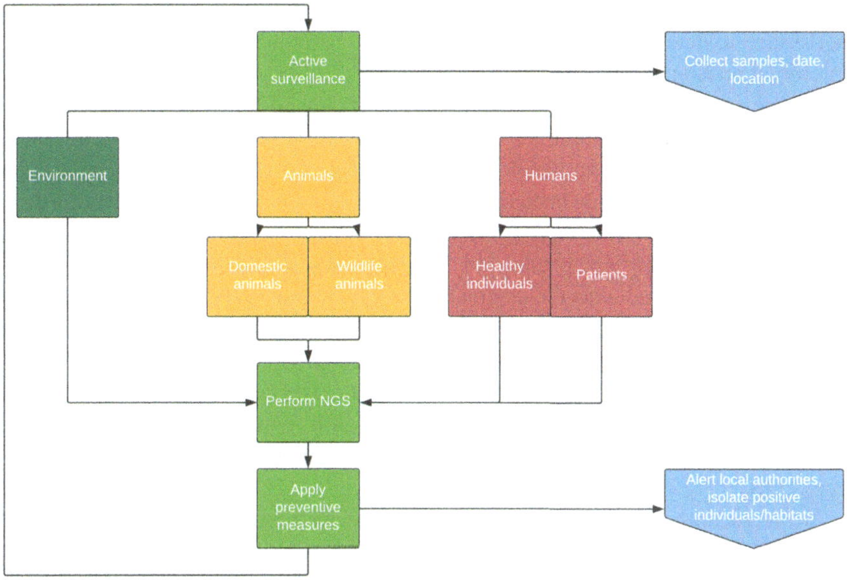

Fig. 2.3 Model for outbreak and infection prevention surveillance

environment. Figure 2.3 shows a model for early outbreak and infection prevention surveillance response that is needed due to detect spilling of new emerging infectious diseases from non-human reservoirs to humans.

Recently, a *K. pneumoniae* clone (ST348) in a horse was found, which had previously been isolated from humans in Portugal and a few other countries [26–31]. The allele differences provided by the cgMLST analysis suggested there was a genetic link, although an epidemiological link could not be found [32]. This indicated that either this particular clone is circulating in humans and horses in Portugal or there was a transfer of this particular isolate from a person to the horse during hospitalisation. In any case, this study demonstrated the importance of identifying and controlling this type of hospital-acquired infections, in both the human and veterinarian hospital settings, in order to avoid antimicrobial resistance dissemination [32].

2.3.4 *Antimicrobial Resistance Characterisation Through NGS*

Before NGS, finding new mechanisms of antimicrobial resistance was demanding since it involved different laborious techniques (e.g. hybridisation, cloning or primer-walking sequencing) until the gene and/or mutation responsible for resistance could be detected. With the entire genome sequenced through NGS, we can, by homology, identify potential new mechanisms. Further experiments can then be performed to determine if these genes are indeed responsible for the observed antimicrobial resistance pattern [33].

In one study, a possible *in vivo* selection of a clinical *Klebsiella oxytoca* isolate showing increased minimum inhibitory concentrations to ceftazidime was described [33]. The patient had been treated with ceftazidime (4 g/day) for a septic episode caused by multiple bacterial species, including *K. oxytoca*, but after 11 days of treatment, another *K. oxytoca* was isolated from a pus sample drained from his wound. The wound isolate showed increased resistance to ceftazidime (MIC $\geq$64 mg/L) compared with the original *K. oxytoca* isolate. WGS revealed the presence of a novel bla_{OXY-2} allele, termed $bla_{OXY-2-15}$, with a two amino acid deletion at Ambler positions 168 and 169 compared to $bla_{OXY-2-2}$. This report showed the risk of *in vivo* selection of ceftazidime-resistant *K. oxytoca* isolates after prolonged ceftazidime treatment and it was the first description of a *K. oxytoca* isolate conferring resistance to ceftazidime by a two amino acid deletion in the omega loop of $bla_{OXY-2-2}$ [33].

More recently, a novel *nim* gene was found in three metronidazole-resistant *Prevotella bivia* strains, the *nimK* gene, which was located on a mobile genetic element [34]. For decades, metronidazole has been the antibiotic of choice when dealing with anaerobic infections. However, metronidazole-resistant bacteria have been reported [35]. The *nimK* gene was associated with an IS*1380* family transposase on a mobile genetic element that also contained a gene encoding an efflux small MDR (SMR) transporter associated with a *crp/fnr* regulator. This was the first description of the presence of a novel *nim* gene in metronidazole-resistant *P. bivia* clinical isolates [34]. The detection of MGEs harbouring *nim* and other relevant genes among anaerobic bacteria is worrying, because these elements may cause a rapid emergence of resistance to the most commonly used antibiotics in anaerobic infections.

2.4 Metagenomics

Metagenomics is here defined as the process of determining the complete DNA or RNA sequence(s), either after reverse transcription to cDNA or directly, of microorganisms and/or viruses from a complex sample that contains several microorganisms and/or viruses. This complex sample can contain the microorganisms' genomes and other genetic elements (i.e. plasmids or phages) that can be present within the organisms' cell or are freely floating in the sample. Sequencing of DNA, cDNA or RNA within a sample can be based on the amplification of (a) specific sequence(s) (amplicon-based or targeted metagenomics) or on the entire genomes (shotgun metagenomics). We will focus this section on the use of metagenomics for three specific purposes that are currently under optimisation and implementation at the UMCG.

2.4.1 Amplicon-Based Metagenomics of the 16S–23S rRNA Encoding Region

The conventional culturing method has long been considered the gold standard for bacterial identification. However, it can take days to weeks to successfully culture bacteria, as some clinically relevant bacteria are slow-growing, difficult to grow, fastidious or sometimes even non-culturable [36, 37]. The 16S *rRNA* gene has proven to be a useful molecular target since it is present in all bacteria, either as a single copy or in multiple copies, and it is highly conserved over time [38]. Consequenttly, until recently, most microbiome studies used this amplicon-based metagenomic approach to investigate the microbial communities of different body sites, and vast literature has been published using this technique.

Nonetheless, this method does not always allow the identification of bacteria to the species level due to high sequence similarities between some species [1]. To overcome this problem, Sabat and colleagues [39] developed an innovative approach based on the sequencing of the 16S–23S *rRNA* encoding region (~4.5 kb). The method proved to be superior to other commonly used identification methods and enabled concurrent identification of several pathogens in clinical samples that were negative by culture and PCR [39]. In order to further improve this method, an in-house database was developed, which, combined with a *de novo* assembly and BLAST (Basic Local Alignment Search Tool) approach, significantly reduced the time needed for analysis [40].

2.4.2 Shotgun Metagenomics for the Identification and Typing of Microbial Pathogens

Several molecular detection techniques have been implemented in the diagnostic laboratory, but these are generally geared towards specific pathogens (e.g. specific RT-PCR or microarrays) and even when unbiased molecular approaches are used, such as 16S/18S *rRNA* gene sequencing, these do not provide all the information that can be obtained by culturing, e.g., antimicrobial susceptibility and molecular typing information [41]. For this reason, the use of shotgun metagenomics as a single method that could provide rapid identification and characterisation of clinically relevant pathogens directly from a sample was evaluated [41]. As the complexity of data analysis is a challenge encountered in shotgun metagenomics, a comparison of a diverse set of bioinformatics tools (commercial and non-commercial) was performed to investigate their performance in taxonomic classification, antimicrobial resistance gene detection and typing [41]. Based on the results obtained, the authors concluded that the tools and databases used for taxonomic classification and antimicrobial resistance had a key impact on the results, suggesting that efforts need to be directed towards standardisation of the methods if shotgun metagenomics is to be used routinely in clinical microbiology.

A study was also conducted to optimise a shotgun metagenomics workflow for the identification and typing of Dengue viruses (DENV), a positive-stranded RNA virus, directly from clinical samples [42]. DENV infection continues to be one of the most prevalent arboviral diseases in tropical and subtropical regions [43]. Nevertheless, to date, there is no fully successful vaccine or specific treatment for DENV [44]. It is therefore essential to monitor circulating DENV. Genotyping is mostly based on sequencing parts of the genes coding for the structural proteins through Sanger sequencing of the E region [45] or the CprM region [46, 47]. However, these methods have poor resolution and do not allow for the detection of recombinant events and detection of escape mutants [42]. A shotgun metagenomics approach was used successfully to sequence whole genomes of DENV directly from clinical samples, without the need for prior sequence-specific amplification steps. The method enabled the identification of intra-host DENV diversity (quasi-species), detection of multiple DENV serotypes in a single sample and generation of phylogenetic trees to understand the dynamics of DENV. Results were obtained within 3 days and the associated reagent costs were low enough to be suitable for a clinical setting.

2.4.3 Shotgun Metagenomics to Characterise the Gut Microbiome/Resistome

In Europe, more than 80% of the total antibiotic consumption in the human sector is prescribed in the community, and the rate of prescription increases with age leading to collateral damage and antibiotic pressure. Although the importance of a healthy gut microbiome and the consequences of dysbiosis have been extensively reported, less is known about the resistome present in the healthy population [48]. A study was conducted at the UMCG to describe the gut resistome of healthy middle-aged people in Northern Netherlands, by using samples from the *Lifelines* cohort [49]. A total of 60 samples were sequenced, in which several resistance genes were identified, and among them the tetracycline resistance genes were predominant. No extended-spectrum β-lactamases (ESBLs) or carbapenemases were found [48]. This study highlighted the importance of monitoring healthy people to identify potential sources of antimicrobial resistance genes and implement effective control measures.

Another study characterised the human intestinal microbiota in faecal samples from STEC-infected patients. The objective was to investigate possible changes in the composition of the intestinal microbiota in samples from STEC-infected patients compared to healthy and healed controls [50]. The stool samples collected from the STEC infected patients had a lower abundance of Bifidobacteriales and Clostridiales members in comparison to controls where these microorganisms predominated. This was the first evidence that changes occur in the intestinal microbiota of patients with STEC infection and it highlighted the importance of metagenomics for the culture-independent diagnosis of infection, as it was able to detect genomic traits associated with STEC in stool samples from infected subjects [50].

2.5 Clinical Impact of NGS

The above-mentioned examples show the power of NGS in the clinical microbiology laboratory. In addition, there is an increasing interest in NGS for clinical microbiology both by laboratories and companies. This becomes clear when looking at the enormous increase in the number of symposia and capacity building workshops related to these topics, and in the available software (commercial and non-commercial) to be used for analyses of shotgun metagenomics and WGS data. We anticipate that the use of NGS in medical microbiology laboratories will further increase over the next years, not only for the characterisation and surveillance of pathogens, the investigation of outbreaks, infection prevention and the detection of novel resistance genes but also for the application of metagenomic approaches in clinical samples for (routine) molecular diagnostics. The latter will have a significant impact on the diagnosis of infectious diseases, on understanding host-pathogen interactions [12], and on the correlation between genotype (provided by NGS) and phenotype [51]. However, the NGS workflow needs further improvement, especially for shotgun metagenomics, to shorten the turnaround time and further reduce costs [1], and to ensure the quality and reproducibilityof the results (including validation and external proficiency testing). Although these and other challenges need to be tackled, we are convinced that NGS will become a powerful tool within the clinical microbiology laboratory and will lead to a personalised approach for diagnosing and monitoring treatment of infectious diseases.

References

1. Deurenberg R, Bathoorn W, Chlebowicz MA, Couto N, Ferdous M, Garcia-Cobos S et al (2017) Application of next generation sequencing in clinical microbiology and infection prevention. J Biotechnol 243:16–24
2. Padmanabhan R, Mishra AK, Raoult D, Fournier PE (2013) Genomics and metagenomics in medical microbiology. J Microbiol Methods 95:415–424
3. Rossen JWA, Friedrich AW, Moran-Gilad J (2018) Practical issues in implementing whole-genome-sequencing in routine diagnostic microbiology. Clin Microbiol Infect 24:355–360
4. EFSA (2014) Use of whole genome sequencing (WGS) of food-borne pathogens for public health protection. EFSA Scientific Colloquium Summ Rep 20
5. Reuter S, Elington MJ, Cartwright EJP, Köser CU, Török ME, Gouliouris T et al (2013) Rapid bacterial whole-genome sequencing to enhance diagnostic and public health microbiology. JAMA Intern Med 173(15):1397–1404
6. Gardy JL, Loman NJ (2018) Towards a genomics-informed, real-time, global pathogen surveillance system. Nat Rev Gen 19:9–20

7. Illumina: iSeq™ 100 Sequencing System (2017) https://www.illumina.com/content/dam/illumina-marketing/documents/products/datasheets/iseq100-sequencing-system-spec-sheet-770-2017-020.pdf. Accessed 24 Jul 2018
8. Thomsen MCF, Ahrenfeldt J, Cisneros JLB, Jurtz V, Larsen MV, Hasman H et al (2016) A bacterial analysis platform: an integrated system for analysing bacterial whole genome sequencing data for clinical diagnostics and surveillance. PLoS One 11(6):e0157718
9. Visconti A, Martin TC, Falchi M (2018) YAMP: a containerized workflow enabling reproducibility in metagenomics research. GigaScience 7(7):giy072
10. Fricke WF, Rasko DA (2014) Bacterial genome sequencing in the clinic: bioinformatic challenges and solutions. Nat Rev Genet 15:49–55
11. Jennings LJ, Arcila ME, Corless C, Kamel-Reid S, Lubin IM, Pfeifer J et al (2017) Guidelines for validation of next-generation sequencing-based oncology panels: a joint consensus recommendation of the association for molecular pathology and college of American pathologists. J Mol Diagn 19:341–365
12. Schlaberg R, Chiu CY, Miller S, Procop GW, Weinstock G, the Professional Practice Committee and Committee on Laboratory Practices of the American Society for Microbiology, and the Microbiology Resource Committee of the College of American Pathologists (2017) Validation of metagenomic next-generation sequencing tests for universal pathogen detection. Arch Pathol Lab Med 41(6):776–786
13. Quainoo S, Coolen JPM, van Hijum SAFT, Huynen MA, Melchers WJG, van Schaik W et al (2017) Whole-genome sequencing of bacterial pathogens: the future of nosocomial outbreak analysis. Clin Microbiol Rev 30:1015–1063
14. Zhou K, Lokate M, Deurenberg RH, Arends J, Lo-Ten Foe J, Grundmann H et al (2015) Characterization of a CTX-M-15 producing *Klebsiella pneumoniae* outbreak strain assigned to a novel sequence type (1427). Front Microbiol 6:1250
15. Zhou X, Chlebowicz MA, Bathoorn E, Rosema S, Couto N, Lokate M et al (2018) Elucidating vancomycin-resistant *Enterococcus faecium* outbreaks: the role of clonal spread and movement of mobile genetic elements. J Antimicrob Chemoth 73(12):3259–3267
16. Fleres G, Couto N, Lokate M, van der Sluis LWM, Ginevra C, Jarraud S et al (2018) Detection of *Legionella anisa* in water from hospital dental chair units and molecular characterization by whole-genome sequencing. Microorganisms 6(3):pii:E71
17. Vaccaro L, Izquierdo F, Magnet A, Hurtado C, Salinas M, Gomes T et al (2016) First case of legionnaire's disease caused by *Legionella anisa* in Spain and the limitations on the diagnosis of *Legionella* non-pneumophila infections. PLoS One 11(7):e0159726
18. Bornstein N, Mercatello A, Marmet D, Surgot M, Deveaux Y, Fleurette J (1989) Pleural infection caused by *Legionella anisa*. J Clin Microbiol 27(9):2100–2101
19. Schönning C, Jernberg C, Klingenberg D, Andersson S, Pääjärvi A, Alm E et al (2017) Legionellosis acquired through a dental unit: a case study. J Hosp Infect 96(1):89–92
20. Zhou K, Lokate M, Deurenberg RH, Tepper M, Arends JP, Raangs EGC et al (2016) Use of whole-genome sequencing to trace, control and characterize the regional expansion of extended-spectrum β-lactamase producing ST15 *Klebsiella pneumoniae*. Sci Rep 6:20840
21. Ferdous M, Friedrich AW, Grundmann H, de Boer RF, Croughs PD, Islam MA et al (2016) Molecular characterization and phylogeny of Shiga toxin-producing *Escherichia coli* isolates obtained from two Dutch regions using whole genome sequencing. Clin Microbiol Infect 22(7):642.e1–642.e9
22. Paton JC, Paton AW (1998) Pathogenesis and diagnosis of Shiga toxin-producing *Escherichia coli* infections. Clin Microbiol Rev 11:450–479
23. Mellmann A, Bielaszewska M, Köck R, Friedrich AW, Fruth A, Middendorf B et al (2008) Analysis of collection of hemolytic uremic syndrome associated enterohemorrhagic *Escherichia coli*. Emerg Infect Dis 14:1287–1290
24. Zhou X, García-Cobos S, Ruijs GJHM, Kampinga GA, Arends JP, Borst DM et al (2017) Epidemiology of extended-spectrum β-lactamase-producing *E coli* and vancomycin-resistant enterococci in the northern Dutch–German cross-border region. Front Microbiol 8:1914

25. Muller J, Voss A, Köck R, Sinha B, Rossen JW, Kaase M et al (2015) Cross-border comparison of the Dutch and German guidelines on multidrug-resistant Gram-negative microorganisms. Antimicrob Res Infect Control 4:7
26. Rodrigues C, Machado E, Ramos H, Peixe L, Novais Â (2014) Expansion of ESBL-producing *Klebsiella pneumoniae* in hospitalized patients: a successful story of international clones (ST15, ST147, ST336) and epidemic plasmids (IncR, IncFIIK). Int J Med Microbiol 304:1100–1108
27. Rodrigues C, Bavlovič J, Machado E, Amorim J, Peixe L, Novais Â (2016) KPC-3-producing *Klebsiella pneumoniae* in Portugal linked to previously circulating non-CG258 lineages and uncommon genetic platforms (Tn4401d-IncFIA and Tn4401d-IncN). Front Microbiol 7:1000
28. Rodrigues C, Mendes A, Sima F, Bavlovič J, Machado E, Novais Â, Peixe L (2017) Long-term care facility (LTCF) residents colonized with multidrug-resistant (MDR) *Klebsiella pneumoniae* lineages frequently causing infections in Portuguese clinical institutions. Infect Control Hosp Epidemiol 38:1127–1130
29. Baraniak A, Izdebski R, Fiett J, Sadowy E, Adler A, Kazma M et al (2013) Comparative population analysis of *Klebsiella pneumoniae* strains with extended-spectrum β-lactamases colonizing patients in rehabilitation centers in four countries. Antimicrob Agents Chemother 57:1992–1997
30. Mshana SE, Hain T, Domann E, Lyamuya EF, Chakraborty T, Imirzalioglu C (2013) Predominance of *Klebsiella pneumoniae* ST14 carrying CTX-M-15 causing neonatal sepsis in Tanzania. BMC Infect Dis 13:466
31. Vubil D, Figueiredo R, Reis T, Canha C, Boaventura L, da Silva G (2016) Outbreak of KPC-3-producing ST15 and ST348 *Klebsiella pneumoniae* in a Portuguese hospital. Epidemiol Infect 143:595–599
32. Trigo da Roza F, Couto N, Carneiro C, Cunha E, Rosa T, Magalhães M et al (2019) Commonality of Multidrug-Resistant *Klebsiella pneumoniae* ST348 Isolates in Horses and Humans in Portugal. Front Microbiol 10:1657
33. Nijhuis RH, Oueslati S, Zhou K, Bosboom RW, Rossen JW, Naas T (2015) OXY-2-15, a novel variant showing increased ceftazidime hydrolytic activity. J Antimicrob Chemother 70:1429–1433
34. Veloo ACM, Chlebowicz M, Winter HLJ, Bathoorn D, Rossen JWA (2018) Three metronidazole-resistant *Prevotella bivia* strains harbour a mobile element, encoding a novel *nim* gene, *nimK*, and an efflux small MDR transporter. J Antimicrob Chemother 73(10):2687–2690
35. Hartmeyer GN, Sóki J, Nagy E, Justesen US (2012) Multidrug-resistant *Bacteroides fragilis* group on the rise in Europe? J Med Microbiol 61:1784–1788
36. Didelot X, Bowden R, Wilson DJ, Peto TEA, Crook DW (2012) Transforming clinical microbiology with bacterial genome sequencing. Nat Rev Gen 13:601–612
37. Salipante SJ, Sengupta DJ, Rosenthal C, Costa G, Spangler J, Sims EH et al (2013) Rapid 16S rRNA next-generation sequencing of polymicrobial clinical samples for diagnosis of complex bacterial infections. PLoS One 8:e65226
38. Petti CA (2007) Detection and identification of microorganisms by gene amplification and sequencing. Clin Infect Dis 44:1108–1114
39. Sabat AJ, van Zanten E, Akkerboom V, Wisselink G, van Slochteren K, de Boer RF et al (2017) Targeted next-generation sequencing of the 16S-23S rRNA region for culture-independent bacterial identification – increased discrimination of closely related species. Sci Rep 7(1):3434
40. Peker N, Garcia-Croes S, Dijkhuizen B, Wiersma HH, van Zanten E, Wisselink G et al (2019) A comparison of three different bioinformatics analyses of the 16S-23S rDNA region for bacterial identification. Front Microbiol 10:620
41. Couto N, Schuele L, Raangs EC, Machado M, Mendes CI, Jesus TF et al (2018) Critical steps in clinical shotgun metagenomics for the concomitant detection and typing of microbial pathogens. Sci Rep 8:13767
42. Lizarazo E, Couto N, Tami A, Vincenti-Gonzalez M, Raangs EC, Velasco Z, Bethencourt S et al (2019) Applied shotgun metagenomics approach for the genetic characterization of dengue viruses. J Biotechnol X 2:100009

43. Bhatt S, Gething PW, Brady OJ, Messina JP, Farlow AW, Moyes CL et al (2013) The global distribution and burden of dengue. Nature 496:504–507
44. Lizarazo E, Couto N, Vincenti-Gonzalez M, Raangs EC, Jaenisch T, Friedrich AW et al (2018) Complete coding sequences of five dengue virus type 2 clinical isolates from Venezuela obtained through shotgun metagenomics. Genome Announc 6:e00545–e00518
45. Wang E, Ni H, Xu R, Barrett ADT, Watowich SJ, Gubler DJ, Weaver SC (2000) Evolutionary relationships of endemic/epidemic and sylvatic dengue viruses. J Virol 74:3227–3234
46. Avilés G, Rowe J, Meissner J, Manzur Caffarena JC, Enria D, Jeor S (2002) Phylogenetic relationships of dengue-1 viruses from Argentina and Paraguay. Arch Virol 147:2075–2087
47. Kukreti H, Chaudhary A, Rautela RS, Anand R, Mittal V, Chhabra M et al (2008) Emergence of an independent lineage of dengue virus type 1 (DENV-1) and its co-circulation with predominant DENV-3 during the 2006 dengue fever outbreak in Delhi. Int J Infect Dis 12(5):542–549
48. García-Cobos S, Rabbers T, Bossers A, Schmitt H, Harmsen HJM, Sinha B, et al (2018) A snapshot of the gut resistome of a middle-aged healthy population in the northern Netherlands (LifeLines). Oral communication (O0350) presented at the 28th European Congress of Clinical Microbiology and Infectious Diseases, Madrid, Spain (available at https://www.escmid.org/escmid_publications/escmid_elibrary/material/?mid=64330)
49. Stolk RP, Rosmalen JG, Postma DS, de Boer RA, Navis G, Slaets JP et al (2008) Universal risk factors for multifactorial diseases: LifeLines: a three-generation population-based study. Eur J Epidemiol 23(1):67–74
50. Gigliucci F, von Meijenfeldt FAB, Knijn A, Michelacci V, Scavia G, Minelli F et al (2018) Metagenomic characterization of the human intestinal microbiota in fecal samples from STEC-infected patients. Front Cell Infect Microbiol 8:25
51. Motro Y, Moran-Gilad J (2017) Next-generation sequencing applications in clinical bacteriology. Biomol Detect Quantif 14:1–6

Chapter 3
WGS for Bacterial Identification and Susceptibility Testing in the Clinical Lab

Sophia Vourli, Fanourios Kontos, and Spyridon Pournaras

3.1 Issues to Consider for the Incorporation of WGS in the Routine Workflow of the Clinical Lab

There are two general approaches for the application of NGS in bacterial genomics: Targeted Next Generation Sequencing (NGS) by sequencing of specific amplicons and Whole Genome Sequencing (WGS). In the first approach, specific genomic regions are enriched by PCR amplification and subsequent selective sequencing. This is the strategy of choice when known segments of the genome are studied. The second approach is used when the organism or the genomic region under investigation is unknown. WGS can be applied to isolated colonies or directly on the clinical specimen (culture independent pathogen detection, usually referred to as "shotgun metagenomics" or Whole-genome metagenomics). Herein we will use the term WGS for sequencing isolates from cultures, and metagenomics when discussing culture-independent sequencing. NGS has several limitations, first of all being the high cost and the time-to-results, if considering incorporating it in the routine of the clinical laboratories. Despite the limitations, the plethora of advantages and potential applications of WGS together with the technological advances that help to circumvent the obstacles, lead to its increasing adoption in the daily workflow of the clinical laboratory. The use of NGS in clinical microbiology laboratories is currently limited, but as standardisation of protocols, automation and data analysis pipelines evolve, its role in bacterial identification and susceptibility testing is expected to widen.

S. Vourli (✉) · F. Kontos (✉) · S. Pournaras (✉)
Laboratory of Clinical Microbiology, Attikon University Hospital, Medical School, National and Kapodistrian University of Athens, Athens, Greece
e-mail: svourli@med.uoa.gr; spournaras@med.uoa.gr

J. Moran-Gilad, Y. Yagel (eds.), *Application and Integration of Omics-powered Diagnostics in Clinical and Public Health Microbiology*,
https://doi.org/10.1007/978-3-030-62155-1_3

Cost and turnaround time are both crucial issues for all clinical laboratories considering implementation of WGS for diagnostic purposes. Estimating the cost of any assay, particularly of a complex one like WGS that comprises many steps, is challenging. For example, the reagents' cost may vary in different settings and countries. Furthermore, the best cost efficiency is succeeded when performing full capacity runs, so the cost is depending on the laboratory's sample throughput. This means that WGS is more cost-effective in reference laboratories, where samples are collected and run in large numbers, but this further increases the time-to-results and limits the use of the provided information for diagnostic or infection control purposes. The choice of the appropriate platform that fits better to each laboratory needs can reduce WGS cost.

Notably, the cost of WGS is progressively decreasing and it is expected that at least large microbiology labs, will be able to implement this technology in their routine workflow in the future. A recent study estimates that the cost of WGS for 16–20 isolates would be 200 euros per isolate and the turnaround time is 2.5–3 days, without considering the time needed for data analysis [1]. As for the turnaround time, recently developed benchtop platforms produce high-quality WGS for many bacterial species in a timeframe relevant for patient treatment or real-time infection control (less than 24 hours).

Furthermore, the development of platforms such as MinION, which produce long reads in significantly shorter runtimes than platforms as Illumina, may help to overcome these limitations [2]. In a laboratory-based study at Cambridge University Hospitals NHS Foundation Trust, which compared WGS with standard diagnostic microbiology, WGS results were concordant with standard phenotypic results. Additionally, with the use of WGS, resistance was attributed to the presence of carbapenemases or other resistance mechanisms [3]. The aim of this study was the comparison of WGS with standard clinical microbiology methods for the investigation of nosocomial outbreaks caused by multidrug-resistant bacteria, the detection of antimicrobial resistance determinants and the typing of other clinically important pathogens. Many more studies show the ability of rapid and accurate WGS [4–8].

Another important question to be answered is which platform fits better to a clinical lab. Different platforms use different sequencing chemistries that lead to differences in throughput, read length, error rate, genome coverage, and cost and run time. Currently, there are two high-throughput sequencing technologies: second generation and third generation sequencing. The main difference between them is the length of reads. Third generation sequencing results in longer read length, thus poising the main shortcomings of second generation sequencing, amplification artifacts and bias. Illumina and Thermo Fisher Scientific (IonTorrent) platforms generate short length reads, while Pacific Biosciences and Oxford Nanopore platforms produce long reads. Second generation sequencing (short reads) platforms produce bacterial draft genomes, suitable for diagnostics and infectious disease surveillance purposes. The lower throughput instruments are also well suited for targeted amplicon sequencing, such as detecting antimicrobial resistance determinants or 16S sequencing. Other applications include RNA sequencing to study the expression of genes and metagenomics sequencing. Third generation sequencing (long reads) platforms are useful for sequencing complex genomic regions such as repeats and are suitable for *de novo* genome assemblies. A detailed comparison of currently available platforms and their advantages and disadvantages is presented in detail in a recent review [9]. As interest in NGS technologies grows every day, new platforms with improved characteristics are developed in a quick pace.

Any clinical laboratory that considers adopting WGS in its workflow must address the validation and quality control of both sample managing (wet-lab) and data analysis (dry-lab) assays. Validation of WGS presents several particularities comparing to the validation of standard microbiological assays. An example of this uniqueness is the complexity of data produced by WGS (identification, resistance and virulence determinants, phylogenetic data are generated with one assay). Currently, there are several clinical NGS guidance initiatives by various organisations [College of American Pathologists: NGS Inspection Checklist (2012), Clinical and Laboratory Standards Institute: MM09 Nucleic Acid Sequencing Methods in Diagnostic Laboratory Medicine (2014), American Academy of Microbiology/ASM: Colloquium on Applications of Clinical Microbial Next-Generation Sequencing (April, 2015), Food and Drug Administration: Draft Guidance: Infectious Disease Next Generation Sequencing Based Diagnostic Devices: Microbial Identification and Detection of Antimicrobial Resistance and Virulence Markers (May 2016)]. The Next Generation Sequencing-Standardisation of Clinical Testing (Next-StoCT) has published detailed guidelines for the validation of the assay, quality control and proficiency testing of NGS [10]. Furthermore, guidelines for validating NGS bioinformatics pipelines are also published [11, 12]. Quality control and validation procedures of NGS assays are analysed in several reports [10, 13, 14]. Two recent reviews summarise all previously published data [1, 15].

The validation process must include all characteristics that determine the assay performance, such as accuracy, precision, repeatability (within-run precision) and reproducibility (between-run precision), analytic sensitivity, analytic specificity, reportable range and reference range [10, 16]. The repeatability can be ascertained by sequencing and analysing the same samples by the same operator and the same instrument and bioinformatics tool in replicates. Similarly, the reproducibility can be determined by sequencing the same samples by different operators, on different runs, and different instruments and bioinformatics tools. The precision and accuracy can be established by comparing NGS results with results of gold-standard methods such as PFGE in the case of typing, and PCR-based techniques for identification of genes or pathogens. The challenge in comparing WGS with already established methods is that, in some cases like bacterial typing for example, WGS has higher discriminatory power than any other existing typing method, thus there is no appropriate "gold-standard" method for comparison. External quality assurance (EQA) and proficiency testing (PT) are fundamental for all assays performed in clinical microbiology laboratories and are equally vital for routine implementation of NGS assays [17]. Performing external quality assurance for WGS is especially demanding because all steps of the assay (DNA extraction, library preparation steps, sequencing reactions, bioinformatics analysis) must be quality assured [15]. Proficiency testing is also essential for laboratories that intend to perform WGS for diagnostic purposes. Both the College of American Pathologists (CAP) and Global Microbial Identifier (GMI) have developed NGS Proficiency Tests schemes.

3.2 Bacterial Identification by WGS in the Clinical Laboratory

Pathogen identification is perhaps the most important duty of the clinical microbiology laboratory. Bacterial infection diagnosis and patient management and outcome heavily rely on accurate and timely pathogen detection and identification by the clinical lab. Correct identification not only drives an adequate antibacterial therapy but also offers suggestions about disease progression and outcome, thus guiding therapeutic interventions.

Currently, most laboratories rely on conventional, culture-based techniques for bacterial detection and identification. Culture of most bacterial species is both cheap, fast and, given the progress in automation of bacterial culture, with reduced labour required. The time-to-result for most samples is around 24–48 hours, except for special samples like blood cultures that require a two-step incubation (blood culture bottle incubation and, as soon as it turns positive, subculture on culture media). Another important factor is that the majority of the personnel already working in clinical laboratories is well trained in standard microbiology techniques, with only a minority already familiar with NGS techniques and, more importantly, in NGS data analysis. Furthermore, very few clinicians know how to interpret NGS results into clinically relevant information. Hence, it seems that culture-based classical microbiology methods still hold their ground in the clinical laboratory workflow. WGS is a new, completely different approach for pathogen detection and identification that offers a plethora of new possibilities and perspectives in the fields of clinical microbiology and infection diagnosis. Recognising the advantages of WGS, many clinical microbiology laboratories are gradually exiting the "comfort zone" of conventional microbiology and consider adopting WGS.

There are two options regarding the implementation of WGS for bacterial detection and identification: i) on isolated bacterial colonies, after subculture or directly from the primary culture plate and ii) directly on the clinical specimen (culture-independent method). The first approach is the most often used until now and better standardised, but the isolation and subculture time significantly increases the time to results. The second one is gaining ground because pathogen identification is quicker and there are fewer risks for contamination of the primary culture or alterations of the bacterial characteristics due to the subculture. There are already several published reports using both approaches that support the feasibility of the integration of NGS for bacterial identification in the clinical laboratory workflow. A characteristic example is the application of WGS in every pathogen isolated in a routine clinical laboratory during 1 day [18], using the automated on-instrument *de novo* assembly workflow (MiSeq Reporter version 2.0; Illumina) and default software settings. The authors handled 130 samples, including several mixed samples, and identified more than 30 bacterial species by WGS. WGS identification was concordant with conventional methods for 115 samples and failed to produce high-confidence organism identifications for 15 samples. Interestingly, the authors of this study pointed out the notable absence of several organisms in the reference genome database. It has been previously shown that it is possible to generate a sequencing library directly from *Escherichia coli* colonies [8]. More recently, this finding was used to perform rapid single-colony WGS for the detection and identification of various pathogens [4]. The researchers developed and validated a simple protocol to enable WGS directly from a single bacterial colony. They correctly identified 17 bacterial pathogens, showing that incorporation of WGS in routine bacterial identification was feasible. For some pathogens, even current WGS turnaround time is advantageous compared to standard methods. As highlighted in one study [9], turnaround time for full characterisation of Shiga toxin-producing *Escherichia coli* at the Centers for Disease Control and Prevention is routinely between 1 and 3 weeks, because several precise and demanding tests must be performed (phenotypic assays for species identification, PCR for virulence profiling, broth microdilution assays for antimicrobial susceptibility testing and agglutination assays for serotyping). All of these assays can be tucked in one assay (WGS) that will produce all of the above information in days instead of weeks.

Moreover, there are cases where culture-independent metagenomics undoubtedly offers an advantage over culture based methods. Detection and identification of pathogens directly from positive blood culture bottles and cerebrospinal fluid are obvious examples. Direct pathogen identification from CSF samples is especially valuable, as several laboratory-confirmed meningococcal disease cases fail to yield a viable invasive isolate, primarily due to the use of antibiotics. Other examples are the detection of slow-growing bacteria, like *Mycobacterium tuberculosis* complex and difficult to culture or uncultured bacterial species like *Chlamydia trachomatis*. Progress in culture-independent NGS as whole-genome metagenomics is already suggesting feasible scenarios in which WGS for pathogen detection directly from clinical specimens may be applied. However, culture-independent whole-genome metagenomics has many limitations, the most important being the presence of high amounts of host DNA in some samples like blood and the presence of commensal bacterial flora, that can hamper NGS assays.

Whole-genome metagenomics for the identification of pathogens directly from positive blood culture bottles is of obvious value, but in this case, the inhibitors present in the primary sample along with the high amount of human DNA may prohibit the recovery of sufficient good-quality pathogen DNA. Several recent studies present various techniques for optimisation of whole genome metagenomics directly from clinical samples, including bacterial DNA enrichment methods. In one such study [19], the researchers succeeded in reducing human DNA and extracting bacterial DNA directly from positive aerobic and anaerobic BACTEC blood culture bottles, by using simple techniques (differential centrifugation and DNA extraction with commercial kits). They subsequently performed whole-genome metagenomics using two platforms (Illumina and MinION) and assessed pathogen recovery and prediction of species and antibiotic susceptibility of 44 Gram-negative and 54 *Staphylococcus* species. Interestingly, according to this study whole genome metagenomics performed better than MALDI-TOF when one pathogen was present in the blood culture bottle, which is the case in the majority of bacteraemia cases. The authors concluded that whole-genome metagenomics offers the potential for an end-to-end diagnostic solution and may replace the multiple clinical workflows that are currently used to support the microbial identification and drug susceptibility testing.

NGS has been applied in several instances for the diagnosis of bacterial infections of the central nervous system. Another application was WGS testing of *Neisseria meningitidis* performed directly from blood and CSF specimens [20]. In this study, a target-specific RNA oligonucleotide bait library was used for the enrichment of bacterial DNA and the Agilent SureSelectXT kit for generating draft meningococcal genomes. Of the ten specimens, eight produced genomes of acceptable quality. The authors considered that half of the non-culture invasive meningococcal CSF disease specimens received by the Meningococcal Reference Unit in Manchester could yield acceptable genomic data. Another proof of concept for whole-genome metagenomics -based bacterial detection directly from CSF samples is the identification of *Leptospira santarosai* in the CSF of an adolescent patient with immune deficiency [21]. This patient had negative results in all the clinically validated assays for leptospirosis and did not receive any empirical antimicrobial agents because of low suspicion for bacterial meningitis. The patient's CSF and serum samples were analysed with Illumina MiSeq platform using an unbiased NGS protocol, resulting in the detection of *Leptospira santarosai* with a turnaround time of 48 hours; the patient was adequately treated and the outcome was favourable.

Another interesting scenario, [7], is the application of whole-genome metagenomics to detect and identify microbial pathogens directly from urine samples. In this study, the authors evaluated WGS performed directly on urine samples using a benchtop sequencer, compared this strategy with conventional bacteriology and WGS of the bacteria yielded by culture, and developed a new fast bioinformatic tool for data analysis. The resulting direct whole-genome metagenomics results were highly concordant with those of cultured isolates in terms of species identification, clonality and resistance genes' identification. A noteworthy finding in this study was the detection with direct whole-genome metagenomics of bacteria known to be urinary tract pathogens that were not detected by conventional culture. The authors noticed an increased number of resistance genes detected by direct sequencing, a fact that they attributed to the presence of natural microbiota in the urethra. By filtering sequencing results and excluding genes with low coverage, all "excess" resistance genes not observed in cultured isolates were removed. On the other hand, direct sequencing did not miss any resistance genes. Turnaround time starting from a urine sample and using the fast bioinformatics pipeline developed in this study was less than 24 hours, significantly less than conventional identification and susceptibility testing turnaround time of 72 hours.

Whole-genome metagenomics for the detection of fastidious pathogens directly from clinical samples can also be of great value. *Chlamydia trachomatis*, for example, is an obligate intracellular pathogen and *in vitro* culture is laborious and time-consuming. For this reason, methods that allow sequencing directly from *C. trachomatis* positive samples are especially attractive. A number of studies [22–25] report various enrichment methods in order to obtain high-quality *C. trachomatis* DNA, suitable for sequencing directly from clinical samples. Christiansen et al. reported a method for *C. trachomatis* sequencing directly from vaginal swabs and urine samples after enrichment with a custom capture RNA bait set, that captures all known diversity amongst *C. trachomatis* genomes. Their method showed increased sensitivity by >ten-fold, comparing to previously reported methods. Rapid whole-genome sequencing of *C. trachomatis* directly from clinical samples was allowed and the authors pointed out the potential to adapt this method to other intracellular or fastidious pathogens.

An exciting aspect is the implementation of whole-genome metagenomics into routine diagnostic of highly dangerous pathogens, like *Bacillus anthracis*, *Francisella tularensis*, *Yersinia pestis* etc. Although such an application seemed unfeasible, mostly because of the impractical size of the equipment for BSL-3 or BSL-4 labs, this option is now realistic with platforms like minION [26].

3.3 Susceptibility Testing by Whole Genome Sequencing

The determination of the antimicrobial resistance profile of pathogens is fundamental for the management of infections and all clinical laboratories must be able to undertake this action. Notably, the determination of the Minimum Inhibitory Concentration (MIC) of antibacterials, apart from categorical results (susceptible/resistant), is also essential. Currently, clinical laboratories perform routine susceptibility testing by conventional methods. Conventional AST methods are well-known, robust and certified, and any new technology has to compete with current reference standards. However, they also have limitations, such as long turnaround time and correlation with clinical outcomes. For these reasons, alternative AST methods like MALDI-TOF and PCR-based detection have been developed. Although these alternative methods confront mainly the problem of the slow turnaround time of conventional AST, they also have limitations that limit their use for full routine AST. PCR-based methods only detect a limited panel of known resistance genes and cannot identify new or rare resistance mechanisms.

On the other hand, WGS offers enormous possibilities in the field of determination of the resistance profile of pathogens. Pathogen identification, detection of all resistance markers (i.e. characterisation of the "resistome") and detection of virulence determinants ("virulome") occur at the same time, thus allowing a complete pathogen characterisation and detection of unknown features that may better guide patient management, but also infection control and resistance containment. It can also identify features related to antibiotic resistance in the genome that standard methods miss. For example, a vancomycin dependent *E. faecium* (i.e. requiring the presence of vancomycin for its growth) isolated from routine blood culture was detected by WGS [27]. Those advantages are the reason why WGS AST, although not widely used at present, is being gradually introduced in the clinical laboratory workflow. Several published reports show that antibiotic resistance data obtained by WGS reliably predict antibiotic resistance phenotype when compared with standard phenotypic methods, with sensitivity and specificity over 95%. Remarkably, many of them propose the implementation of WGS-AST as the primary susceptibility testing method, followed by standard phenotypic susceptibility testing. Stoesser et al. [28], used WGS to predict resistance phenotypes of *E. coli* and *K. pneumoniae* clinical isolates from bacteraemias. Whole-genome data were compared to phenotypic results obtained by the BD Phoenix system. The sensitivity of genotype for predicting resistance across all antibiotics for both species was 0.96 (95% CI: 0.94–0.98) and the specificity was 0.97 (95% CI: 0.95–0.98). Very major and major error rates, at 1.2% and 2.1%, respectively, were within the accepted 1.5% and 3% FDA limits. In another study, 335 clinical isolates of *S. Sonnei* were subjected to WGS and conventional AST (agar dilution) and results were compared [29]. Databases used to detect antibiotic resistance genes were the Comprehensive Antimicrobial Resistance Database (CARD) and Resfinder. Only fifteen isolates showed discrepancy for one of the ten antibiotics tested. Interestingly, all 15 discrepant results concerned isolates that were phenotypically susceptible to specific antimicrobials and were predicted to be resistant by WGS due to the detection of a resistance determinant which was not expressed or expressed poorly. In another study, bloodstream isolates of common Gram-negative bacteria from neutropenic patients were subjected to WGS by Illumina MiSeq and results were compared with the routine method used in the clinical laboratory (e-test and automated methods), using broth microdilution assay as the gold standard [30]. In order to analyse WGS results, the researchers developed a customised database of AMR protein sequences by merging the data of ARDB and CARD. The interesting finding of this study was that WGS was equal or superior to conventional methods in predicting antimicrobial resistance when both approaches were compared to the broth microdilution method. This study is an example of the shortcomings of many conventional methods that are routinely used in clinical microbiology laboratories. An example of rapid WGS-based antimicrobial-resistance prediction is presented in this published report [31]. The authors developed a user-friendly software ('Mykrobe predictor') that can generate antibiotic resistance reports from raw sequence data very quickly (3 minutes). They implemented their method to *S. aureus* and *M. tuberculosis* sequence data, compared the results with standard phenotypic methods and found comparable error rates. They also demonstrated that this method works with Oxford Nanopore Technologies MinION and produces high-quality results in 7 hours. An interesting example of the correlation of the variability in MIC values of clinically relevant antibiotics with resistome data of *Pseudomonas aeruginosa* was presented in a resistome-wide association study [32]. Among the novel mutations identified in this study, 29 were variants of the *oprD* gene associated with variation in meropenem MIC. Many other studies show the same high sensitivity and specificity values of

WGS versus standard susceptibility testing, concluding that WGS can be a viable alternative for predicting resistance to antibacterial agents [33–43].

Nevertheless, the genomic prediction of resistance for clinical purposes requires caution. First, in contrast to phenotypic testing, as all molecular methods, it detects resistance markers that predict resistance, but it does not provide information on susceptibility. Furthermore, the absence of a resistance gene does not necessarily exclude resistance to that antibiotic, as new resistance genes that are not included in the AR gene database might have been missed. Moreover, there is doubt as to whether hetero-resistance detection, a very important feature of certain pathogens/antibacterials combinations, can be detected by WGS. Additionally, the vast volume of data that is produced by WGS makes it heavily reliant on AR gene data resources for interpretation and quality control. Additionally, WGS-based AST demands thorough quality assurance and quality control together with in depth clinical evaluation and cost-effectiveness analyses. Because of these limitations, in the EUCAST Subcommittee report [44] it was stated that "for most bacteria, the available evidence for using WGS as a tool to infer antimicrobial susceptibility (i.e. to rule-in as well as to rule-out resistance) accurately is either poor or non-existent. More focused studies and additional funding resources are needed as a priority to improve knowledge". In this report, systematic errors and limitations of WGS-AST by pathogen are thoroughly analysed.

Antibiotic resistance gene data resources must be continuously updated and curated to remain comprehensive. Several literature reviews summarised available antibiotic resistance gene data resources [45, 46]. Novel user-friendly databases are continuously developed. The "Mykrobe Predictor" mentioned earlier [31] was recently developed for prediction of a WGS-AST for *Staphylococcus aureus* and *Mycobacterium tuberculosis*. The software had sensitivity and specificity of 99.1% and 99.6% for 12 antibiotics against *S. aureus*. The authors also addressed the problem of minor resistance alleles. The SEAR (Search Engine for Antimicrobial Resistance) [47] is another recently created tool targeting horizontally-acquired resistance determinants. This software includes gene dosage and sequence variation assessment, but the most interesting feature is the possibility to identify resistance determinants from metagenomics data. There are many more examples of recently developed or updated bioinformatics tools [48–53] which indicate the intense research efforts in trying to incorporate WGS-AST into the everyday practice of the clinical microbiology lab.

3.4 WGS for the Identification and Susceptibility Testing of Mycobacteria

Mycobacteria are among the clinically most important slow-growing, fastidious pathogens. As some of the species-level identification methods and drug susceptibility testing (DST) are laborious, time-consuming and sometimes lack sensitivity and specificity, the role of WGS utilisation can be of paramount significance to patient care. The next paragraphs will discuss the current experience with using NGS techniques on mycobacteria, as a representative of the potential of NGS to change the clinical viewpoint on identification and susceptibility testing.

Tuberculosis (TB) is caused by bacteria belonging to members of the *Mycobacterium tuberculosis* complex (MTBC) and, more rarely, by *Mycobacterium canettii* [54]. In 2016 the World Health Organization [55] estimated 10.4 million new TB cases caused by *Mycobacterium tuberculosis*. Moreover, 1.3 million TB deaths occurred among HIV-negative and additional 374,000 deaths among HIV-positive people, thus making TB the leading cause of death by a single pathogen. One major challenge that has to be overcome is the resistance to anti-tuberculous drugs [56]. Multidrug-resistant tuberculosis (MDR-TB), is defined as bacteria resistant to both rifampicin and isoniazid, is hardly curable by the standard four-drug regimen and requires administration of second-line drugs [57]. In 2016, a globally estimated 4.1% of new cases and 19% of previously treated cases were MDR or rifampicin-resistant TB but isoniazid-susceptible (RR-TB) [55]. The emergence of resistance to additional drugs is not scarce and there are reports of extensively drug-resistant tuberculosis (XDR-TB; MDR-TB bacteria also exhibiting resistance to fluoroquinolones and injectable second-line drugs). The average proportion of XDR-TB cases has been estimated to be 6.2% [55]. The presence of drug-resistant TB strains underlines the necessity for fast and accurate resistance detection to allow effective treatment and restriction of transmissions.

The gold standard for diagnosing drug-resistant *M. tuberculosis* is still the phenotypic drug susceptibility testing (DST, CLSI 2011). However, DST is technically challenging, requires prolonged time due to the slow growth of the hydrophobic TB bacteria, needs expensive laboratory facilities that are not available in most high-burden countries and are not yet standardised for all anti-TB compounds [58–61]. Furthermore, strict laboratory safety precautions are required when handling TB cultures. Lastly, the shipment of samples for long distances to referral laboratories may be difficult and, when considering international shipment to quality assurance laboratories, also very expensive [57].

The emergence of drug resistance in MTB arises from single nucleotide polymorphisms (SNPs), insertions/deletions (indels) and, more rarely, large deletions in genes encoding drug targets or drug-converting enzymes [62]. Unlike other pathogenic bacteria, resistance is not due to horizontal transfer of genes via mobile genetic elements [63].

In the published literature, 286 mutations have been reported to be associated with resistance to rifampicin, isoniazid, ofloxacin/levofloxacin, moxifloxacin, amikacin, kanamycin, capreomycin, streptomycin, ethionamide/prothionamide and pyrazinamide with confidence for predicting resistance to be high, moderate or minimal [64].

Several conventional molecular assays have been used for years and were also recently endorsed by WHO [55], such as the line Probe Assays (LPA), MTBDRplus (Hain, Lifescience 2012) for detection of isoniazid and rifampicin resistance and MTBDRsl (Hain, Lifescience, 2015) for resistance to fluoroquinolones and second-line injectable agents. More recently, WHO has also approved another molecular assay, the Cepheid Xpert MTB/RIF test (Cepheid, USA), which simultaneously discriminates MTBC from other acid-fast bacteria and detects resistance to rifampicin [55]. Even though these assays are rapid and convenient, they can detect only limited numbers of loci that mediate drug resistance.

These limitations of the conventional genotypic assays do not exist with WGS that can be applied on either cultured TB isolates or directly on the clinical specimen and has the potential to identify resistance to any drug in a single assay. In more detail, a single WGS procedure can incorporate all loci, detect all types of mutations and distinguish changes resulting in resistance from silent mutations. Another advantage of WGS is that sequencing data may be stored and revisited in the future when new resistance loci arise in the literature. Finally, the implementation of WGS reduced handling and shipping of highly infectious, drug-resistant MTB isolates.

WGS has been used in many studies to explore resistance patterns against anti-TB drugs [65–69], investigate dynamics of transmission [70–72] and outbreaks [73, 74], exhibiting extremely high discriminatory power. Furthermore, WGS has been used to elucidate the genetic basis of drug resistance in MTB, such as novel mutations and insertions/deletions related to resistance and also mutations compensating for fitness cost [75–77].

Few data exist on the performance of WGS for the detection of drug resistance derived from prospective studies. In a multi-national surveillance study of Zingol et al. [78], sequencing applied on MTB isolates from 7094 patients, in comparison with DST, exhibited sensitivity for resistance detection against rifampicin, isoniazid, ofloxacin, moxifloxacin, pyrazinamide, kanamycin, amikacin and capreomycin of 91%, 86%, 85%, 88%, 54%, 79%, 90% and 81%, respectively. In another prospective study, conducted in the UK on 777 MTB isolates, WGS compared to DST for resistance detection against isoniazid, rifampicin, ethambutol and pyrazinamide showed a sensitivity of 93.1%, 100%, 100% and 81.8%, respectively [79] and high specificity, >98.5% for all four drugs.

Coll et al. [80] performed WGS on 792 isolates from six countries, using the phenotypic tests as reference and found sensitivity/specificity values in predicting resistance against the first and second-line anti-TB drugs of 96.2/98.1% for rifampicin; 92.8/100% for isoniazid; 88.7/81.7% for ethambutol; 87.1/89.7% for streptomycin; 70.9/93.9% for pyrazinamide; 85.5/94.9% for ofloxacin; 82.9/98.3% for amikacin.

In a similar study performed by Walker et al. [65] on 2099 isolates from different countries, sensitivities/specificities were: 91.7/99.2% for rifampicin; 85.2/ 98.4% for isoniazid; 82.3/95.1% for ethambutol; 81.6/99.1% for streptomycin; 24.0/99.9% for pyrazinamide; 45.5/100% for ofloxacin; 88.1/99.5% for amikacin. Chattergie et al. [81] analysed 74 isolates from Mumbai, India: resistance to rifampicin and isoniazid was predicted with sensitivity 100% and specificity 94%; to ethambutol resistance prediction sensitivity was 100% and specificity 78%; to streptomycin, sensitivity was 85% and specificity 100%. Finally, Feliciano et al. [61], tested a small collection of 30 isolates from patients in Brazil and Mozambique that harboured MDR-TB: respective sensitivity/specificity were for rifampicin, 87.5/92.3%; isoniazid, 95.6/100%; streptomycin, 85.7/93.3%; ethambutol, 100/77.2%.

Culturing before DNA extraction for WGS requires prolonged time when testing slow-growing bacteria, such as *M. tuberculosis*. For the sake of gaining time, Bjorn-Mortensen et al. (2015) tested a protocol of extracting DNA for WGS directly from frozen glycerol stocks; libraries and sequencing results were comparable with those derived using revived bacteria after subculturing the glycerol stocks.

Votintseva et al. [82] applied a low-cost DNA extraction in 1 mL of an early positive mycobacterial growth indicator tube (MGIT) (median culture age, 4 days): WGS results could be available 3 days after a MGIT culture flagged positive; 98% of the samples were correctly identified as MTB and successfully mapped to the H37Rv reference MTB genome with sequence coverage >90%.

Brown et al. [83] designed a method using biotinylated RNA baits specific for *M. tuberculosis* DNA, in order to capture by whole-genome metagenomics full bacterial genomes directly from infected sputum specimens without culture: *M. tuberculosis* whole genome metagenomics data were successfully extracted directly from all 24 smear- and culture-positive sputum samples; 20 of which were of high quality (>20X depth and > 90% of the covered genome). Comparing whole-genome metagenomics results with those of conventional molecular assays and cultures, high levels of concordance were observed. The process could be completed, even from low-grade smear-positive samples, within 96 hours. Disadvantages of sequencing were its high cost (approximately $350/sample), the skills needed and the instrumentation that is currently not available in most laboratories.

An important challenge of TB diagnosis, therapy and control is also represented by infections caused simultaneously by different strains [84]. This can be due to >1 distinct strains with different genomes (a single transmission event) or to super-infection with >1 clonal variants (multiple transmissions) during a single episode of disease [85]. Preforming WGS results in faster detection and has the highest sensitivity and discriminatory power for the detection of mixed infections, which otherwise cannot be discriminated by other assays.

The depth of coverage (number of reads covering individual nucleotides) indicates the quality of the sequence data: the high number of reads assures the correct nucleotide call. No consensus exists on the number of reads needed for the analysis of MTB resistance mutations. High coverage may be needed for samples that contain >1 MTB strain (mixed infections) or when resistant and susceptible bacilli simultaneously exist in the same sample (heteroresistance).

Another possible barrier for implementing WGS methods in routine settings could be the lack of expertise in bioinformatics among the clinical microbiology personnel: data generated are highly complex, necessitating user-friendly comprehensive and validated workflows [86]. The most common bioinformatics platforms are PhyResSE [87], http://www.phyresse.org/, TB Profiler [80], https://github.com/jodyphelan/TBProfiler, Mykrobe predictor [31], http://www.mykrobe.com/products/predictor/ and TGS-TB [88] https://gph.niid.go.jp/tgs-tb/. These platforms are freely available, user-friendly, can run in a common PC and accept raw data derived directly from the WGS instrument (FASTQ files). While the few available studies exhibited high sensitivity and specificity using such platforms for isoniazid and rifampicin resistance, a substantial variation seems to exist with the remaining first- and second-line drugs.

In particular, Macedo et al. [86] simultaneously evaluated the performance of the four most common free online WGS-based platforms (PhyResSE, Mykrobe Prediktor, TB profiler, TGS-TB,) to predict resistance against first- and second-line anti-TB drugs, using a collection of MDR-TB bacterial strains. Overall, the sensitivity of resistance prediction ranged from 84.2% (Mykrobe predictor) to 95.2% (TB profiler) and specificity was higher and homogeneous (varied from 94.0% using TGS-TB to 100.0% using Mykrobe predictor; [86]). TB profiler and TGS-TB were shown to be the best-ranked platforms for resistance prediction against almost all anti-TB drugs, with sensitivity/specificity >90%, being highly promising for implementation in the routine practice of clinical microbiology laboratories.

In another study, van Beek et al. [89] tested 211 *M. tuberculosis* isolates for first-line drugs resistance both by phenotypic DST and WGS. The authors analysed the results of WGS using five software platforms (KvarQ, Mykrobe Prediktor, PhyResSE, TB profiler, and TGS-TB). The time needed and costs of reagents were compared for both approaches. The sensitivity of the five softwares for the prediction of any resistance among *M. tuberculosis* strains was 74–80% and the specificity >95% for all platforms, except for TGS-TB, that had lower specificity (81.6%; [89]).

More recent studies underline the need to standardise the databases for the interpretation of genotype-phenotype correlations based on clinical grounds [66, 90]. In a large collaborative project established by academic institutions, public health agencies and non-governmental organisations, the world-wide consortium "CRyPTIC" (Comprehensive Resistance Prediction for Tuberculosis, www.crypticproject.org/) was developed. This is a global collaboration of TB research institutions aiming to achieve improved, faster and targeted treatment of MDR-TB infections by applying genetic resistance prediction.

On the contrary, three main limitations hamper the utility of genotypic AST in comparison with phenotypic assays [44]:

(a) Systematic errors due to low limit of detection by WGS;
(b) systematic errors caused by poorly established breakpoints for phenotypic DST used as a reference method for the validation of WGS-based AST;
(c) poor understanding of the genotypic basis of phenotypic resistance.

Another challenge for clinical laboratories is to correctly identify mycobacterial species other than MTB complex and *M. leprae* (non-tuberculous mycobacteria, NTM). There are >180 NTM species published and available online http://www.bacterio.net/mycobacterium.html, while novel species are continuously described [91]. NTM represent emerging pathogens that infect both immunocompromised and competent patients. Human NTM disease is classified into the following clinical syndromes: chronic pulmonary disease, lymphadenitis, cutaneous disease and disseminated disease. Of these, chronic pulmonary disease is clinically the most common entity [92]. Given the increasing incidence of NTM infections [93], the correct identification of clinical NTM isolates is crucial because the clinical relevance of NTM species is different [94].

The more traditional phenotypic and biochemical assays and high-performance liquid chromatography analysis of mycolic acids, applied for the identification of mycobacteria are laborious, time-consuming, need experienced laboratory personnel and are thus being replaced by molecular methods. Such assays include line probe hybridisation, PCR-RFLP analysis, real-Time PCR, DNA sequencing, and MALDI-TOF [95].

Currently, the identification of NTM relies mostly on commercial DNA probe systems that provide excellent identification but only for few species. Strains that are either identified only at the *Mycobacterium* genus level or identified incorrectly due to cross-reactions of particular probes have to be identified accurately by genetic sequencing [96].

The gene encoding 16S *rRNA*, commonly allows unambiguous identification of the vast majority of species and exhibits, in contrast with other genetic targets, only a small limited of micro heterogeneity [96].

For the correct identification of NTM bacilli, sequencing of the 16S *RNA* gene represented the reference method [94]. However, single-target sequencing cannot accurately differentiate all species; for higher degrees of discrimination, to the subspecies level, sequencing several genes such as the 16S–23S *rRNA* internal transcribed spacer (ITS) region, the heat shock protein of 65 KD (*hsp*65) and the beta RNA polymerase subunit (*rpoB*) may be necessary [96, 97].

The current limitations of the genotypic tests and single-target sequencing can be easily overcome by WGS of the cultured bacterium. Several studies from the UK, USA and Italy showed that WGS might be more rapid and cost-effective for identification of mycobacteria than the more traditional assays [67, 79, 98–101]. In a prospective study performed by Quan et al. [79], 96% of 1902 mycobacteria tested by WGS were correctly identified at the species level.

WGS has also been used to discover and describe new mycobacterial species, as the role of the traditional DNA/DNA hybridisation (DDH) is controversial (although still considered the gold standard to define whether two closely related strains belong to the same species or not; [96]. A robust WGS method for estimating genomic relatedness, the average nucleotide identity (ANI), was very recently acknowledged by the Committee of Systematic Bacteriology and is now available. In this approach, two strains definitely represent different species when the ANI between their genomes is <95–96% (Kim et al. 2014). The availability of WGS, which is prerequisite for ANI calculation, is gradually becoming available in many countries [96].

Lastly, WGS can be reliably performed on *M. leprae* DNA extracted from biopsies, in order to discriminate cases of relapse from and reinfections, being powerful for evaluating outcomes of different therapeutic schemes and following disease transmission [102].

3.5 Conclusions

WGS allows simultaneous bacterial identification, AST and virulence markers detection. This provides us with a complete and in-depth picture of pathogens that no other method is producing and permits associations and suggestions that were never feasible before. Although currently only a small number of clinical microbiology laboratories are ready to incorporate WGS for bacterial identification and AST in their routine workflow, it is evident that the advantages offered in particular cases, such as fastidious and slow-growing pathogen identification and AST, will lead more and more laboratories to adopt it. Furthermore, in the near future, new clinically relevant species may be discovered, new resistance genes and their role in resistance and clinical effectiveness of antibacterial will be determined and genetic features that allow particular bacterial clones to spread and persist will be deeply studied, making WGS a "must-have" technology for the updating clinical laboratory.

References

1. Rossen JWA, Friedrich AW, Moran-Gilad J (2018) Practical issues in implementing whole-genome-sequencing in routine diagnostic microbiology. Clin Microbiol Infect 24:355–360
2. Judge K, Harris SR, Reuter S, Parkhill J, Peacock SJ (2015) Early insights into the potential of the Oxford Nanopore MinION for the detection of antimicrobial resistance genes. J Antimicrob Chemother 70:2775–2778
3. Reuter S, Ellington MJ, Cartwright EJP et al (2014) Europe PMC Funders Group. Rapid bacterial whole-genome sequencing to enhance diagnostic and public health microbiology. JAMA Intern Med 173:1397–1404
4. Köser CU, Fraser LJ, Ioannou A et al (2014) Rapid single-colony whole-genome sequencing of bacterial pathogens. J Antimicrob Chemother 69:1275–1281
5. Cummings CA, Bormann Chung CA, Fang R et al (2010) Accurate, rapid and high-throughput detection of strain-specific polymorphisms in Bacillus anthracis and Yersinia pestis by next-generation sequencing. Investig Genet 1:1–14
6. Bertelli C, Greub G (2013) Rapid bacterial genome sequencing: methods and applications in clinical microbiology. Clin Microbiol Infect 19:803–813
7. Hasman H, Saputra D, Sicheritz-Ponten T, Lund O, Svendsen CA, Frimodt-Moller N, Aarestrup FM (2014) Rapid whole-genome sequencing for detection and characterization of microorganisms directly from clinical samples. J Clin Microbiol 52:139–146
8. Adey A, Morrison HG, Asan et al (2010) Rapid, low-input, low-bias construction of shotgun fragment libraries by high-density in vitro transposition. Genome Biol 11:R119
9. Besser J, Carleton HA, Gerner-Smidt P, Lindsey RL, Trees E (2018) Next-generation sequencing technologies and their application to the study and control of bacterial infections. Clin Microbiol Infect 24:335–341
10. Gargis AS, Kalman L, Berry MW et al (2013) Assuring the quality of next-generation sequencing in clinical laboratory practice. Nat Biotechnol 30(11):1033–1036. https://doi.org/10.1038/nbt.2403.Assuring
11. Roy S, Coldren C, Karunamurthy A et al (2018) Standards and guidelines for validating next-generation sequencing bioinformatics pipelines: a joint recommendation of the Association for Molecular Pathology and the College of American Pathologists. J Mol Diagn 20:4–27

12. Gargis AS, Kalman L, Bick DP et al (2015) Good laboratory practice for clinical next-generation sequencing informatics pipelines. Nat Biotechnol 33:689–693
13. Lefterova MI, Suarez CJ, Banaei N, Pinsky BA (2015) Next-generation sequencing for infectious disease diagnosis and management: a report of the Association for Molecular Pathology. J Mol Diagn 17:623–634
14. Pont-Kingdon G, Gedge F, Wooderchak-Donahue W, Schrijver I, Weck KE, Kant JA, Oglesbee D, Bayrak-Toydemir P, Lyon E (2012) Design and analytical validation of clinical DNA sequencing assays. Arch Pathol Lab Med 136:41–46
15. Gargis A, Kalman L, Lubin IM (2016) Assuring the quality of next-generation sequencing in clinical laboratory practice. J Clin Microbiol 54:2857–2865
16. Rehm HL, Bale SJ, Bayrak-toydemir P, Jonathan S, Brown KK, Deignan JL, Friez MJ, Birgit H (2013) ACMG clinical laboratory standards for next-generation sequencing. Genet Med 15:733–747
17. Moran-Gilad J, Sintchenko V, Pedersen SK, Wolfgang WJ, Pettengill J, Strain E, Hendriksen RS (2015) Proficiency testing for bacterial whole genome sequencing: an end-user survey of current capabilities, requirements and priorities. BMC Infect Dis 15:1–10
18. Long SW, Williams D, Valson C, Cantu CC, Cernoch P, Musser JM, Olsen RJ (2013) A genomic day in the life of a clinical microbiology laboratory. J Clin Microbiol 51:1272–1277
19. Anson LW, Chau K, Sanderson N et al (2018) DNA extraction from primary liquid blood cultures for bloodstream infection diagnosis using whole genome sequencing. J Med Microbiol 67:347–357
20. Clark SA, Doyle R, Lucidarme J, Borrow R, Breuer J (2018) Targeted DNA enrichment and whole genome sequencing of Neisseria meningitidis directly from clinical specimens. Int J Med Microbiol 308:256–262
21. Wilson MR, Naccache SN, Samayoa E et al (2014) Actionable diagnosis of neuroleptospirosis by next-generation sequencing. N Engl J Med 370:2408–2417
22. Seth-Smith HMB, Harris SR, Skilton RJ et al (2013) Whole-genome sequences of chlamydia trachomatis directly from clinical samples without culture. Genome Res 23:855–866
23. Christiansen MT, Brown AC, Kundu S et al (2014) Whole-genome enrichment and sequencing of chlamydia trachomatis directly from clinical samples. BMC Infect Dis 14:1–11
24. Andersson P, Klein M, Lilliebridge RA, Giffard PM (2013) Sequences of multiple bacterial genomes and a chlamydia trachomatis genotype from direct sequencing of DNA derived from a vaginal swab diagnostic specimen. Clin Microbiol Infect 19:E405–E408
25. Joseph SJ, Li B, Ghonasgi T, Haase CP, Qin ZS, Dean D, Read TD (2014) Direct amplification, sequencing and profiling of chlamydia trachomatis strains in single and mixed infection clinical samples. PLoS One. https://doi.org/10.1371/journal.pone.0099290
26. Wołkowicz T (2017) The utility and perspectives of NGS-based methods in BSL-3 and BSL-4 laboratory – sequencing and analysis strategies. Brief Funct Genomics 17(6):471–476.
27. Mitchell SL, Mattei LM, Alby K (2017) Whole genome characterization of a naturally occurring vancomycin-dependent Enterococcus faecium from a patient with bacteremia. Infect Genet Evol 52:96–99
28. Stoesser N, Batty EM, Eyre DW, Morgan M, Wyllie DH, Del Ojo EC, Johnson JR, Walker AS, Peto TEA, Crook DW (2013) Predicting antimicrobial susceptibilities for Escherichia coli and Klebsiella pneumoniae isolates using whole genomic sequence data. J Antimicrob Chemother 68:2234–2244
29. Sadouki Z, Day MR, Doumith M, Chattaway MA, Dallman TJ, Hopkins KL, Elson R, Woodford N, Godbole G, Jenkins C (2017) Comparison of phenotypic and WGS-derived antimicrobial resistance profiles of Shigella sonnei isolated from cases of diarrhoeal disease in England and Wales, 2015. J Antimicrob Chemother 72:2496–2502
30. Shelburne SA, Kim J, Munita JM et al (2017) Whole-genome sequencing accurately identifies resistance to extended-spectrum β-lactams for major gram-negative bacterial pathogens. Clin Infect Dis 65:738–745

31. Bradley P, Gordon NC, Walker TM et al (2015) Rapid antibiotic-resistance predictions from genome sequence data for Staphylococcus aureus and Mycobacterium tuberculosis. Nat Commun. https://doi.org/10.1038/ncomms10063
32. Jaillard M, van Belkum A, Cady KC et al (2017) Correlation between phenotypic antibiotic susceptibility and the resistome in Pseudomonas aeruginosa. Int J Antimicrob Agents 50:210–218
33. Zankari E, Hasman H, Kaas RS, Seyfarth AM, Agersø Y, Lund O, Larsen MV, Aarestrup FM (2013) Genotyping using whole-genome sequencing is a realistic alternative to surveillance based on phenotypic antimicrobial susceptibility testing. J Antimicrob Chemother 68:771–777
34. Zankari E, Hasman H, Cosentino S, Vestergaard M, Rasmussen S, Lund O, Aarestrup FM, Larsen MV (2012) Identification of acquired antimicrobial resistance genes. J Antimicrob Chemother 67:2640–2644
35. Neuert S, Nair S, Day MR et al (2018) Prediction of phenotypic antimicrobial resistance profiles from whole genome sequences of non-typhoidal Salmonella enterica. Front Microbiol 9:1–11
36. Gordon NC, Price JR, Cole K et al (2014) Prediction of staphylococcus aureus antimicrobial resistance by whole-genome sequencing. J Clin Microbiol 52:1182–1191
37. Tyson GH, McDermott PF, Li C et al (2015) WGS accurately predicts antimicrobial resistance in Escherichia coli. J Antimicrob Chemother 70:2763–2769
38. Mcdermott PF, Tyson GH, Kabera C, Chen Y, Li C, Folster JP, Ayers SL, Lam C, Tate HP (2016) Whole-genome sequencing for detecting antimicrobial resistance in. Antimicrob Agents Chemother 60:5515–5520
39. Holden MTG, Hsu L, Kurt K et al (2013) A genomic portrait of the emerfences, evolution, and global spread of methicillin-resistant Staphylococcus aureus. Genome Res 23:653–664
40. Eyre DW, Golubchik T, Gordon NC et al (2012) A pilot study of rapid benchtop sequencing of Staphylococcus aureus and Clostridium difficile for outbreak detection and surveillance. BMJ Open 2:1–9
41. Hazen TH, Zhao L, Boutin MA, Stancil A, Robinson G, Harris AD, Rasko DA, Johnson JK (2014) Comparative genomics of an IncA/C multidrug resistance plasmid from Escherichia coli and Klebsiella isolates from intensive care unit patients and the utility of whole-genome sequencing in health care settings. Antimicrob Agents Chemother 58:4814–4825
42. Kos VN, Déraspe M, McLaughlin RE, Whiteaker JD, Roy PH, Alm RA, Corbeil J, Gardner H (2015) The resistome of Pseudomonas seudomonas aeruginosa in relationship to phenotypic susceptibility. Antimicrob Agents Chemother 59:427–436
43. Luo Y, Luo R, Ding H, Ren X, Luo H, Zhang Y, Ye L, Cui S (2017) Characterization of carbapenem-resistant *Escherichia coli* isolates through the whole-genome sequencing analysis. Microb Drug Resist 24(2). https://doi.org/10.1089/mdr.2017.0079
44. Ellington MJ, Ekelund O, Aarestrup FM et al (2017) The role of whole genome sequencing in antimicrobial susceptibility testing of bacteria: report from the EUCAST subcommittee. Clin Microbiol Infect 23:2–22
45. Xavier BB, Das AJ, Cochrane G, De Ganck S, Kumar-Singh S, Aarestrup FM, Goossens H, Malhotra-Kumar S (2016) Consolidating and exploring antibiotic resistance gene data resources. J Clin Microbiol 54:851–859
46. McArthur AG, Tsang KK (2017) Antimicrobial resistance surveillance in the genomic age. Ann N Y Acad Sci 1388:78–91
47. Rowe W, Baker KS, Verner-Jeffreys D, Baker-Austin C, Ryan JJ, Maskell D, Pearce G (2015) Search engine for antimicrobial resistance: a cloud compatible pipeline and web interface for rapidly detecting antimicrobial resistance genes directly from sequence data. PLoS One. https://doi.org/10.1371/journal.pone.0133492
48. Gupta SK, Padmanabhan BR, Diene SM, Lopez-Rojas R, Kempf M, Landraud L, Rolain JM (2014) ARG-annot, a new bioinformatic tool to discover antibiotic resistance genes in bacterial genomes. Antimicrob Agents Chemother 58:212–220

49. de Man TJB, Limbago BM (2016) SSTAR, a stand-alone easy-to-use antimicrobial resistance gene predictor. mSphere 1:1–10
50. Davis JJ, Boisvert S, Brettin T et al (2016) Antimicrobial resistance prediction in PATRIC and RAST. Sci Rep 6:1–12
51. Brittnacher MJ, Heltshe SL, Hayden HS, Radey MC, Weiss EJ, Damman CJ, Zisman TL, Suskind DL, Miller SI (2016) GUTSS: an alignment-free sequence comparison method for use in human intestinal microbiome and fecal microbiota transplantation analysis. PLoS One 11:1–16
52. Jia B, Raphenya AR, Alcock B et al (2017) CARD 2017: expansion and model-centric curation of the comprehensive antibiotic resistance database. Nucleic Acids Res 45:D566–D573
53. Lakin SM, Dean C, Noyes NR et al (2017) MEGARes: an antimicrobial resistance database for high throughput sequencing. Nucleic Acids Res 45:D574–D580
54. Galagan JE (2014) Genomic insights into tuberculosis. Nat Rev Genet 15:307–320
55. WHO (2017) Global tunerculosis report 2017. WHO, Geneva
56. Lange C, Chesov D, Heyckendorf J, Leung CC, Udwadia Z, Dheda K (2018) Drug-resistant tuberculosis: an update on disease burden, diagnosis and treatment. Respirology. https://doi.org/10.1111/resp.13304
57. McNerney R, Zignol M, Clark TG (2018) Use of whole genome sequencing in surveillance of drug resistant tuberculosis. Expert Rev Anti-Infect Ther 16:433–442
58. Piersimoni C, Olivieri A, Benacchio L, Scarparo C (2006) MINIREVIEW current perspectives on drug susceptibility testing of Mycobacterium tuberculosis complex : the automated nonradiometric systems. Society 44:20–28
59. Lange C, Abubakar I, Alffenaar JWC et al (2014) Management of patients with multidrugresistant/extensively drug-resistant tuberculosis in Europe: a TBNET consensus statement. Eur Respir J 44:23–63
60. Domínguez J, Boettger EC, Cirillo D et al (2016) Clinical implications of molecular drug resistance testing for Mycobacterium tuberculosis: a TBNET/RESIST-TB consensus statement. Int J Tuberc Lung Dis 20:24–42
61. Feliciano CS, Namburete EI, Rodrigues Plaça J, Peronni K, Dippenaar A, Warren RM, Silva WA, Bollela VR (2018) Accuracy of whole genome sequencing versus phenotypic (MGIT) and commercial molecular tests for detection of drug-resistant Mycobacterium tuberculosis isolated from patients in Brazil and Mozambique. Tuberculosis 110:59–67
62. Hameed HMA, Islam MM, Chhotaray C et al (2018) Molecular targets related drug resistance mechanisms in MDR-, XDR-, and TDR-*Mycobacterium tuberculosis* strains. Front Cell Infect Microbiol 8:114
63. Gillespie S (2002) Evolution of drug resistance in Mycobacterium tuberculosis: clinical and molecular perspective. Antimicrob Agents Chemother 46:267–274
64. Miotto P, Tessema B, Tagliani E et al (2017) A standardised method for interpreting the association between mutations and phenotypic drug resistance in Mycobacterium tuberculosis. Eur Respir J 50(6):1701354. https://doi.org/10.1183/13993003.01354-2017
65. Walker TM, Kohl TA, Omar SV et al (2015) Whole-genome sequencing for prediction of Mycobacterium tuberculosis drug susceptibility and resistance: a retrospective cohort study. Lancet Infect Dis 15:1193–1202
66. Witney AA, Gould KA, Arnold A et al (2015) Clinical application of whole-genome sequencing to inform treatment for multidrug-resistant tuberculosis cases. J Clin Microbiol 53:1473–1483
67. Pankhurst LJ, del Ojo EC, Votintseva AA et al (2016) Rapid, comprehensive, and affordable mycobacterial diagnosis with whole-genome sequencing: a prospective study. Lancet Respir Med 4:49–58
68. Manson AL, Cohen KA, Abeel T, Desjardins CA, Cho N, Gabrielian A, Gomez J, Jodals AM, Joloba M (2017) HHS Public Access 49:395–402
69. Coll F, Phelan J, Hill-Cawthorne GA et al (2018) Genome-wide analysis of multi- and extensively drug-resistant Mycobacterium tuberculosis. Nat Genet 50:307–316

70. Bryant JM, Schürch AC, van Deutekom H et al (2013) Inferring patient to patient transmission of Mycobacterium tuberculosis from whole genome sequencing data. BMC Infect Dis 13:1–12
71. Glynn JR, Guerra-Assunção JA, Houben RMGJ et al (2015) Whole genome sequencing shows a low proportion of tuberculosis disease is attributable to known close contacts in rural Malawi. PLoS One 10:1–12
72. Guthrie JL, Delli Pizzi A, Roth D, Kong C, Jorgensen D, Rodrigues M, Tang P, Cook VJ, Johnston J, Gardy JL (2018) Genotyping and whole genome sequencing to identify tuberculosis transmission to Pediatric patients in British Columbia, Canada, 2005–2014. J Infect Dis. https://doi.org/10.1093/infdis/jiy278
73. Holden KL, Bradley CW, Curran ET, Pollard C, Smith G, Holden E, Glynn P, Garvey MI (2018) Unmasking leading to a healthcare worker Mycobacterium tuberculosis transmission. J Hosp Infect:1–7
74. Gardy JL, Johnston JC, Sui SJH et al (2011) Whole-genome sequencing and social-network analysis of a tuberculosis outbreak. N Engl J Med 364:730–739
75. Casali N, Nikolayevskyy V, Balabanova Y et al (2014) Europe PMC Funders Group. Evolution and transmission of drug resistant tuberculosis in a Russian population. Nat Genet 46:279–286
76. Zeng X, Kwok JS-L, Yang KY, Leung KS-S, Shi M, Yang Z, Yam W-C, Tsui SK-W (2018) Whole genome sequencing data of 1110 Mycobacterium tuberculosis isolates identifies insertions and deletions associated with drug resistance. BMC Genomics 19:365
77. Zhang H, Li D, Zhao L et al (2013) Genome sequencing of 161 Mycobacterium tuberculosis isolates from China identifies genes and intergenic regions associated with drug resistance. Nat Genet 45:1255–1260
78. Zignol M, Cabibbe AM, Dean AS et al (2018) Genetic sequencing for surveillance of drug resistance in tuberculosis in highly endemic countries: a multi-country population-based surveillance study. Lancet Infect Dis 18:675–683
79. Quan TP, Bawa Z, Foster D et al (2017) Evaluation of whole genome sequencing for mycobacterial species identification and drug susceptibility testing in a clinical setting: a large-scale prospective assessment of performance against line-probe assays and phenotyping. J Clin Microbiol 56:JCM.01480-17
80. Coll F, McNerney R, Preston MD et al (2015) Rapid determination of anti-tuberculosis drug resistance from whole-genome sequences. Genome Med 7:1–10
81. Chatterjee A, Nilgiriwala K, Saranath D, Rodrigues C, Mistry N (2017) Whole genome sequencing of clinical strains of Mycobacterium tuberculosis from Mumbai, India: a potential tool for determining drug-resistance and strain lineage. Tuberculosis 107:63–72
82. Votintseva AA, Pankhurst LJ, Anson LW et al (2015) Mycobacterial DNA extraction for whole-genome sequencing from early positive liquid (MGIT) cultures. J Clin Microbiol 53:1137–1143
83. Brown AC, Bryant JM, Einer-Jensen K et al (2015) Rapid whole-genome sequencing of mycobacterium tuberculosis isolates directly from clinical samples. J Clin Microbiol 53:2230–2237
84. Cohen T, van Helden PD, Wilson D, Colijn C, McLaughlin MM, Abubakar I, Warren RM (2012) Mixed-strain Mycobacterium tuberculosis infections and the implications for tuberculosis treatment and control. Clin Microbiol Rev 25:708–719
85. McIvor A, Koornhof H, Kana BD (2017) Relapse, re-infection and mixed infections in tuberculosis disease. Pathog Dis 75:1–16
86. Macedo R, Nunes A, Portugal I, Duarte S, Vieira L, Gomes JP (2018) Dissecting whole-genome sequencing-based online tools for predicting resistance in Mycobacterium tuberculosis: can we use them for clinical decision guidance? Tuberculosis 110:44–51
87. Feuerriegel S, Schleusener V, Beckert P, Kohl TA, Miotto P, Cirillo DM, Cabibbe AM, Niemann S, Fellenberg K (2015) PhyResSE: a web tool delineating Mycobacterium tuberculosis antibiotic resistance and lineage from whole-genome sequencing data. J Clin Microbiol 53:1908–1914

88. Sekizuka T, Yamashita A, Murase Y, Iwamoto T, Mitarai S, Kato S, Kuroda M (2015) TGS-TB: Total genotyping solution for Mycobacterium tuberculosis using short-read whole-genome sequencing. PLoS One 10:1–12
89. van Beek J, Haanperä M, Smit PW, Mentula S, Soini H (2018) Evaluation of whole genome sequencing and software tools for drug susceptibility testing of Mycobacterium tuberculosis. Clin Microbiol Infect 25(1):82–86. https://doi.org/10.1016/j.cmi.2018.03.041
90. Cirillo DM, Miotto P, Tortoli E (2017) Evolution of phenotypic and molecular drug susceptibility testing. Adv Exp Med Biol 1019:221–246. https://doi.org/10.1007/978-3-319-64371-7_12
91. Gcebe N, Rutten VPMG, Van Pittius NG, Naicker B, Michel AL (2018) Mycobacterium komaniense sp. Nov., a rapidly growing nontuberculous Mycobacterium species detected in South Africa. Int J Syst Evol Microbiol 68(5). https://doi.org/10.1099/ijsem.0.002707
92. Griffith DE, Aksamit T, Brown-Elliott BA et al (2007) An official ATS/IDSA statement: diagnosis, treatment, and prevention of nontuberculous mycobacterial diseases. Am J Respir Crit Care Med 175:367–416
93. Prevots DR, Marras TK (2015) Epidemiology of human pulmonary infection with non-tuberculous mycobacteria: a review. Clin Chest Med 36:13–34
94. van Ingen J (2015) Microbiological diagnosis of nontuberculous mycobacterial pulmonary disease. Clin Chest Med 36:43–54
95. Somoskovi A, Salfinger M (2014) Nontuberculous mycobacteria in respiratory infections: advances in diagnosis and identification. Clin Lab Med 34:271–295
96. Tortoli E (2014) Microbiological features and clinical relevance of new species of the genus Mycobacterium. Clin Microbiol Rev 27:727–752
97. Koh W-J (2017) Nontuberculous mycobacteria—overview. Microbiol Spectr 5(1). https://doi.org/10.1128/microbiolspec.tnmi7-0024-2016
98. Olaru ID, Patel H, Kranzer K, Perera N (2018) Turnaround time of whole genome sequencing for mycobacterial identification and drug susceptibility testing in routine practice. Clin Microbiol Infect 24:659.e5–659.e7
99. Shea J, Halse TA, Lapierre P, Shudt M, Kohlerschmidt D, Van Roey P, Limberger R, Taylor J, Escuyer V, Musser KA (2017) Comprehensive whole-genome sequencing and reporting of drug resistance profiles on clinical cases of Mycobacterium tuberculosis in New York. J Clin Microbiol 55:1871–1882
100. Votintseva AA, Bradley P, Pankhurst L et al (2017) Same-day diagnostic and surveillance data for tuberculosis via whole-genome sequencing of direct respiratory samples. J Clin Microbiol 55:1285–1298
101. Cabibbe AM, Trovato A, De Filippo MR et al (2018) Early View Countrywide implementation of whole genome sequencing : an opportunity to improve tuberculosis management, surveillance and contact tracing in low incidence countries. Eur Respir J. https://doi.org/10.1183/13993003.00387-2018
102. Stefani MMA, Avanzi C, Bührer-Sékula S et al (2017) Whole genome sequencing distinguishes between relapse and reinfection in recurrent leprosy cases. PLoS Negl Trop Dis 11:1–13

Chapter 4
Whole-Genome Sequencing for Bacterial Virulence Assessment

Florian Tagini, Trestan Pillonel, and Gilbert Greub

4.1 Introduction

In recent years, whole-genome sequencing (WGS) is increasingly being considered a technique that could change clinical microbiology [1, 2]. In addition to microbial typing and prediction of antibiotic susceptibility, one of the major clinical application of bacterial genomics is the detection of clinically relevant virulence factors and virulence prediction. In this chapter, we will thus explore what this technique really implies [3]. Before discussing virulence factors, the terms "virulence" and "pathogenicity" need to be defined.

For virulence, the definition used in this chapter is "the relative capacity of a micro-organism to cause damage to a host" as proposed by Casadevall & Pirofski [4]. Conversely, the pathogenicity is the general capacity of a microorganism to cause damage to a host, and depends on both the pathogen and the host-response [4]. Pathogenicity is to be opposed to commensalism, where the interaction results in no clear benefits or damages for any of the involved organism. Of notes, the damage-response framework of pathogens is not restricted to the direct effects of a micro-organism on a host [4]. For instance, immunological molecular mimicry or oncogenesis can cause damage to a host and are not directly related to the entry of

F. Tagini · T. Pillonel
Institute of Microbiology, Department of Laboratory Medicine, Lausanne University Hospital and University of Lausanne, Lausanne, Switzerland

G. Greub (✉)
Institute of Microbiology, Department of Laboratory Medicine, Lausanne University Hospital and University of Lausanne, Lausanne, Switzerland

Division of Infectious Diseases, Department of Medicine, Lausanne University Hospital, Lausanne, Switzerland
e-mail: gilbert.greub@chuv.ch

J. Moran-Gilad, Y. Yagel (eds.), *Application and Integration of Omics-powered Diagnostics in Clinical and Public Health Microbiology*,
https://doi.org/10.1007/978-3-030-62155-1_4

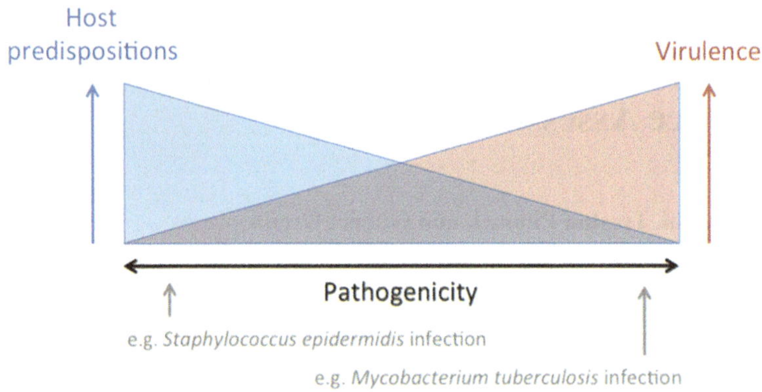

Fig. 4.1 Pathogenicity as the result of the host-pathogen interaction. In this model, highly virulent bacteria can be pathogenic regardless of the host status (e.g. *Mycobacterium tuberculosis*), while other bacteria are generally considered as less virulent than most pathogen and would be pathogenic only in specific situations (e.g. *Staphylococcus epidermidis* is pathogenic only when an intravenous catheter is in place or when patients are immunosuppressed)

a given bacterial isolate [4]. Thus, bacterial proteins implicated in such pathogenesis represent virulence factors. In this chapter, we will mainly focus on the direct damages that can be caused by a pathogen to a host and we will use a simplified model to define pathogenicity and virulence (Fig. 4.1), where a given bacteria, upon the presence and expression of virulence factors and according to host's susceptibility (e.g. immune status, epithelial breach, genetic predisposition), can be pathogenic, i.e. causing tissue lesions or organ damage.

The virulence of a bacterial strain depends on the presence and expression of virulence factors and is solely dependent on the strain characteristics. A virulence factor can be defined as a determining factor (i.e. gene product) that would help improve the survival of a bacterium within the host or by causing more cellular and tissue damage. Several classes of virulence factors should be recognised, including (1) toxins, (2) effectors of secretion systems, (3) adhesive factors, (4) invasive factors, (5) resistance to reactive oxygen and nitrogen species, (6) immune system escape, and (7) nutrient uptake. Antibiotic resistance determinants, although they may contribute to the pathogenesis (e.g. in a patient treated with antibiotics), form a special class of genes and are not discussed in this chapter.

4.1.1 Toxins

Bacterial toxin is a general term to describe a diverse set of virulence factors that are generally secreted by the bacterium and cause damage to host cells. It consists of several subcategories with different modes of actions: (i) pore-forming toxins, (ii) adenylate or guanylate cyclase-affecting toxins, (iii) protein synthesis-inhibiting toxins, (iv) surfactant-like toxins, (v) superantigens, and (vi) neurotoxins [5].

(i) Pore-forming toxins have one of the most straightforward mechanisms. Indeed, these molecules are able to form pores in host cells, which causes influx and efflux of ions, small molecules and proteins and eventually leading to cell death [6]. For instance, bacterial-mediated haemolysis, unraveled early on in the history of microbiology (19th and 20th centuries), was shown to be due to pore-forming toxins, such as the listeriolysin O of *Listeria monocytogenes*, streptolysins O and S of *Streptococcus pyogenes*, or the staphylococcal alpha- and gamma-toxin [7–12]. Another example of pore-forming toxins is the *Staphylococcus aureus* Panton-Valentine Leucocidin (PVL, LukSF) (or other leucocidins such as LukGH or LukDE), which can directly lyse human leukocytes [10].

(ii) Adenylate and guanylate cyclase–affecting toxins are a particular class of toxins, found for instance in enteropathogens such as *Escherichia coli* and *Vibrio cholerae,* or in respiratory pathogens such as *Bordetella pertussis*, the causative agent of pertussis [5]. These toxins penetrate the host cell and lead to an increased production of cyclic AMP or cyclic GMP, and eventually to the up-regulation of ion channels and to an increased volume of mucosal secretion (water follows the osmolality and is attracted into the lumen) [13–20]. Diarrhoea, emesis and cough are, respectively, the resulting symptoms, which are thought to favor the bacterial spread to other hosts. Interestingly, these toxins may have a broader spectrum of action. For instance, *B. pertussis* toxins could also inhibit phagocytosis, cytokine production and oxidative burst [5].

(iii) Protein synthesis–inhibiting toxins dramatically contribute to the pathogenicity of several bacteria, eventually leading to host cell death. For instance, the diphtheria toxin, encoded by a lysogenic bacteriophage of *Corynebacterium diphtheriae*, ADP-ribosylates the elongation factor 2 of the host cell and thus prevents protein synthesis [21]. This toxin causes local damages at the site of infection, the respiratory tract, with the formation of characteristic pseudomembranes and also systemic damages, such as heart and other end-organ injuries [22, 23]. Another classical example is the Shiga toxin producing *Escherichia coli* (STEC) and the role of Shiga toxins in the pathogenesis of haemolytic and uremic syndromes [24].

(iv) Surfactant-like toxins constitute a particular class of toxins with amphipathic properties capable to disrupt lipid bilayers of the host membrane. Best exemplified by phenol-soluble modulins of staphylococci, they have a broad spectrum of actions, such as host cell lysis, pro-inflammatory stimulation, and contribution to biofilm formation [25, 26]

(v) Superantigens, produced mainly by *S. aureus* or *Streptococcus pyogenes,* are a specific class of toxins that can bind both the lymphocytic T-cell receptors and the Major Histocompatibility Complex (MHC). Thus, it activates up to 20% of lymphocytes in a non-specific manner, ultimately leading to an inflammatory cytokine storm in the host and potentially to cardiovascular collapse due to an increased vascular permeability and death due to shock and multi-organ failure [27].

(vi) Neurotoxins, such as the botulinum or tetanus toxins produced by *Clostridium botulinum* and *Clostridium tetani*, respectively, are a separate category of toxin, able to modulate the transmissions of nerves impulses [28, 29].

4.1.2 Secretion Systems and Their Effectors

Although most toxins are secreted by various bacterial secretion systems, some specific secretion systems have an important role in secreting the so called "effectors" that are able to induce damage in the target cell, and could be considered a distinct class of virulence factors. First, the type III secretion system (T3SS) is found in several Gram-negative pathogens, such as *Salmonella* spp., *Shigella, Pseudomonas* spp. and *Yersinia* spp. [30]. In addition, it is also encoded by intracellular pathogens, such as *Chlamydia trachomatis, Waddlia chondrophila* and plays a central role in their pathogenesis [31]. T3SS assemble into a needle-like apparatus conserved across distant bacterial species. It is able to secrete its effectors into the target cell, which may affect a broad range of cellular functions, such actin cytoskeletal dynamics, gene expression and post-translational modifications, signal transduction pathways, and vesicle transport and endocytic trafficking [32]. Second, the type-four secretion system is an important virulence factor of Gram negative and Gram positive bacteria involved in various cellular processes including conjugative horizontal gene transfers and contact-independent DNA uptake, as well as secretion of toxins or effector proteins in the target cell [33]. Third, the type VI secretion system is also particularly interesting for bacterial virulence: in the context of polymicrobial infection, it helps pathogens to compete with other bacteria and to colonize a niche; for intracellular bacteria such as *Burkholderia* spp. and *Francisella tularensis*, it can also specifically mediate virulence (e.g. phagosomal escape for *F. tularensis*) [34, 35]. Finally, the type VII secretion systems (T7SS) is a key virulence factor of *M. tuberculosis* and other mycobacteria [36]. It can also be found in many *Actinobacteria* and in *Firmicutes* (with a type-VII-like secretion system) [37]. The most famous *M. tuberculosis* effectors are Esx-A (ESAT-6) and Esx-B (CFP-10). They are involved in modulation of the T-cell response, phagosomal escape and exhibit some direct effect on the membrane of host cells [38].

4.1.3 Adhesive Factors

Adherence to various surfaces (e.g. the endothelium or any prosthetic device), can be an important determinant of bacterial pathogenesis [39, 40]. A broad range of proteins or protein assemblies can promote bacterial attachment and are classically called adhesins (one protein) or pili (large protein assemblies). For instance, several pathogens associated with endocarditis, such as *Bartonella henselae*, *Eikenella corrodens* and *Cardiobacterium hominis* have been shown to encode adhesins [41–43].

Regarding pili, an example is the type IV pili of *Neisseria meningitidis* that allows attachment to the epithelium, invasion into the bloodstream, attachment to the brain microvascular endothelium and crossing of the blood-brain barrier to cause meningitis [44, 45]. Furthermore many pathogens are able to produce biofilms, which are matrices of hydrated extracellular polymeric substances, composed mainly of polysaccharides, proteins, nucleic acids and lipids formation [46]. It promotes the mechanical attachment of the micro-organisms and the large three-dimensional structure of some adhesins, reduces the susceptibility of bacteria to various stresses and to antibiotic therapies [47] as well as their engulfment by phagocytic cells.

4.1.4 Invasive Factors

Some virulence factors can help the bacteria invade tissues and promote their systemic dissemination. For instance, streptokinase and staphylokinase activate plasminogen into plasmin, which can then break down fibrin clots. These proteases are involved in tissue spreading by destroying the extracellular matrix and fibrin fibers that holds cells together [48, 49]. Many other bacterial proteases can degrade the extracellular matrix or even surprisingly the DNA nets of neutrophils and participate in bacterial invasion [50–53].

4.1.5 Resistance to Reactive Oxygen and Nitrogen Species

Many genes are involved in the resistance to stresses that bacteria encounter within the host [54]. For instance, genes involved in resistance to reactive oxygen or nitrogen species can affect bacterial survival. The *S. aureus* catalase enables the breakdown of hydrogen peroxide and thus was thought to improve the survival of bacteria to the killing by neutrophils [55]. However, it was later shown that catalase-negative *S. aureus* infection can retain virulence, highlighting the plethora of bacterial compensatory "virulence" mechanisms [56, 57].

4.1.6 Immune System Escape

To increase their survival, bacteria have developed a broad range of molecular means to subvert both the innate and adaptive immune systems of the host. First, many bacteria are able to prevent phagocytosis, the most straightforward way to clear bacteria. For instance, the production of a polysaccharide capsule (e.g. for *Streptococcus pneumoniae*) can prevent bacterial opsonisation by the complement system or by immunoglobulins [58]. Furthermore, bacteria, and particularly intracellular bacteria, have developed many different ways to escape the endosome–/

phagosome- lysosome pathway, by manipulating the host cellular pathways [59–61]. Some pathogen strategies aim at degrading host chemokines involved in the inflammatory response to attract neutrophiles, like interleukin-8, with specific proteases (e.g. SpyCEP of *S. pyogenes*) [62]. Bacterial proteases can also cleave immunoglobulins, such as IgA1 or IgG, which promote bacterial attachment to mucosal surfaces and survival, respectively [63, 64]. Conversely, bacteria can also recruit regulatory molecules. *Neisseria meningitidis* recruits factor H, which prevents the activation of the complement [65]. Lipopolysaccharide, besides being an important immune system stimulating factor (that can eventually lead to septic shock), is also known to contribute to the resistance to complement of *K. pneumoniae* [66].

4.1.7 Nutrient Uptake

The fight for nutrients is a complex interplay between the host and the pathogen. Iron, an inorganic ion, is required for many eukaryotic and bacterial processes and is involved in virulence and pathogenicity [67]. Bacteria have developed many ways to circumvent every iron-sequestration strategies of the host. For instance, bacteria can acquire iron, which is bound to transferrin, lactoferrin or hemoglobin. Furthermore, iron tightly regulates many virulence factors (e.g. the diphtheria toxin) [68]. For intracellular bacteria, the acquisition of nutrients is also critical for their survival [61, 69]. In addition to iron and other nutrients, the acquisition of cholesterol, for instance, is made after manipulation of the host cell machinery [70].

From a genomic perspective, bacteria – including pathogenic ones – generally present highly plastic genomes. We should differentiate the core-genome, consisting of conserved genes between all bacteria of a species or any other clade (core genes), from the accessory genome, which includes all the variable genetic elements of a given species or clade. Horizontal gene transfers, mediated by bacteriophages, plasmids or recombination, are important drivers of bacterial evolution and pathogen adaptation [71]. Virulence determinants can thus be either encoded in all bacterial strains of a given species or sporadically occur in some virulent strains. For instance, several *S. aureus* toxins, such as the alpha-toxin and some phenol-soluble proteins [10], are present in every *S. aureus* isolates whereas the PVL toxin is only encoded by the genome of some more virulent *S. aureus* strains. Core-genome sizes vary a lot across different bacterial species [72–74]. Good knowledge of the genomics of the pathogen is thus required for a successful identification and interpretation of virulence markers [3]. If a virulence factor is encoded by a core-gene, it generally means that the species identification itself implies the presence of that trait (if strictly present in the core-genome). However, when assessing virulence factors we should not focus only on the accessory genes since variants (such as single-nucleotide polymorphisms (SNPs), deletion or insertion events) may occur in core genes, leading potentially to loss or gain of function that may increase or decrease the virulence of a given strain.

Most of the knowledge on virulence factors has been gathered thanks to *in vitro* experiments or animal models of infection. However, the overall contribution to pathogenicity of virulence factors is often less clear in humans. Indeed, significant differences between humans and mice may impact the interpretations of animal experiments and extrapolation of obtained results to humans. Furthermore, the presence of genes encoding virulence factors is usually not sufficient by itself to increase the virulence of a bacteria. Indeed, transcriptional modulation and expression of the protein, which depend on complex regulatory networks, are truly determining the virulence end-phenotype. These regulators depend on various mechanisms, such as the activity of two-component systems [75], expression of regulatory non-coding RNAs [76] or sigma factors [77].

By providing a detailed map of the virulence factors encoded in a bacterial strain, WGS could bring new useful predicting tools for clinical microbiologists. In this chapter, we will review the main current and foreseen clinical applications where WGS has or could have an added value. In addition, we will discuss the main technical approaches and limitations of WGS as well as the validation and interpretation of the results.

4.2 Clinical Applications

For WGS analyses, requests to identify known virulence factors may be driven by various reasons. A critical assessment of the benefits for the patient or public health is thus important. Therefore, one should always wonder: will the analysis have any impact on patient's treatment or on any other management aspects (for instance, by undertaking specific infection control measures)? If the answer is negative, then the utility of the analysis is probably limited to the research field. Thus, there are two main motives for virulence determinants detection: (a) VFs detection for personalised treatment and/or clinical management, and (b) epidemiological surveillance of virulent clones.

In conventional clinical microbiology workflows, the virulence properties of bacterial isolates are rarely characterised [1]. Indeed, the identification of the species brings up usually the possibility to infer the general pathogenic potential of an isolate. For instance, the identification of *Listeria monocytogenes* in the cerebrospinal fluid of a patient proves meningitis. Based soley on the identification of the bacterium, we assume that the strain is pathogenic in that situation, and that a set of genes involved in virulence and pathogenesis is present. Knowing whether the strain encodes some accessory virulence genes is unlikely to change patient management. However, in this particular example, WGS could still be useful for typing and public health surveillance. Furthermore, as the ultimate typing method, WGS can also provide a rapid taxonomic assignment of an isolate by classifying it to a particularly virulent clade (e.g. species, subspecies or serovar). Indeed, for many pathogens, several clonal complexes have been associated to increased virulence or poorer prognosis (e.g. for *S. aureus*, *E. coli*, and many more) [78, 79]. As many virulence genes are acquired through horizontal gene transfer, typing analyses based on whole genomes data can add additional assessment of virulence factor content, allowing to monitor the spread of established virulence factors.

E. coli, is another well-illustrating example. Due to its highly plastic genome and in the context of the host-pathogen interaction, *Escherichia coli* presents variable phenotypes ranging from commensal interactions to very invasive diseases. For instance, upon the acquisition of specific virulence genes (Table 4.1), *E. coli* can present very specific pathogenic features, classified into pathotypes [80]. However, the pathotype definition seems to lose relevance with the rising number of virulence factor combinations and virulence phenotypes [80]. For instance, the recently described Shiga-toxin producing enteroaggregative *E.coli* (STEAEC) is a hybrid between STEC and EAEC [81]. In addition, the Shigella B13 carrying the locus enterocyte effacement (LEE) pathogenicity island is more closely related to *E. alberti* than to *E. coli* [82]. Therefore, there is a clear added value to detect virulence factors in order (a) to monitor and predict the emergence of new pathotype combinations, and (b) to identify horizontal transfers events of known virulence factors.

In addition to *E. coli* and *S. aureus*, the detection of virulence factors can also be recommended in some other specific cases. This concerns mostly bacterial species that are known to exhibit a high virulence variability. For instance, *C. diphtheriae* can cause the clinical disease diphtheria when encoding *tox* gene and expressing the diphtheria toxin (cf. section 1). Specific PCRs are available to detect the *tox* gene but are usually only available at national reference centers (together with toxin production assays). Toxigenic *C. diphtheriae* infections require specific isolation strategies and specific patient's follow-up to monitor potential cardiac toxicity. Ruling out the presence of the toxin may take time because of referral to a reference center for testing [83]. Local WGS analysis in this context can be more time- and cost-effective than the standard procedures [83].

Another example of extensively studied toxin is the Panton-Valentin leucocidin (PVL) of *S. aureus*. For recurrent skin and soft tissue infections, the detection of the toxin was shown to be useful. Indeed, specific decolonisation strategies may be introduced [84]. Although considered to be a marker of invasiveness, its association with pneumonia, bacteraemia and musculoskeletal infections was shown to be questionable [85]. Conversely, PVL-positive strains are associated with skin and soft tissue infections and are more likely to be treated surgically. A large set of virulence factors is encoded in *S. aureus* genomes but their identification is currently unlikely to lead to the modification of the antibiotic therapy (Table 4.1) [86]. Indeed, the choice of therapy is currently driven mainly by the antibiotic susceptibility of the strain and by clinical presentation rather than by the presence of the PVL or other virulence factors [85]. For instance, in the presence of severe necrotising pneumonia, clindamycin or linezolid, both inhibitors of the ribosomes and protein synthesis, can be introduced [87]. Interestingly, a genome-wide association study of methicillin-resistant *S. aureus* could predict isolates toxicity from their genomic sequences, looking at SNPs, insertion and deletions events [88]. However, there is a need for more genotype-outcome associations, before virulence determinants can be integrated into the clinical management of *S. aureus* infections [89].

Table 4.1 Selected virulence factors

Pathogen	Virulence determinants	Utility / potential role	Comments
Escherichia coli/ Shigella spp.	LEE PAI; *eae, bfpA*	EPEC defining region.	Part of the accessory genome; the search of these genes allows pathotyping of the isolates.
	Plasmid pINV; *ipaH* and *ipa* genes	EIEC/*Shigella* defining plasmid.	
	ST and LT toxins; *est, elt*	ETEC defining genes.	
	Shiga toxins; *stx1, stx2*	EHEC defining genes.	
	Plasmid pAA; *aggR, aatA, aaiC*	Associated with EAEC phenotype.	
	Afa/Dr. adhesins encoding genes	DAEC defining genes.	
Klebsiella pneumoniae	*rmpA/A2* (regulator of mucoid phenotype)	Association with community-acquired pyogenic liver abscesses and other hypervirulent phenotypes.	Accessory genes; some of them are present on plasmids.
	Yersiniabactin and aerobactin loci		
	Colibactin toxin locus		
	pld1 (phospholipase D family protein)		
Staphylococcus aureus	Panton-Valentine toxin; *lukSF*	Mainly useful for recurrent SSTIs; various associations with outcome.	Accessory gene (cf. text).
	Exfoliative toxin A, B and D; *eta, etb, etd*	Association with the pathogenesis of bullous impetigo and staphylococcal scaled-skin syndrome.	Although part of the accessory genome, the pathogenicity depends also on the gene expression.
	Staphylococcal enterotoxins and enterotoxin-like toxins; *se(a-e), se(g-j), se(r-t), sel(k-q), sel(u-w)*, *tst* (TSST-1)	Associated with infective endocarditis, food poisoning and toxic shock syndrome.	
	Alpha-hemolysin; *hla/hly*	Its expression was associated with the outcome of bacteraemia and its activity with ventilator-associated pneumonia.	Core gene.
	Phenol soluble modulins; *psm* genes	Expression associated with SSTI.	Core genes (usually). Their expression influence virulence.
	Fibronectin-binding protein; *fnbA* and *fnbB*	Associated with cardiac device infections.	*fnbA* is a core-gene and *fnbB* is part of the accessory genome.
	Clumping-factor A and B (fibrinogen binding, platelet activation); *clfA*, *clfB*	Association with infective endocarditis.	Accessory genes.
	Collagen adhesin; *cna*	Associated with invasive infections and infective endocarditis.	Accessory genes.

(continued)

Table 4.1 (continued)

Pathogen	Virulence determinants	Utility / potential role	Comments
	Biofilm formation; *ica* genes	Associated with invasive infections.	Accessory genes.
	Staphylokinase, chemotaxis inhibitory protein, staphylococcal complement inhibitor; *sak, chp, scn*	Associated with bacteremia (all) and with mortality (staphylokinase).	Accessory genes.
	Accessory gene regulator; *agr*	Gene polymorphisms and expression were associated with persistent bacteremia and mortality.	Core genes; involved in the regulation of many virulence genes.
Streptococcus pyogenes	Superantigens; *speA, speC, speH, speI, speJ, speL/M, ssa*, SMEZ	Associated with the pathogenesis of toxic shock syndrome (and scarlet fever).	Accessory genes.
	Cysteine protease; *speB*	Its expression was associated with tissue invasion.	Core gene.
	Streptolysin S and O; *sagA, sagB, sagC, slo*	Expression associated with invasive diseases.	Core genes.
	Antiphagocytic protein (M protein); *emm*	Required for full virulence in animal models; Utility for reverse-compatibility with *emm*-typing.	Core gene.
	Hyaluronic acid capsule synthesis; *hasA, hasB, hasC*	The production of the capsule was associated with more invasive phenotypes.	Accessory genes.
	Interleukin-8 protease; *spyCEP*	Expression associated with invasive diseases.	Core gene.
	Streptodornase D; *sda1*	Expression associated with invasive diseases.	Accessory gene.
Corynebacterium diphtheriae	Diphtheria toxin; *tox*	Clear role in the pathogenesis of diphtheria.	The absence of the gene rules out a toxigenic strain. If detected, the occurrence of a non-toxigenig *tox*-bearing strain cannot be excluded.

Data gathered and adapted from Robins-Browne et al., Tagini et al., Giulieri et al., Ye et al., Catalán-Nájera et al., Cole et al. and Turner et al. [3, 80, 89, 124–126]. EPEC, enteropathogenic *E. coli*; EIEC, enteroinvasive *E. coli*; ETEC, enterotoxigenic E. coli; EHEC, enterohaemorrhagic *E. coli*; EAEC, enteroaggregative *E. coli*; DAEC, diffusely adhering *E. coli;* LEE PAI, Locus of enterocyte effacement pathogenicity island; pINV and pAA, virulence plasmids of EIEC and EAEC, respectively; SSTI, skin and soft tissue infection

The characterisation of a strain in the presence of a severe clinical presentation could be beneficial for the patient or for the public health measures to prevent transmission that may occur. In the context of toxic shock syndrome, sequencing of the *S. aureus* or *S. pyogenes* strain can potentially bring useful information. For instance, when dealing with clustered cases of severe infection, WGS may be not only useful to investigate the genetic distance between organisms but also to characterise the hypervirulent emerging clone [72]. However, the management of toxic shock syndrome should be independent of WGS results. Indeed, clinical recognition of the syndrome, cardiovascular resuscitation, removal of the source (foreign body removal or surgery), antibiotic treatment and adjunctive therapies, such as clindamycin or intravenous immunoglobulins, should be introduced in every patient presenting these syndromes without delay [90].

Overall, when facing an outbreak or for epidemiological surveillance, WGS can be used to quickly characterise VFs encoded in the genome of epidemic clones or the emergence of a new virulent strain, which is particularly relevant from a public health or hospital hygiene perspectives. Furthermore, the characterisation of putative bioterrorist agents can be relevant for reference centers [91]. WGS-based clinical management will probably rise in the coming years.

4.3 Methods and Procedures

To implement WGS in clinical microbiology or public health laboratories, the genomic platform needs to fulfill national and international standards in laboratory medicine (for instance, the ISO 15189:2012 certification) [92]. Therefore, each part of the workflow should benefit from standard operating procedures (SOPs), each lab member from competency assessments and a strict monitoring of laboratory supplies should be done [93]. To ensure good quality results, several specific quality control checkpoints as well as the use of internal and external quality controls are required [94] (Table 4.2). External quality control programmes (proficiency testing) are currently developed for WGS in microbiology, but most of them have initially been aimed at assessing the performance of WGS in outbreak investigation, SNP calling and antibiotic resistance gene detection [95]. Furthermore, each proposed analysis should undergo validation studies, through research and development projects, to assess how the technique performs in the laboratory setting. Such accreditation has been completed by our bacterial genomics diagnostic laboratory in 2018 for typing and resistance genes prediction of any bacterial strain as well as for a few virulence factors of *S. aureus*, since analysis of "virulome" information is still in the process of routine implementation.

Table 4.2 General quality metrics and controls for an Illumina-based workflow

Category	Metrics	Comments
DNA extraction	DNA concentration	Hard criteria (e.g. >1 ng/ µl).
	DNA purity	Hard criteria (e.g. A260/280 ratio ≥ 0.7).
Library preparation	Library size concentration	Hard criteria (e.g. ≥1 nM).
	Library size distribution	Hard criteria (e.g. fragments sizes should be comprised between <0.3 kb and > 3 kb).
Sequencing	PhiX control	Hard criteria; assessment of error rates (e.g. <4.9%).
	Per cent of bases with quality score > Q30 FOR the run (%Q30)	Hard criteria; it should be defined according to the platform characteristics.
Reads QC	Number of reads passing the QC	This metric should provide sufficient theoretical coverage (number of reads * mean read length / expected size of the genome).
Assemblies	Assembly length	Soft criteria; should correspond to the expected genome size.
	Number of contigs	Soft criteria; it should correspond to expected fragmentation level.
	GC content	Soft criteria; it should correspond to the expected GC content.
	N50	Soft criteria; it is a metric of the assembly fragmentation and should be a warning sign when lower than an expected cutoff.
	L50	Soft criteria; number of contigs larger or equal than N50. It should be a warning when higher than an expected cutoff.
	Sequencing depth	Mean sequencing depth should be above a cutoff (e.g. hard cutoff of 60x; soft cutoff of 100x).
Detection of the VFs	Percentage of detected VF in a control strain	Hard criteria; the reason for the false negative should be determined before any interpretation of the run.
Detection of the variants	Number of called mutations of variant in a control strain	Hard criteria; if above a defined cutoff, the reason should be determined.

As for all WGS analysis, it is also recommended to use one control strain (e.g. for each run) that should pass each of the QC checkpoints and would be a positive control for the complete workflow [94]. In addition, external quality controls should also be performed on a regular basis. The use of a no-template control can also be added to control for de-multiplexing errors of Illumina. Hard criteria mean that if these steps do not pass the cutoff, the analysis should be repeated. When a value does not reach a soft criterion, this needs to be highlighted and should be critically interpreted. N50, size of the contig lying at the 50% of the total assembly length when contigs are ordered by sizes; L50, the smallest number of contigs to reach the N50

4.3.1 Sequencing

Whole-genome sequencing is usually performed from pure bacterial culture. Standard procedures are increasingly available for this technique [96]. The choice of sequencer should be done to match the laboratory settings, budget and desired output [92, 97]. The main technical features of available sequencers was already discussed elsewhere [98]. In recent years, the Illumina company (San Diego, USA) took a large proportion of the sequencing market share [99]. Short read technologies can answer many clinical questions, including in terms of virulence assessment. Long read technologies, such as Nanopore Sequencing (Oxford Nanopore, GB) or Pacific Biosciences (Menlo Park, USA), have the advantage of improving the genome assembly by facilitating the resolution of repeats. Solving the genome structure provides additional useful information, such as the genomic localisation of virulence factors (ex: on a plasmid or a chromosomal location), which could indicate the potential for transmissibility of virulence factors between strains. Furthermore, the Nanopore sequencing technology can acquire data in real-time, which can speed up the time-to-diagnosis; as soon as a defined number of reads has been acquired, a preliminary interpretation could be theoretically done, potentially reducing the turnaround time.

4.3.2 Software and Pipelines

Raw sequencing reads should first be quality controlled and trimmed for remaining adapters (when using an Illumina sequencer) or low-quality nucleotides (for instance, using Trimmomatic [100]). Then, three main approaches can be used (Fig. 4.2):

(i) Virulence factors can be searched in the assembled genome (e.g. assembled using SPAdes [101]). A reference database of known virulence factors is used to identify homologs in the newly sequenced genome. The detection of virulence factors can be made with any sequence similarity search tool such as BLAST (NCBI). Searches can be performed using nucleotide or amino acid sequences. The use of amino acid sequences is less stringent than nucleotide sequences and allows the detection of more distantly related homologs. The presence of repeats in bacterial genomes impairs the assembly of the complete genome from short reads. Most assemblies are split in multiple fragments (what is commonly referred as "contigs") If a gene is split over two different contigs, it might be missed by homology search tools; particularly if results are filtered based on query coverage, which is actually recommended to avoid detecting only fragments sharing high sequence identity. An example of a specific tool is Kleborate, which is dedicated to the assessment of *Klebsiella spp.* virulence factors directly from genome assemblies [102].

(ii) The second approach bypass the assembly step and search for known virulence factors in reads, reducing the probability of false negative results associated with gapped assemblies. Various tools can be used to map reads (e.g BWA, Samtools, Diamond) on a chosen reference genome or a set of virulence factors [103–105]. This allows the detection of specific genetic variants of virulence genes. Alternatively, specialised homology search tools for big datasets

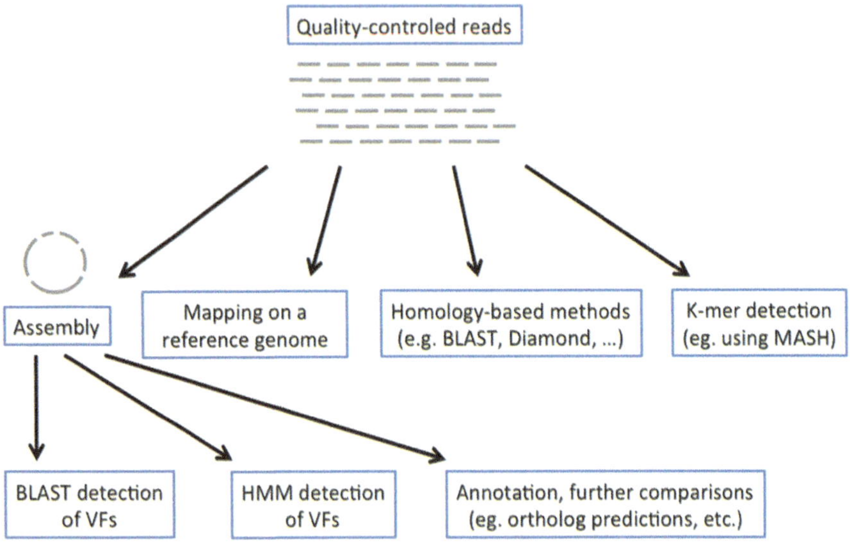

Fig. 4.2 General overview of the possible bioinformatic alternatives to detect virulence factors (VFs). Several analyses are proposed by online pipelines. This is a simplified view; approaches can be combined. BLAST analyses can also be performed after gene annotation (for instance, pipelines can BLAST predicted amino acid sequences on the reference database of virulence factors). HMM, hidden Markov model

(e.g Diamond) can be used to perform direct searches of virulence factors in reads.

(iii) Methods based on the detection of *k*-mer associated with virulence, combine both the advantage of detecting variants and genes [106]. As a reminder, *k*-mers are *in silico* fragments of *k* size of a DNA sequence (i.e. reads). The general principle of this technique is to look for exact matches, allowing to check for the presence of specific genetic variants and of any sequence of interest. Nevertheless, this approach will fail to identify virulence factor genes that diverge too much from the reference sequence. When dealing with poorly characterised pathogens, Hidden Markov Models (HMM) can be used to detect amino acid sequences sharing low sequence identity but sharing likely a similar functional domain and by extension, functions. For instance, a curated database of protein domains associated to antibiotic resistance, called ResFAM, was recently developed [107]. Similarly, dedicated curated databases for virulence domain, when created, would be useful in the same manner. For secretion systems, a tool using HMM called Macsyfinder was developed [108]. However, the use of HMM may be limited to the research field, since its principal advantage is to identify domains encoded in distant bacterial genomes.

To ensure reproducibility and to fulfill laboratory medicine standards, the use of robust and versioned bioinformatics pipeline is required. Depending on the bioinformatics workforce available at a given setting, clinical microbiologists may prefer commercial, *in-house* software or web-based pipelines. Commercial software has the advantage of being developed and validated by the company. For instance, the Ridom SeqSphere and software, which can perform a broad range of WGS analyses, can also detect a set of virulence genes and variants developed for a DNA-microarray detecting *Staphylococcus aureus* virulence and resistance genes [109]. However, commercial software tools are usually black boxes, which prevent the understanding and correct interpretation of inherent technical limits. Conversely, *in-house* software or open-source software allow continuous developments and provide a larger flexibility when the analysis needs to be adapted to a certain case, to specific set of virulence factors. As many software tools have dependencies, dedicated tools allow the creation of stable informatic environments. Furthermore, the whole pipeline can be contained in virtual machines. One example of developed pipeline is MetaGenLab pipeline (docker container available https://hub.docker.com/r/metagenlab/diag_pipelines), which is a versioned snakemake pipeline, calling software using a conda environment to perform typing, antibiotic resistance and virulence analyses (development open source version available on GitHub (https://github.com/metagenlab). Finally, online resources such as VFanalyzer and PATRIC are briefly discussed below. Concerns about online platforms include data protection of the patients and, traceability, versioning, and reproducibility of the results.

4.3.3 Databases

A good reference is necessary to make reliable identifications of virulence factors. Several databases have been designed specifically for virulence factors or contain specific sections associated to virulence. The most widely used database of virulence factors for human pathogens is the curated Virulence Factor Database (VFDB) [110, 111]. VFDB is associated with a web-resource (named VFanalyzer) to submit assembled sequences for online analysis [112]. Victor is another manually curated database of virulence factors integrating data from bacterial, viral and eukaryotic pathogens [113]. The Pathosystems Resource Integration Center (PATRIC) is a large database integrating various data, including genomics, transcriptomics, protein–protein interactions, three-dimensional protein structures and sequence typing data as well as associated metadata [114]. PATRIC integrates data from both VFDB and Victors databases as well as additional manually curated virulence factors [115]. Sequencing reads can be submitted to the PATRIC platform for analysis [116]. Finally, PHI-base is a database containing curated genes involved in host-pathogen interactions [117]. Initially focusing on plant pathogens, it also contains approximately 30% of data on bacteria of medical and environmental importance [117].

Surprisingly, there is very little overlap between the VFs indexed in those four databases (Fig. 4.3). It means that each dataset presents very specific VFs. Focusing on *S. aureus* VFs, we observed that only VFDB (as well as PATRIC since it integrates VFDB data) presented the classical and main VFs that could be expected for this pathogen (Table 4.1). For Victors and PHI-base, many of listed VFs were

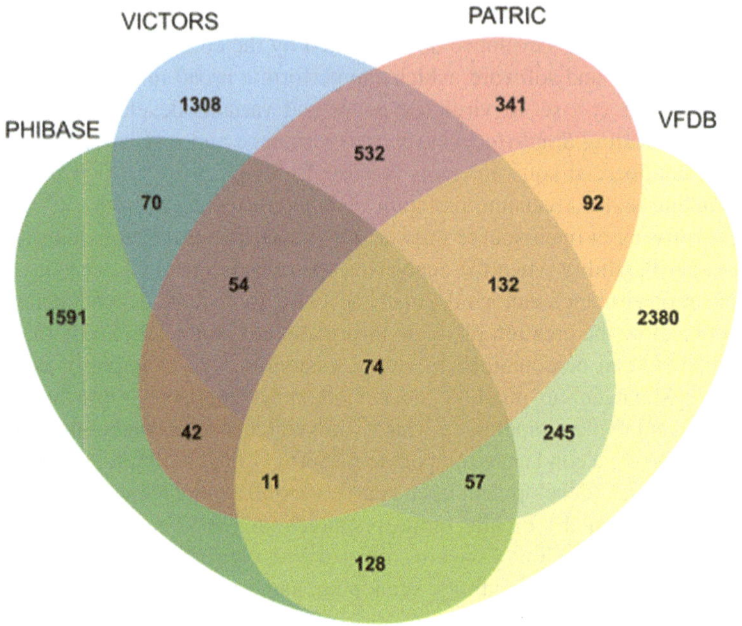

Fig. 4.3 Comparison of the content of VFDB, VICTORS, PHI-Base and PATRIC_VF (February 2019). Protein sequences were clustered at 90% of amino acid sequence identity with silix (https://bmcbioinformatics.biomedcentral.com/articles/10.1186/1471-2105-12-116). PATRIC also integrate VFDB and VICTORS VFs and those were discarded from the analysis

identified from large screens for mutants presenting attenuated in virulence as compared to wild-type strains (as reported for instance by Mei et al. [118]). As of 2019, VFDB or PATRIC seem to be the most suited database to investigate the virulence of human pathogens.

4.3.4 *Ontologies*

There is a need for standardised vocabularies (or ontologies) to properly describe virulence factors and their interactions with their host. Ontologies provide a denomination reference and should be resistant to homonymy. VFDB uses a standardised classification system and PHI-base uses standardised terms to annotate VFs, but there are currently no standards that are used by those web resources. A lot of work is still needed to setup an ontology for pathogen-host interactions and virulence factors that could be used to harmonise the information stored in curated VF databases [119].

4.3.5 Further Developments

In the previous section, we discussed methods focusing on WGS from pure bacterial cultures. However, the development of culture-independent approaches, such as shotgun metagenomics, is very promising. Indeed, the detection of specific reads associated with virulence factors in various specimen types, could be sufficient in order to take measures to improve patient management. This approach has already provided promising results for pathogen detection in the context of antibiotic treatments or for the diagnosis of fastidious bacteria [120, 121]. Going further than assessing the presence and absence of genes associated to virulence, the development of dedicated diagnostic tools and databases allowing to effectively characterise SNPs (in coding and non-coding sequences) as well as insertion / deletions events, would be a major advance in the understanding of virulence. In addition, conditional gene annotation, taking into account the presence of other genetic features (mutations or other genes), could help determine the virulence of a gene. Combination with other omics technologies, such as transcriptomics and proteomics could be the next revolution in clinical microbiology. For instance, the development of SWATH-MS for proteomics analyses showed promising results for *M. tuberculosis* infection [122]. It is also likely that particular VF could also be detected and integrated in the clinical interpretation.

WGS opens up the possibility to perform large-scale studies (correlation studies) in order to identify putative variant or genes with prognosis value [106]. Furthermore, a progressive integration of WGS data in clinical score or even genomic status of the host could be the next step needed to reach a good predictive score, always aiming for more personalised medicine. Predictive data could set for instance the indication for a dedicated treatment, follow-up or management.

4.4 Interpretation, Validation and Impact

As for any microbiological analysis, the interpretation of the results should take into account pre-analytical and analytical variables. Several quality scores should be assessed in order to validate the analysis (Table 4.2). Hard criteria (e.g. if below, the analysis should be repeated) and soft criteria may be used (e.g. if borderline with a quality metrics, results can still be interpreted depending on the rest of the metrics and on the performed analysis). Once the analysis has been validated technically, a critical interpretation must be done by the clinical microbiologist, who should be able to integrate (i) technical aspects and understand the limits of the test, and (ii) the clinical and microbiological aspects, such as the isolation site, the suspected disease, the biology of the bacterial species, etc. All these data should finally be transmitted to the clinician requesting the analysis using a standardised report. The format of these reports can be complex to design, as it is required to present in a highly summarised way the main patient's data, the main genomic findings and the interpretation. To help design such a report, back-and-forth discussions between the technical team, the clinical microbiologists and the physicians should be done [123]. As WGS data also includes typing analysis and the identification of antibiotic resistant determinant, generic reports must have the possibility to integrate all these data in a comprehensive manner.

The training in the interpretation of WGS analyses will be a challenge in the coming years for clinical microbiologists, particularly in the context of a rapidly evolving technology. Clinical microbiologists will also have to teach this to a variety of medical personal such as medical students, infectious diseases specialists, epidemiologists and any person involved in the management of patients with severe infectious disease (e.g. intensive-care specialists, …).

4.5 Future Perspectives

As of 2019, WGS appears to be a game-changing technology for clinical microbiology because it allows the broad detection of any DNA sequence, regardless of the availability of a specific diagnostic test (e.g. PCRs). As such, sequencing platforms definitely open up the possibility to detect specific virulence factors in a competitive turnaround time if the strain needs to be sent to a national reference center. Furthermore, WGS allows the epidemiological surveillance of the emergence of virulent clones, therefore possibly preventing the spread of these at early stages.

However, virulence assessment using WGS has not yet revealed its full potential. Indeed, for many situations, the technique is limited by its poor predictive value in terms of patient outcome, which depends on the expression level of virulence factors as well as on the host's susceptibility. Many developments are foreseen thanks to various methods, including the identification of new targets, their implication in clinical scores, and the combinations with other omics techniques, such as transcriptomics and proteomics, which could be the next developmental steps. On the genomic side, not only the detection and characterisation of virulence factors will be further developed, but also the detection of specific variants in regulatory mechanisms associated to increased virulence.

Conflict of Interest The authors declare no conflict of interest.

References

1. Didelot X, Bowden R, Wilson DJ, Peto TEA, Crook DW (2012) Transforming clinical microbiology with bacterial genome sequencing. Nat Rev Genet 13(9):601–612
2. Bertelli C, Greub G (2013) Rapid bacterial genome sequencing: methods and applications in clinical microbiology. Clin Microbiol Infect 19(9):803–813
3. Tagini F, Greub G (2017) Bacterial genome sequencing in clinical microbiology: a pathogen-oriented review. Eur J Clin Microbiol Infect Dis 36(11):2007–2020

4. Casadevall A, Pirofski L (2003) The damage-response framework of microbial pathogenesis. Nat Rev Microbiol 1(1):17–24
5. Rudkin JK, McLoughlin RM, Preston A, Massey RC (2017) Bacterial toxins: offensive, defensive, or something else altogether? PLoS Pathog 13(9):e1006452
6. Peraro MD, van der Goot FG (2015) Pore-forming toxins: ancient, but never really out of fashion. Nat Rev Microbiol 14:77
7. Macfarlane MG (1950) The biochemistry of bacterial toxins. 5. Variation in haemolytic activity of immunologically distinct lecithinases towards erythrocytes from different species. Biochem J 47(3):270–279
8. Henkel JS, Baldwin MR, Barbieri JT (2010) Toxins from Bacteria. EXS 100:1–29
9. Berube BJ, Bubeck Wardenburg J (2013) Staphylococcus aureus α-toxin: nearly a century of intrigue. Toxins 5(6):1140–1166
10. Otto M (2014) Staphylococcus aureus toxins. Curr Opin Microbiol 0:32–37
11. Kayal S, Charbit A (2006) Listeriolysin O: a key protein of Listeria monocytogenes with multiple functions. FEMS Microbiol Rev 30(4):514–529
12. Alouf JE (1980) Streptococcal toxins (streptolysin O, streptolysin S, erythrogenic toxin). Pharmacol Ther 11(3):661–717
13. Sharp GW, Hynie S (1971) Stimulation of intestinal adenyl cyclase by cholera toxin. Nature 229(5282):266–269
14. Chen LC, Rohde JE, Sharp GW (1971) Intestinal adenyl-cyclase activity in human cholera. Lancet Lond Engl 1(7706):939–941
15. Cohen MS, Chang P (2018) Insights into the biogenesis, function, and regulation of ADP-ribosylation. Nat Chem Biol 14(3):236–243
16. Currie MG, Fok KF, Kato J, Moore RJ, Hamra FK, Duffin KL et al (1992) Guanylin: an endogenous activator of intestinal guanylate cyclase. Proc Natl Acad Sci U S A 89(3):947–951
17. Kaper JB, Nataro JP, Mobley HL (2004) Pathogenic Escherichia coli. Nat Rev Microbiol 2(2):123–140
18. Schulz S, Green CK, Yuen PS, Garbers DL (1990) Guanylyl cyclase is a heat stable enterotoxin receptor. Cell 63(5):941–948
19. Melvin JA, Scheller EV, Miller JF, Cotter PA (2014) Bordetella pertussis pathogenesis: current and future challenges. Nat Rev Microbiol 12(4):274–288
20. Graf R, Codina J, Birnbaumer L (1992) Peptide inhibitors of ADP-ribosylation by pertussis toxin are substrates with affinities comparable to those of the trimeric GTP-binding proteins. Mol Pharmacol 42(5):760–764
21. Simon NC, Aktories K, Barbieri JT (2014) Novel bacterial ADP-ribosylating toxins: structure and function. Nat Rev Microbiol 12(9):599–611
22. Boyer NH, Weinstein L (1948) Diphtheritic myocarditis. N Engl J Med 239(24):913–919
23. Kneen R, Nguyen MD, Solomon T, Pham NG, Parry CM, Nguyen TTH et al (2004) Clinical features and predictors of diphtheritic cardiomyopathy in Vietnamese children. Clin Infect Dis Off Publ Infect Dis Soc Am 39(11):1591–1598
24. Melton-Celsa AR. Shiga Toxin (Stx) classification, structure, and function. Microbiol Spectr [Internet]. 2014 [cited 2019 Feb 5];2(2). Available from: https://www.ncbi.nlm.nih.gov/pmc/articles/PMC4270005/
25. Li S, Huang H, Rao X, Chen W, Wang Z, Hu X (2014) Phenol-soluble modulins: novel virulence-associated peptides of staphylococci. Future Microbiol 9(2):203–216
26. Cheung GYC, Joo H-S, Chatterjee SS, Otto M (2014) Phenol-soluble modulins – critical determinants of staphylococcal virulence. FEMS Microbiol Rev 38(4):698–719
27. Spaulding AR, Salgado-Pabón W, Kohler PL, Horswill AR, Leung DYM, Schlievert PM (2013) Staphylococcal and streptococcal superantigen exotoxins. Clin Microbiol Rev 26(3):422–447
28. Rossetto O, Scorzeto M, Megighian A, Montecucco C (2013) Tetanus neurotoxin. Toxicon 66:59–63
29. Rossetto O, Pirazzini M, Montecucco C (2014) Botulinum neurotoxins: genetic, structural and mechanistic insights. Nat Rev Microbiol 12(8):535–549

30. Galán JE, Collmer A (1999) Type III secretion machines: bacterial devices for protein delivery into host cells. Science 284(5418):1322–1328
31. Stephens RS, Kalman S, Lammel C, Fan J, Marathe R, Aravind L et al (1998) Genome sequence of an obligate intracellular pathogen of humans: chlamydia trachomatis. Science 282(5389):754–759
32. Deng W, Marshall NC, Rowland JL, McCoy JM, Worrall LJ, Santos AS et al (2017) Assembly, structure, function and regulation of type III secretion systems. Nat Rev Microbiol 15(6):323–337
33. Grohmann E, Christie PJ, Waksman G, Backert S (2018) Type IV secretion in gram-negative and gram-positive bacteria. Mol Microbiol 107(4):455–471
34. Russell AB, Peterson SB, Mougous JD (2014) Type VI secretion system effectors: poisons with a purpose. Nat Rev Microbiol 12(2):137–148
35. Brodmann M, Dreier RF, Broz P, Basler M (2017) Francisella requires dynamic type VI secretion system and ClpB to deliver effectors for phagosomal escape. Nat Commun 16(8):15853
36. Simeone R, Bottai D, Frigui W, Majlessi L, Brosch R (2015) ESX/type VII secretion systems of mycobacteria: Insights into evolution, pathogenicity and protection. Tuberculosis 95(Supplement 1):S150–S154
37. Bottai D, Gröschel MI, Brosch R (2017) Type VII secretion systems in gram-positive bacteria. Curr Top Microbiol Immunol 404:235–265
38. Renshaw PS, Panagiotidou P, Whelan A, Gordon SV, Hewinson RG, Williamson RA et al (2002) Conclusive evidence that the major T-cell antigens of the Mycobacterium tuberculosis complex ESAT-6 and CFP-10 form a tight, 1:1 complex and characterization of the structural properties of ESAT-6, CFP-10, and the ESAT-6*CFP-10 complex. Implications for pathogenesis and virulence. J Biol Chem 277(24):21598–21603
39. Okaro U, Addisu A, Casanas B, Anderson B (2017) Bartonella species, an emerging cause of blood-culture-negative endocarditis. Clin Microbiol Rev 30(3):709–746
40. Oliveira WF, Silva PMS, Silva RCS, Silva GMM, Machado G, Coelho LCBB et al (2018) Staphylococcus aureus and Staphylococcus epidermidis infections on implants. J Hosp Infect 98(2):111–117
41. Tagini F, Pillonel T, Asner S, Prod'hom G, Greub G (2016) Draft genome sequence of a cardiobacterium hominis strain isolated from blood cultures of a patient with infective endocarditis. Genome Announc 4(5)
42. Riess T, Raddatz G, Linke D, Schäfer A, Kempf VAJ (2007) Analysis of Bartonella adhesin a expression reveals differences between various B. henselae strains. Infect Immun 75(1):35–43
43. Yumoto H, Azakami H, Nakae H, Matsuo T, Ebisu S (1996) Cloning, sequencing and expression of an Eikenella corrodens gene encoding a component protein of the lectin-like adhesin complex. Gene 183(1–2):115–121
44. Merz AJ, Enns CA, So M (1999) Type IV pili of pathogenic Neisseriae elicit cortical plaque formation in epithelial cells. Mol Microbiol 32(6):1316–1332
45. Kolappan S, Coureuil M, Yu X, Nassif X, Egelman EH, Craig L (2016) Structure of the Neisseria meningitidis type IV pilus. Nat Commun 7:13015
46. Flemming H-C, Wingender J (2010) The biofilm matrix. Nat Rev Microbiol 8(9):623–633
47. Flemming H-C, Wingender J, Szewzyk U, Steinberg P, Rice SA, Kjelleberg S (2016) Biofilms: an emergent form of bacterial life. Nat Rev Microbiol 14(9):563–575
48. Peetermans M, Vanassche T, Liesenborghs L, Claes J, Vande Velde G, Kwiecinksi J et al (2014) Plasminogen activation by staphylokinase enhances local spreading of S. aureus in skin infections. BMC Microbiol 14:310
49. McArthur JD, Cook SM, Venturini C, Walker MJ (2012) The role of streptokinase as a virulence determinant of Streptococcus pyogenes--potential for therapeutic targeting. Curr Drug Targets 13(3):297–307
50. Elgaml A, Miyoshi S-I (2017) Regulation systems of protease and hemolysin production in Vibrio vulnificus. Microbiol Immunol 61(1):1–11

51. Hyde JA (2017) Borrelia burgdorferi keeps moving and carries on: a review of Borrelial dissemination and invasion. Front Immunol [Internet]. [cited 2019 Feb 11];8. Available from: https://www.frontiersin.org/articles/10.3389/fimmu.2017.00114/full
52. Boehm M, Simson D, Escher U, Schmidt A-M, Bereswill S, Tegtmeyer N et al (2018) Function of serine protease HtrA in the lifecycle of the foodborne pathogen campylobacter jejuni. Eur J Microbiol Immunol 8(3):70–77
53. Storisteanu DML, Pocock JM, Cowburn AS, Juss JK, Nadesalingam A, Nizet V et al (2017) Evasion of neutrophil extracellular traps by respiratory pathogens. Am J Respir Cell Mol Biol 56(4):423–431
54. Delany I, Seib KL (2012) Stress response in microbiology. Horizon Scientific Press, Requena JM. 450 p
55. Mandell GL (1975) Catalase, superoxide dismutase, and virulence of Staphylococcus aureus. In vitro and in vivo studies with emphasis on staphylococcal--leukocyte interaction. J Clin Invest 55(3):561–566
56. Messina CGM, Reeves EP, Roes J, Segal AW (2002) Catalase negative Staphylococcus aureus retain virulence in mouse model of chronic granulomatous disease. FEBS Lett 518(1–3):107–110
57. Berenger B, Chen J, Bernier A-M, Bernard K (2016) Draft whole-genome sequence of a catalase-negative Staphylococcus aureus subsp. aureus (sequence type 25) Strain isolated from a patient with endocarditis and septic arthritis. Genome Announc 4(6)
58. Hyams C, Camberlein E, Cohen JM, Bax K, Brown JS (2010) The <em>Streptococcus</em> <em>pneumoniae</em> capsule inhibits complement activity and neutrophil phagocytosis by multiple mechanisms. Infect Immun 78(2):704
59. Weiss G, Schaible UE (2015) Macrophage defense mechanisms against intracellular bacteria. Immunol Rev 264(1):182–203
60. Ray K, Marteyn B, Sansonetti PJ, Tang CM (2009) Life on the inside: the intracellular lifestyle of cytosolic bacteria. Nat Rev Microbiol 7(5):333–340
61. Mitchell G, Chen C, Portnoy DA (2016) Strategies used by bacteria to grow in macrophages. Microbiol Spectr 4(3)
62. Edwards RJ, Taylor GW, Ferguson M, Murray S, Rendell N, Wrigley A et al (2005) Specific C-terminal cleavage and inactivation of interleukin-8 by invasive disease isolates of Streptococcus pyogenes. J Infect Dis 192(5):783–790
63. Weiser JN, Bae D, Fasching C, Scamurra RW, Ratner AJ, Janoff EN (2003) Antibody-enhanced pneumococcal adherence requires IgA1 protease. Proc Natl Acad Sci U S A 100(7):4215–4220
64. von Pawel-Rammingen U, Johansson BP, Björck L (2002) IdeS, a novel streptococcal cysteine proteinase with unique specificity for immunoglobulin G. EMBO J 21(7):1607–1615
65. McNeil LK, Zagursky RJ, Lin SL, Murphy E, Zlotnick GW, Hoiseth SK et al (2013) Role of factor H binding protein in Neisseria meningitidis virulence and its potential as a vaccine candidate to broadly protect against meningococcal disease. Microbiol Mol Biol Rev MMBR 77(2):234–252
66. Merino S, Camprubí S, Albertí S, Benedí VJ, Tomás JM (1992) Mechanisms of Klebsiella pneumoniae resistance to complement-mediated killing. Infect Immun 60(6):2529–2535
67. Sheldon JR, Laakso HA, Heinrichs DE (2016) Iron acquisition strategies of bacterial pathogens. Microbiol Spectr. 4(2)
68. Tao X, Schiering N, Zeng HY, Ringe D, Murphy JR (1994) Iron, DtxR, and the regulation of diphtheria toxin expression. Mol Microbiol 14(2):191–197
69. Cornejo E, Schlaermann P, Mukherjee S (2017) How to rewire the host cell: a home improvement guide for intracellular bacteria. J Cell Biol 216(12):3931–3948
70. Samanta D, Mulye M, Clemente TM, Justis AV, Gilk SD (2017) Manipulation of host cholesterol by obligate intracellular bacteria. Front Cell Infect Microbiol 7:165
71. Juhas M (2015) Horizontal gene transfer in human pathogens. Crit Rev Microbiol 41(1):101–108

72. Tagini F, Aubert B, Troillet N, Pillonel T, Praz G, Crisinel PA et al (2017) Importance of whole genome sequencing for the assessment of outbreaks in diagnostic laboratories: analysis of a case series of invasive Streptococcus pyogenes infections. Eur J Clin Microbiol Infect Dis 36(7):1173–1180
73. Rasko DA, Rosovitz MJ, Myers GSA, Mongodin EF, Fricke WF, Gajer P et al (2008) The Pangenome structure of Escherichia coli: comparative genomic analysis of E. coli commensal and pathogenic isolates. J Bacteriol 190(20):6881–6893
74. Namouchi A, Didelot X, Schöck U, Gicquel B, Rocha EPC (2012) After the bottleneck: genome-wide diversification of the Mycobacterium tuberculosis complex by mutation, recombination, and natural selection. Genome Res 22(4):721–734
75. Churchward G (2007) The two faces of Janus: virulence gene regulation by CovR/S in group A streptococci. Mol Microbiol 64(1):34–41
76. Quereda JJ, Cossart P (2017) Regulating bacterial virulence with RNA. Ann Rev Microbiol 71:263–280
77. Kazmierczak MJ, Wiedmann M, Boor KJ (2005) Alternative sigma factors and their roles in bacterial virulence. Microbiol Mol Biol Rev MMBR 69(4):527–543
78. Xiong YQ, Fowler VG, Yeaman MR, Perdreau-Remington F, Kreiswirth BN, Bayer AS (2009) Phenotypic and genotypic characteristics of persistent methicillin-resistant Staphylococcus aureus bacteremia in vitro and in an experimental endocarditis model. J Infect Dis 199(2):201–208
79. Chattaway MA, Day M, Mtwale J, White E, Rogers J, Day M et al (2017) Clonality, virulence and antimicrobial resistance of enteroaggregative Escherichia coli from Mirzapur, Bangladesh. J Med Microbiol 66(10):1429–1435
80. Robins-Browne RM, Holt KE, Ingle DJ, Hocking DM, Yang J, Tauschek M (2016) Are Escherichia coli Pathotypes still relevant in the era of whole-genome sequencing? Front Cell Infect Microbiol 6:141
81. Clements A, Young JC, Constantinou N, Frankel G (2012) Infection strategies of enteric pathogenic Escherichia coli. Gut Microbes 3(2):71–87
82. Walters LL, Raterman EL, Grys TE, Welch RA (2012) Atypical Shigella boydii 13 encodes virulence factors seen in attaching and effacing Escherichia coli. FEMS Microbiol Lett 328(1):20–25
83. Tagini F, Pillonel T, Croxatto A, Bertelli C, Koutsokera A, Lovis A et al (2018) Distinct genomic features characterize two clades of Corynebacterium diphtheriae: proposal of Corynebacterium diphtheriae Subsp. diphtheriae Subsp. nov. and Corynebacterium diphtheriae Subsp. lausannense Subsp. nov. Front Microbiol 9:1743
84. Saeed K, Gould I, Esposito S, Ahmad-Saeed N, Ahmed SS, Alp E et al (2018) Panton–valentine leukocidin-positive Staphylococcus aureus: a position statement from the International Society of Chemotherapy. Int J Antimicrob Agents 51(1):16–25
85. Shallcross LJ, Fragaszy E, Johnson AM, Hayward AC (2013) The role of the Panton-valentine leucocidin toxin in staphylococcal disease: a systematic review and meta-analysis. Lancet Infect Dis 13(1):43–54
86. Grumann D, Nübel U, Bröker BM (2014) Staphylococcus aureus toxins--their functions and genetics. Infect Genet Evol J Mol Epidemiol Evol Genet Infect Dis 21:583–592
87. Micek ST, Dunne M, Kollef MH (2005) Pleuropulmonary complications of Panton-valentine Leukocidin-positive community-acquired methicillin-resistant Staphylococcus aureus: importance of treatment with antimicrobials inhibiting exotoxin production. Chest 128(4):2732–2738
88. Laabei M, Recker M, Rudkin JK, Aldeljawi M, Gulay Z, Sloan TJ et al (2014) Predicting the virulence of MRSA from its genome sequence. Genome Res 24(5):839–849
89. Giulieri SG, Holmes NE, Stinear TP, Howden BP (2016) Use of bacterial whole-genome sequencing to understand and improve the management of invasive Staphylococcus aureus infections. Expert Rev Anti-Infect Ther 14(11):1023–1036

90. Wilkins AL, Steer AC, Smeesters PR, Curtis N (2017) Toxic shock syndrome – the seven Rs of management and treatment. J Infect 74:S147–S152
91. Gilchrist CA, Turner SD, Riley MF, Petri WA, Hewlett EL (2015) Whole-genome sequencing in outbreak analysis. Clin Microbiol Rev 28(3):541–563
92. Rossen JWA, Friedrich AW, Moran-Gilad J (2018) Practical issues in implementing whole-genome-sequencing in routine diagnostic microbiology. Clin Microbiol Infect 24(4):355–360
93. Carey RB, Bhattacharyya S, Kehl SC, Matukas LM, Pentella MA, Salfinger M et al (2018) Implementing a quality management system in the medical microbiology laboratory. Clin Microbiol Rev 31(3)
94. Kozyreva VK, Truong C-L, Greninger AL, Crandall J, Mukhopadhyay R, Chaturvedi V (2017) Validation and implementation of clinical laboratory improvements act-compliant whole-genome sequencing in the public health microbiology laboratory. J Clin Microbiol 55(8):2502–2520
95. Moran-Gilad J, Sintchenko V, Pedersen SK, Wolfgang WJ, Pettengill J, Strain E et al (2015) Proficiency testing for bacterial whole genome sequencing: an end-user survey of current capabilities, requirements and priorities. BMC Infect Dis 15:174
96. Fricke WF, Rasko DA (2014) Bacterial genome sequencing in the clinic: bioinformatic challenges and solutions. Nat Rev Genet 15(1):49–55
97. Quainoo S, Coolen JPM, van Hijum SAFT, Huynen MA, Melchers WJG, van Schaik W et al (2017) Whole-genome sequencing of bacterial pathogens: the future of nosocomial outbreak analysis. Clin Microbiol Rev 30(4):1015–1063
98. Besser J, Carleton HA, Gerner-Smidt P, Lindsey RL, Trees E (2018) Next-generation sequencing technologies and their application to the study and control of bacterial infections. Clin Microbiol Infect 24(4):335–341
99. CNBC.com AP special to. Illumina manufactures a genetically-designed market disaster [Internet]. 2016 [cited 2017 Oct 20]. Available from: https://www.cnbc.com/2016/10/12/illumina-manufactures-a-genetically-designed-market-disaster.html
100. Bolger AM, Lohse M, Usadel B (2014) Trimmomatic: a flexible trimmer for Illumina sequence data. Bioinformatics 30(15):2114–2120
101. Bankevich A, Nurk S, Antipov D, Gurevich AA, Dvorkin M, Kulikov AS et al (2012) SPAdes: a new genome assembly algorithm and its applications to single-cell sequencing. J Comput Biol 19(5):455–477
102. Lam MMC, Wick RR, Wyres KL, Gorrie CL, Judd LM, Jenney AWJ, et al (2018) Genetic diversity, mobilisation and spread of the yersiniabactin-encoding mobile element ICEKp in Klebsiella pneumoniae populations. Microb Genomics [Internet]. [cited 2019 Feb 13] 4(9). Available from: https://www.ncbi.nlm.nih.gov/pmc/articles/PMC6202445/
103. Buchfink B, Xie C, Huson DH (2014) Fast and sensitive protein alignment using DIAMOND. Nat Methods 12:59
104. Li H, Handsaker B, Wysoker A, Fennell T, Ruan J, Homer N et al (2009) The sequence alignment/map format and SAMtools. Bioinforma Oxf Engl 25(16):2078–2079
105. Li H, Durbin R (2010) Fast and accurate long-read alignment with burrows-wheeler transform. Bioinforma Oxf Engl 26(5):589–595
106. Méric G, Mageiros L, Pensar J, Laabei M, Yahara K, Pascoe B et al (2018) Disease-associated genotypes of the commensal skin bacterium Staphylococcus epidermidis. Nat Commun 9(1):5034
107. Gibson MK, Forsberg KJ, Dantas G (2015) Improved annotation of antibiotic resistance determinants reveals microbial resistomes cluster by ecology. ISME J 9(1):207–216
108. Abby SS, Rocha EPC (1615) Identification of protein secretion systems in bacterial genomes using MacSyFinder. Methods Mol Biol Clifton NJ 2017:1–21
109. Strauß L, Ruffing U, Abdulla S, Alabi A, Akulenko R, Garrine M et al (2016) Detecting Staphylococcus aureus virulence and resistance genes: a comparison of whole-genome sequencing and DNA microarray technology. J Clin Microbiol 54(4):1008–1016

110. Chen L, Yang J, Yu J, Yao Z, Sun L, Shen Y et al (2005) VFDB: a reference database for bacterial virulence factors. Nucleic Acids Res 33(Database issue):D325–D328
111. Chen L, Zheng D, Liu B, Yang J, Jin Q (2016) VFDB 2016: hierarchical and refined dataset for big data analysis—10 years on. Nucleic Acids Res 44(D1):D694–D697
112. Liu B, Zheng D, Jin Q, Chen L, Yang J (2019) VFDB 2019: a comparative pathogenomic platform with an interactive web interface. Nucleic Acids Res 47(D1):D687–D692
113. Sayers S, Li L, Ong E, Deng S, Fu G, Lin Y et al (2019) Victors: a web-based knowledge base of virulence factors in human and animal pathogens. Nucleic Acids Res 47(D1):D693–D700
114. Wattam AR, Abraham D, Dalay O, Disz TL, Driscoll T, Gabbard JL et al (2014) PATRIC, the bacterial bioinformatics database and analysis resource. Nucleic Acids Res 42(Database issue):D581–D591
115. Mao C, Abraham D, Wattam AR, Wilson MJC, Shukla M, Yoo HS et al (2015) Curation, integration and visualization of bacterial virulence factors in PATRIC. Bioinforma Oxf Engl 31(2):252–258
116. Wattam AR, Davis JJ, Assaf R, Boisvert S, Brettin T, Bun C, et al. Improvements to PATRIC, the all-bacterial Bioinformatics Database and Analysis Resource Center. Nucleic Acids Res. 2017;45(D1):D535–42
117. Urban M, Cuzick A, Rutherford K, Irvine A, Pedro H, Pant R et al (2017) PHI-base: a new interface and further additions for the multi-species pathogen–host interactions database. Nucleic Acids Res 45(Database issue):D604–D610
118. Mei J-M, Nourbakhsh F, Ford CW, Holden DW (1997) Identification of Staphylococcus aureus virulence genes in a murine model of bacteraemia using signature-tagged mutagenesis. Mol Microbiol 26(2):399–407
119. Korves T, Colosimo ME (2009) Controlled vocabularies for microbial virulence factors. Trends Microbiol 17(7):279–285
120. Lazarevic V, Gaïa N, Girard M, Leo S, Cherkaoui A, Renzi G et al (2018) When bacterial culture fails, metagenomics can help: a case of chronic hepatic Brucelloma assessed by next-generation sequencing. Front Microbiol 9:1566
121. Wilson MR, Naccache SN, Samayoa E, Biagtan M, Bashir H, Yu G et al (2014) Actionable diagnosis of neuroleptospirosis by next-generation sequencing. N Engl J Med 370(25):2408–2417
122. Schubert OT, Gillet LC, Collins BC, Navarro P, Rosenberger G, Wolski WE et al (2015) Building high-quality assay libraries for targeted analysis of SWATH MS data. Nat Protoc 10(3):426–441
123. Crisan A, McKee G, Munzner T, Gardy JL (2018) Evidence-based design and evaluation of a whole genome sequencing clinical report for the reference microbiology laboratory. PeerJ 6:e4218
124. Catalán-Nájera JC, Garza-Ramos U, Barrios-Camacho H (2017) Hypervirulence and hypermucoviscosity: two different but complementary Klebsiella spp. phenotypes? Virulence 8(7):1111–1123
125. Ye M, Tu J, Jiang J, Bi Y, You W, Zhang Y et al (2016) Clinical and genomic analysis of liver abscess-causing Klebsiella pneumoniae identifies new liver abscess-associated virulence genes. Front Cell Infect Microbiol 6:165
126. Cole JN, Barnett TC, Nizet V, Walker MJ (2011) Molecular insight into invasive group a streptococcal disease. Nat Rev Microbiol 9(10):724–736

Chapter 5
Epidemiological Typing Using WGS

Lieke B. van Alphen, Christian J. H. von Wintersdorff, and Paul H. M. Savelkoul

5.1 Basic Concepts of Bacterial Typing

5.1.1 General Concepts

Bacterial typing is the identification of different strain-types within the same species. Bacterial typing has been used more and more intensively for infection control purposes of identifying outbreaks of a specific strain. With the use of typing, both person-to-person transmission of bacterial strains can be identified, source identification of the outbreak and follow up of effectiveness of infection control measurements can all be determined.

Typing has two levels of interpretation: the actual comparison of bacterial isolates and their identity at the taxonomic level, followed by a interpretation in combination with the clinical and epidemiological data of the patients involved. The latter is crucial to identify transmission, clustering of patients carrying the same strain and eventually analysing an outbreak.

Since the era of molecular techniques, many different methods have been developed and used for strain typing. Each method has its pros and cons and may be either species-specific (e.g. *spa* typing *S. aureus*) or general (e.g. Amplified fragment length polymorphism -AFLP), so the selection of the appropriate technique should be made based on the performance characteristics of the test in combination with the goal of infection control typing.

The classical general performance characteristics for typing methods include typeability, discriminatory power, reproducibility, flexibility, portability and

L. B. van Alphen (✉) · C. J. H. von Wintersdorff · P. H. M. Savelkoul
Department of Medical Microbiology, Care and Public Health Research Institute (CAPHRI), Maastricht University Medical Center, Maastricht, The Netherlands
e-mail: Lieke.van.alphen@mumc.nl

J. Moran-Gilad, Y. Yagel (eds.), *Application and Integration of Omics-powered Diagnostics in Clinical and Public Health Microbiology*,
https://doi.org/10.1007/978-3-030-62155-1_5

concordance (Table 5.1) [1].Typeability indicates the proportion of isolates the method is able to type, while the discriminatory power is a measure of the resolution: how well the method can discriminate between related/unrelated isolates. A method should also be reproducible, flexible, meaning many different species can be typed, and portable, so the results can be compared or transferred between labs. Concordance with epidemiological data and ease of interpretation are necessary to reliably interpret the result in a timeframe that fits with the appropriate intervention actions. In addition, the technical difficulty is important, as not all laboratories might be able or willing to adopt certain techniques.

Next to the classical characteristics of typing methods, the ideal typing method for infection control purposes should generate data that can be stored for comparison into a portable database. Moreover, it should be cheap and fast and preferably applicable to all bacterial species. Unfortunately, there is still no ideal typing method fulfilling all these optimal characteristics.

Table 5.1 summarises the basic characteristics of all commonly used typing methods. Among the genotypic typing methods a rough distinction can be made between band-based methods-(e.g. multi-loci variable-number tandem repeat analysis (MLVA), AFLP, Pulsed-field Gel Electrophoresis (PFGE) and sequence-based

Table 5.1 General typing method characteristics

Discriminatory reach												
Genus	Species	Subsp.	Strain	Accessory	Time	Technical difficulty	Costs	Flexibility[a]	Typeability	Reproducibility	Concordance	Portability
PFGE					48-72h	++	$$	H	+	++	++	-
Rep-PCR					24h	+	$$	H	++	++	++	-
AFLP					12h	+/++	$	H	++	++	++	++
MLST					48h	+	$$	SS	++	++	++	+++
MLVA					12h	+	$	SS	++	+/++[b]	++	-/+++[c]
WGS - SNP analysis					72h	+++	$$$	H/SS[d]	+++	+++	+++	+++
WGS - cgMLST					72h	+++	$$$	H/SS[d]	+++	+++	+++	+++
WGS - wgMLST					72h	+++	$$$	H/SS[d]	+++	+++	+++	+++

Typeability: usefulness of a method to identify a strain type (= % typeable strains)
Discriminatory power: resolution of methods to discriminate independent strains
Flexibility: the number of different species that can be typed by the same method
Portability: method can be easily transferred between labs
Concordance: level of agreement of the typing results to the epidemiological observations
Reproducibility: reproducibility of the typing results independent to time
[a]H highly flexibility; SS: species specific
[b]Depending on band-based analysis (+) or automated fragment size analysis (++)
[c]Depending on band-based analysis (−) or automated fragment size analysis (+++)
[d]Methodology is highly flexible, bioinformatical analysis is species specific

methods (e.g. *spa*, multilocus sequence typing (MLST), whole-genome sequencing). In general, fragment-based methods have shorter turnaround times (except for PFGE), as PCR is the underlying technique and typing results can be generated within a day. This makes these techniques ideal when rapid results are needed, however, the reproducibility of some of these techniques tends to pose a problem and the portability is not as good. Recently, the possibility of using automated band separation on capillary sequencers has increased the reproducibility and portability remarkably. New developments in these methods promise even higher reproducibility, portability and speed, but this needs to be further established.

In the sequence-based methods, the results are more straightforward, increasing the portability and reproducibility. Recent technical and pipeline algorithm developments in the NGS applications and equipment has remarkably decreased the turnaround times and costs. The bioinformatics and data analysis, combined with data storage and management, seems to be the main limiting factor. Rapid advancements in automated data-analysis systems are currently being made to address this problem.

5.1.2 Purpose/Applications of Typing

Choosing the optimal typing technique relies greatly on the application of typing results. Typing techniques used for surveillance purposes such as monitoring the prevalence of multi-drug resistant organisms (MDROs) on a national or global scale, may be different from the ones used for local typing for rapid implementation of infection control measures. In both applications, high discriminatory power is essential, but the speed of results and rapid recognition of spread within a hospital might be more critical. These considerations, in combination with costs and skills/machinery needed to perform specific tests, might lead to choosing different typing methods. It is therefore of utmost importance to recognise all the different aspects of a particular technique, in order to able to choose the optimal technique (or techniques) that is best suited for the problem at hand.

5.1.3 Variability

Interpretation of data in epidemiological typing can be difficult due to inherent variability at several levels within the typing system used. This variability can be at a technical level, a biological level or even an evolutionary level.

Every method used for typing will show a certain degree of technical variability, which can be assessed by repeating measurements of the same biological samples. This technical variability will differ per employed method and can be heavily dependent on the protocols and chemicals used. An important issue to reduce technical variability is the standardisation of the complete typing procedure. Small differences within a given method will introduce variability in the results and may eventually lead to misinterpretation of related/unrelated strains. Use of internal quality/reference markers may be useful to further reduce the effect of small technical differences within a given method.

Understanding or reducing the biological differences between strains of a given species, however, is much more challenging. Biological variation is a natural survival mechanism in all micro-organisms. Biological variation is a regulated and reversible process, wherein specific DNA fragments are changed under certain environmental conditions as a reaction of the micro-organism to these conditions. This may lead to temporarily different genotypic (and phenotypic) results especially with high discriminatory typing techniques. The level of these changes is species-dependent and varies from very sporadic to large parts of the genome. When using a typing method, the genetic variation of a given species should be validated and taken into account, to be able to determine cut-off values for different and identical strains. Basically a cut-off value has to be beyond the biological variation level.

The last important consideration when discussing variability is the evolutionally variation or evolutionary clock. This variation is divided into two timeframes. First, under short term conditions within the host the microorganism may change in minor point mutations, like single nucleotide polymorphisms (SNP), based on its own molecular clock. These changes are well established and unique to each isolate and the cut-off value used for that micro-organism will indicate when it can be designated as a different strain. Second, when a given micro-organism stays within a host for a long period of time, the genetic changes within the micro-organism will be directed by the host immune status selecting a host specific strain (e.g. *Helicobacter pylori*). These species, however, are not part of nosocomial transmission routes and will not play a role in outbreak situations, but when typing these strains, it should be realised that they may have adapted to the host in a unique genomic way.

5.2 Established Typing Methods

5.2.1 Pulsed-Field Gel Electrophoresis (PFGE)

The traditional "gold-standard" molecular typing method for a wide spectrum of clinically relevant pathogens is PFGE, a fragment-based method analysing the complete genome of a bacterium [2]. Briefly, the (intact) genomic DNA is released and cleaved with one or more restriction enzymes that infrequently cut DNA, resulting in large DNA fragments. These fragments are separated on an agarose gel by pulsed-field electrophoresis, where the orientation of the electric field changes periodically. In PFGE a large portion of the genome is assessed and large genomic events, such as recombinations, insertions or deletions of large mobile elements will result in a change in the PFGE pattern due to altered restriction sites. In addition, large plasmids can also be observed on the gel. PFGE used to be the primary approach for outbreak characterisation and the analysis of transmission events, but dependent on the pathogen and the resources of the laboratory, other higher throughput methods, like multilocus sequence typing (MLST), multi-loci variable-number tandem repeat analysis (MLVA) and next-generation sequencing are replacing this method. Notably, it is a relatively inexpensive, but time-consuming method, with relatively high discriminatory power, good epidemiological concordance and excellent reproducibility, and thus it will remain the method of choice in low-resource settings until other methods become less expensive. In addition, standardised protocols (through Pulsenet and Harmony) and quality control panels are available for some bacterial species, increasing the portability and quality control [3, 4].

5.2.2 *Repetitive Extragenetic Palindromic Sequence-Based PCR (REP-PCR)*

One of the first PCR-based typing techniques enabling database storage and pattern comparison is the REP-PCR. The PCR uses specific primers directed against specific repetitive elements on the chromosomal DNA of bacteria for amplification. These repetitive elements include the BOX; REP; ERIC elements. The amount of these elements and their variation are strain-specific, and the resulting patterns can be stored in a database for future comparison with new strains. Although this method was very rapid and sensitive, it also mandated the use of standardised buffers and protocols. The method is still being used by different labs and a commercial system including interpretation and database comparison is available. The discriminatory power of the method is acceptable but not as high as fragment-based methods. The advantage of this method is its flexibility since the method is more or less independent of the bacterial species involved [5].

5.2.3 *Variable Number of Tandem Repeat (VNTR) Typing*

In bacterial genomes, many regions with repetitive nucleotide repeats can be found, both in coding and non-coding regions. These repeats can be very small (a few bases) or extend over longer stretches such as 100 base pairs in length. Loci consisting of several repeats directly adjacent to each other vary in number: a variable number of tandem repeat locus. These repeats are prone to mutation due to slipped strand mispairing during DNA replication. In VNTR typing, differences in the number of repeats are assessed to distinguish isolates using PCR based techniques and separation on an automated sequencer. To increase the discriminatory power, frequently multiple loci are used, in a multiple-locus VNTR analysis (MLVA). This method is used for a wide array of organisms. Pulsenet harmonisation protocols are available for *E. coli* O157 and *S. enterica* serotypes Enteritidis and Typhimurium at the CDC pulsenet website. Other (curated) databases exist for MRSA, *Haemophilus influenza* and many other organisms. Typing by MLVA is fast, unambiguous and discriminatory. The main limitation is that the method is not universal, as primers for each pathogenic species need to be developed. In addition, harmonisation of the allele amplicons is very important for the reproducibility of the data [6].

5.2.4 Amplified Fragment Length Polymorphism Typing (AFLP)

In the AFLP method, genomic DNA fragments resulting from restriction with (one or) two enzymes are specifically ligated to double-stranded adaptors. These restriction fragments with ligated adaptors are then selectively amplified by PCR, using universal primers complementary to the adaptor sequence, the restriction site sequence and a specific number of extra nucleotides (to be able to obtain the most optimal number of fragments). The PCR is performed with highly stringent conditions at the beginning, ensuring efficient primer binding to the complementary nucleotide sequence of the template. In this way, a large number of restriction fragments is amplified in a single reaction. The reaction is usually performed with one fluorescently labelled primer to facilitate separation on an automated DNA sequencer with a subsequent computer-assisted comparison. The most significant advantages of this method are the excellent discriminatory power, the flexibility and the speed: with only a limited set of reagents, using the same procedure, a wide variety of organisms can be typed. Developments in this technique have decreased the turn-around time to 8-12 h, making this a suitable method when rapid results are needed [7, 8].

5.2.5 Single Locus Sequence Typing

Single locus sequence typing (SLST) can be used in organisms which contain genes that show a high rate of variability, usually because the gene product is exposed on the surface and consequently undergoes antigenic variation. In SLST, the gene of interest is amplified by PCR and subsequently sequenced [9]. Sequences can then be compared to an online database to determine a type. For example, typing of group A *Streptococcus* can be performed based on the sequence of the *emm* gene, which encodes the surface M-protein, a major virulence and immunological determinant of this pathogen. For *Campylobacter*, the short variable region of the *flaB* gene, encoding the minor flagellin subunit FlaB, can be used to study *Campylobacter* epidemiology [10]. In these organisms, SLST can be used for epidemiological studies, but to fully discriminate closely related organisms, in outbreak situations for example, it usually has to be combined with another typing method as the single locus sequencing does not give enough discriminatory power in closely related isolates. In general, this method can be performed easily and the results are easy to interpret and portable, however, the resolution and costs are problematic in certain settings.

A particular form of SLST is the *S. aureus* protein A gene (*spa*) typing, in which the polymorphic X region of this gene is amplified and sequenced. This method relies on the identification of the short repeats in this region, which vary in the number of repeats and point mutations within the repeats. Each repeat is assigned a code and the *spa*-type is deduced from the order of the specifically coded repeats. The discriminatory power of *spa* typing is lower than PFGE, but it is swift, cost-effective, easy to use and has a standardised international nomenclature [11]. The major drawback is the fact that it is a single locus technique and can mis-classify isolates. This is illustrated by comparison to other techniques such as MLVA. According to the data generated over several years of performing simultaneous spa and MLVA typing, it was shown that the resolution of MLVA is superior to spa typing and that MLVA is sufficient to characterise MRSA isolates for surveillance [12].

Another example of SLST is ribotyping of the 16S–23S interspacer region (ITS) of *Clostridium*. Even though the method is not universally applicable and therefore not used for many species, it is successful for typing *Clostridium difficile*. With this technique, different strains are characterised based on the length heterogeneity of the ITS region. Currently, more than 400 ribotypes are described for *C. difficile* among which is ribotype O27 which is known for its virulence, high transmission rate and toxin production [13]. The ribotyping method for *C. difficile* is well standardised and, although compared to other methods not the ideal typing method, is still very useful for infection control purposes in health care institutions.

5.2.6 Multilocus Sequence Typing

In multilocus sequence typing (MLST) the genetic relatedness is assessed using multiple genes. Here, multiple loci on the genome are amplified by PCR and subsequently sequenced. As multiple loci are analysed simultaneously, the discriminatory power is higher than SLST. Depending on the varying degrees of genetic drift, which can be gene type and organism-specific, various types of genes can be chosen to be part of the MLST scheme [14]. Schemes using housekeeping genes are most common, as these are stable enough to be amplified by PCR, but show sufficient variation for discrimination. For example, for *E. feacium* an organism with high genome plasticity, resulting in a high degree of DNA banding pattern polymorphisms, a seven gene MLST gave insight into the population structure. However, the limited extent of variation in these genes also limits the discriminatory power, as for *E. feacium*, the discriminatory power of MLST is insufficient for hospital outbreak investigations [15].

Typing systems using polymorphic genes (usually genes under selective pressure, such as virulence genes) have also been developed to increase the discriminatory power. For *C. trachomatis*, several multilocus sequencing typing systems that vary in both the number and type of loci used (housekeeping versus polymorphic genes) have been developed due to the need for genotyping methods with high resolution [16]. Schemes employing combinations of housekeeping and polymorphic genes have been developed as well, as for *Salmonella enterica,* for which MLST that combines two housekeeping genes with two variable flagellin genes has been developed [17]. In MLST, the unique sequences of each loci (alleles) are assigned a number and the combination of alleles or allelic profile results in a sequence type which can then be determined using international online databases. The great advantage of this method is that it leads to unambiguous nomenclature of highly reproducible typing results. The most considerable disadvantages of this method are the time and labour involved, combined with the sometimes high cost of sequencing. In addition, for some pathogens, this method does not have enough discriminatory power for routine use in outbreak settings.

5.2.7 Plasmid Typing

Plasmid fingerprinting (PF) was the first molecular method to be used as a bacterial typing tool and has been used successfully for the analysis of outbreaks of both nosocomial and community-acquired infections [18]. The number and size of the plasmids present are used as the basis for strain differentiation. The differentiation of large plasmids (100–150 kb) can be problematic using this method resulting in a loss of discriminatory power for isolates that contain only large plasmids. For some species, such as staphylococci, restriction enzyme analysis (REA) of plasmid DNA (REAP) is therefore performed to achieve better discriminatory power [18]. A significant limitation is that some strains may not have detectable plasmids, necessitating other techniques. Moreover, bacterial plasmid content is highly subjective to change over time, and as such, PF has limited uses and has been replaced by more recent and robust typing methods.

As opposed to using plasmid content to type bacterial strains, methods to characterise plasmids themselves were later designed due to their pronounced role in virulence, and antibiotic resistance. The most widely used plasmid typing classification is based on the differentiation between incompatibility or Inc-types. This trait determines the ability of plasmids to coexist with other plasmids in the same bacteria stably and is defined by their origin of replication. The most recent detection method is based on PCR detection (PCR-Based Replicon Typing or PBRT), targeting the major plasmid families occurring in Enterobacteriaceae [19]. More recently, this scheme was converted into a more rapid real-time PCR based method [20]. For an extensive collection of resistant Enterobacteriaceae, however, the majority of plasmids were found to belong to only four different plasmid families [21]. In order to achieve higher resolution within Inc-types of plasmids, they may be further subtyped by plasmid multilocus sequence typing (pMLST) [19].

While typing of plasmids by these methods can provide valuable insights on the epidemiology of plasmids in bacterial populations, its use in tracking clonal outbreaks is limited. As with PF, this is because plasmid content is subjective to change over time, usually independent of chromosomal alterations. Therefore it does not accurately reflect bacterial evolution and should only be used as an additional feature for differentiation.

However, with the increasing spread of plasmid-borne resistances such as the extended-spectrum beta-lactamases (ESBLs) and carbapenemases, typing of plasmids is becoming increasingly important. Moreover, the spread of resistance between different strains or even species through plasmid transfer is becoming an increasing concern in healthcare settings. The issue of typing of plasmids by whole genome sequencing (WGS) will be further discussed in Sect. 5.3.4 of this chapter.

5.2.8 Proteomic Alternatives for Typing

Genome-based methods for typing are currently by far the most commonly used technologies. Nevertheless, there are new methodologies on the horizon that have the potential to play a role in future routine typing procedures, although some improvements are still needed. Currently, several proteomic methods are being tested for the applicability of bacterial strain typing. These methods include matrix-assisted laser desorption/ionisation time-of-flight mass spectrometry (MALDI-TOF MS), Fourier Transform infrared spectroscopy (FTIR); Raman spectroscopy; near FTIR; ART FTIR [22–27].

All these methods produce a spectrum/profile based on phenotypic characteristics of the species/strain. Almost all of these techniques can discriminate bacteria at the level of identification of species. Of which MALDI-TOF is the most well-known method. In many labs, MALDI-TOF is routinely used for the identification of cultured bacterial species, which is a significant advantage for the implementation of different applications such as strain typing. [28].

Although most of these proteomic techniques are, in theory, capable of discriminating bacteria at the strain level, there are still substantial differences in practice. Based on strain analysis of these techniques compared to different genotypic methods like PFGE, Randomly amplified polymorphic DNA (RAPD), *spa* typing and MLST it is clear that none of these techniques reaches the discriminatory level needed for routine strain typing [28, 29]. The use of these technologies for typing is further discussed in chapters 9 and 11.

5.3 Whole-Genome Sequencing for Typing

5.3.1 Whole-Genome Sequencing (WGS)

Using next-generation sequencing the whole genome of a bacterium can be sequenced. Depending on the sequencing technique used, the genome needs to be fragmented to a specific size and is then sequenced as short or longer fragments (reads) after which bioinformatics tools are used to sort the sequenced fragments back into genome pieces. For most applications, like SNP-based analysis and cgMLST, the sequence reads are assembled into the whole genomes, although assembly-free SNP analysis using k-SNP or assembly-free cgMLST can be performed as well [30]. Assembly can be performed *de novo*, where the reads are assembled using only the information available in the reads, like a giant puzzle. These *de novo* assemblies most commonly do not result in completely assembled genomes, especially when using short read sequencing, as large stretches of identical sequence can be distributed throughout the genome and presence of plasmids and mobile elements can complicate assembly even further.

WGS has high typeability and flexibility and can be used as a standardised method for many pathogens. The large amount of data generated and the high resolution and discriminatory power that can be achieved makes this a very powerful tool for surveillance and epidemiological investigations [31]. Currently, the main drawbacks are the costs, the computational power necessary for data analysis, and time, as the workflow needs to be improved with regard to turnaround time. Generating typing information within a timeframe where infection control measures can still be implemented is currently difficult to achieve. The availability of web-accessible bioinformatics platforms for data processing and analysis could facilitate the use of WGS for typing in clinical laboratory settings. These are becoming more and more available, as automated extraction of typing information from whole genome sequence data has currently been developed for some bacterial pathogens on several platforms and development for other bacterial species is in progress.

5.3.2 SNP Based Versus Allele-Based Approaches

The common set of genes shared between isolates within the same species is known as the core genome (cg). However, the biological and evolutionary variation within bacteria will give rise to differences in the genomic content between isolates of the same species. These differences can be point mutations like SNPs or deletions and insertions, which occur at a specific rate for each species, that varies widely between microorganisms [32].

Analysis of the genomic data for epidemiological or infection control purposes can be performed in a variety of approaches. For typing purposes using the whole genome sequence data analysis can be performed using SNP based approaches or allele based approaches, like cgMLST or whole-genome (wg)MLST. Typically a cgMLST scheme is a little less discriminatory compared to an SNP based approach, but more suited for prospective analysis. When a wgMLST approach is used in which both cgMLST and the accessory genome are analysed, the discriminatory power can reach that of SNP-based approaches [33]. For some microorganisms, a cgMLST based scheme can perform similar to that of an SNP-based approach, as has been shown for *E. faecium* [34].

The major advantage of an allele-based approach is the possibility of a universal, standardised and expandable public nomenclature, which facilitates easy data sharing. For SNP-based approaches, this is more difficult because of the requirement of a closely related reference sequence for the analysis. To exchange data, the use of the same reference sequence is obligatory for comparison. The choice of reference sequences can have a significant impact on the identification of SNPs, so the reproducibility of a specific SNP analysis is dependent on the availability of all parameters used, like the analysis pipeline and the reference sequence. If a relative distant organism is chosen as a reference, mapping will only occur against the similar regions, and unique or specific variable regions will not be analysed, leading to a loss of resolution [35]. When a closely related reference genome is unavailable, it is reasonable to generate a reference genome consensus by combining both *de novo* and mapping approaches to avoid losing coverage and detection of false SNPs. This approach has been used in the analysis of an outbreak *of Burkholderia cepacia* complex (Bcc) isolates [36].

In SNP-based typing, it is essential to identify regions within the core genome that contain high degrees of sequence diversity due to possible recombination, insertion of mobile genetic elements or other complex genetic events. Inclusion of these regions in the SNP typing could result in technical variability obscuring the actual evolutionary and/or biological variability and subsequently the phylogenetic analysis [37]. A genetic mutation introduces a single SNP whereas a recombination event can result in multiple SNPs or complex rearrangements. Both are, however, detected as a single recombination event in allele-based methods where whole alleles are used for comparison. This could allow a better definition of genetic relationships in bacteria with high recombination rates than using SNP based approaches. However, in allele-based approaches, it is also possible to misinterpret genetic events, especially when large regions flanking alleles have been exchanged.

One of the drawbacks of cgMLST is that comparisons can only be made based on common genes. For microorganisms with a high degree of variability, this means that the number of comparable genes, and thus the resolution, will be lower.

Micro-evolution events and genomic variability during outbreaks will need to be accounted for by validating specific appropriate thresholds for relatedness per species and sometimes even for a specific outbreak clade. Some thresholds of clonality have been proposed [37] but should be considered as guidelines, as specific relatedness thresholds are difficult to establish. These can vary per wg/cgMLST scheme/procedure and need to be validated epidemiologically, thereby taking population genetics of the organism into account. The timeframe of comparison should also be considered. i.e., the longer an outbreak lasts, the more likely some isolates will surpass the proposed relatedness threshold.

5.3.3 Nomenclature of WGS Data

The nomenclature of WGS data is standardised for some micro-organisms in public databases where cgMLST data can be translated into sequence type and clonal complex data. However, this is not always sufficient to demonstrate the clonal distance between isolates. New methods are being suggested to infer more information in the nomenclature, such as including a measure of clonal distance in the nomenclature of SNP-based analysis- a method called 'SNP-address' [38]. The SNP address is build up demonstrating the number of SNP differences between isolates, inferred from phylogenetic analysis of a large number of isolates. This will quickly demonstrate the distance between two isolates based on the SNP address. Recently, a similar nomenclature structure was proposed for cgMLST data within PulseNet International, where a strain nomenclature contained a hierarchy of several levels of similarity [39]. This could increase the amount of information inferred within a name, compared to the currently used sequence type and clonal complex. However, SNP or MLST address is still not widely implemented in scientific literature.

5.3.4 The Accessory Genome

A considerable strength of employing WGS for epidemiological typing is that all genetic data becomes available, enabling the analysis of many additional genetic traits. This may include mobile genetic elements (MGEs) carrying antimicrobial resistance genes, as well as virulence factors.

Hence, the clustering of strains may be based on the presence or absence of accessory genes, in addition to variations in the core genome. The combined analysis of variation in both the core, as well as the accessory genome can provide a superior resolution to establish a comprehensive understanding of the bacterial population epidemiology [33].

MGEs, such as transposons or plasmids, play a significant role in bacterial epidemiology. Virulence or antimicrobial resistance profiles are often associated with these structures, which can be highly mobile or have a broad host range, enabling their spread within a species, as well as between different bacterial species. In order to understand the potential of such MGEs, it is therefore essential to characterise their epidemiological spread.

Currently, WGS surveillance and outbreaks analyses typically revolve around tracing clonal strains through SNPs or cgMLST. However, this approach may not be appropriate when there is a rapid spread of a plasmid, leading to a 'plasmid outbreak' rather than, or even in addition to, a clonal outbreak (see Fig. 5.1) [37, 40].

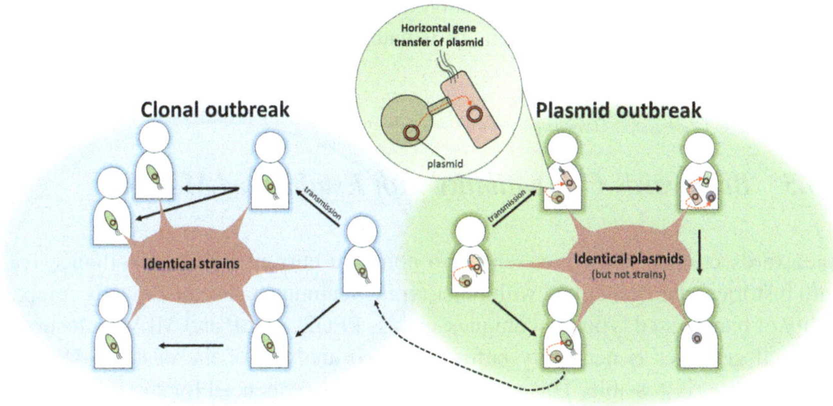

Fig. 5.1 Illustration of a plasmid outbreak (green background) in comparison with a clonal outbreak (blue background). In a clonal outbreak, transmission (black arrow) of one single strain occurs between different subjects. In the event of a plasmid outbreak, transmission of various different strains or even species may occur, which harbour the same specific plasmid. The 'outbreak plasmid' is disseminated between strains or species through horizontal gene transfer and may go unnoticed when screening is performed only at the species or chromosomal level. Specific successful strain-plasmid combinations may subsequently lead to clonal outbreaks (dotted black arrow)

Additionally, resistance may also spread through MGEs such as transposons, which may rearrange the genetic configuration of genomes or plasmids [41]. In order to be able to include such events in epidemiological investigations, in-depth analysis and understanding of reconstructed genomes, as well as plasmids, is required.

At this time, the characterisation of plasmids by WGS is still a challenge. High-throughput sequencing technologies generally produce short reads (a few hundred bases in length) [42]. Consequently, the complete reconstruction of plasmids is often problematic due to commonly occurring repetitive structures in plasmids. While there are several software algorithms available to improve the reconstruction of plasmids from fragmented assemblies [42], these often rely on heuristic techniques, which may not always result in accurate assemblies [43, 44].

Improvements are being made, as newer software packages like PlasFlow [45] and MOB-suite [46] achieve higher accuracy of plasmid identification from short-read sequencing data. Also, a web-accessible database has been developed (pATLAS) to investigate and visualise relationships between plasmids [47]. As an alternative to overcome the problems short-read data generates, long-read sequencing technologies are capable of producing tens- to hundreds of kilobase long reads. This dramatically simplifies the assembly process, making these techniques better suited to reconstruct genomes and plasmids successfully. Compared to short-read technologies, however, these generally suffer from a lower throughput, higher error rate, and higher costs which restrict their use [48].

In order to make the epidemiological tracking of specific mobile genetic elements (e.g., horizontal spread of a plasmid) more feasible, a reference assembly may be constructed using long-read techniques, after which short reads produced by WGS of more extensive collections of isolates or samples can be mapped against this reference (40).

5.3.5 Backwards Compatibility with Established Methods

Backwards compatibility is essential to compare current WGS surveillance data with historical data generated with other typing techniques. For backwards compatibility of band-based typing techniques such as PFGE, AFLP and MLVA, closure of the draft genomes is necessary before *in silico* analysis of the ensuing AFLP or PFGE profiles is possible. This means that genomes sequenced for surveillance purposes should routinely be closed using long-read sequence technologies to ensure a proper transition and comparison from current reference technologies to WGS.

Obtaining sequence-based microbial typing information, such as *spa* typing or MLST typing data from draft genomes, is possible, and methods are freely available online [49]. For *spa* typing of methicillin-resistant *staphylococcus aureus*, WGS data combined with an optimised de novo assembly showed good compatibility with sanger sequencing-based *spa* typing [50]. However, using draft genomes can pose difficulties as well, especially when analysing a multicopy gene for backwards compatibility. Thus, novel bioinformatic tools circumventing these difficulties need to be developed, as has been done, for example, for *Legionella pneumophila* typing [51].

5.3.6 Variability and Comparability of WGS Data

For whole-genome sequencing, efforts have been made to assess the variability and reproducibility of different whole genome sequencing technologies and analysis tools and the implications for the outcome. Phelan et al. assessed the variability between two sequencing platforms and two analysis pipelines for tuberculosis resistance and demonstrated platform-specific variability in coverage of some regions, which could have implications for drug resistance predictions [52]. In another study, high reproducibility and accuracy of bacterial typing using cgMLST for *Staphylococcus aureus* was seen among five laboratories assessing 20 bacterial samples [53]. However, in this study, isolated DNA was distributed, the same sequencing platform was used and both the sequence runs and data analysis were performed according to harmonised protocols. In a recent congress presentation, Carriço and colleagues showed that trimming, the choice of assembler and assembly post-processing could have a significant impact on the cgMLST allele calling process, for an *E. coli* dataset [54]. This could mean that for cgMLST data, sharing of raw data from the same platform and/or the use of the harmonised protocols for sequencing and data analysis could be necessary for reproducible typing results which can be compared between institutes. More research into the variability and comparability of data is necessary.

5.4 Typing for Infection Control

Whole-genome sequencing has become an essential tool in the investigation of outbreaks as tracing and characterisation of an outbreak can be performed with high resolution. Using this technique in real-time in health care settings is an exciting possibility; however, the cost-benefit and feasibility of achieving results within an actionable timeframe remains challenging. This method might not always be timely enough for standard continuous infection control surveillance, where there is a need for reliable typing systems that are easy and quick to handle aiming to screen and exclude clonal relatedness rather than confirm strain relatedness. Recent studies have shown that average sequencing turn-around time after initial culture, identification and susceptibility testing ranged from 4.4 to 5.3 days [55], but shorter times of around 48 h have been reported as well [56]. For actionable results within the hospital setting, this might not be timely enough, so using a faster but less discriminatory method in parallel, such as AFLP, MLVA or resistance-gene specific PCRs could help prevent a delay in outbreak recognition. Preferably, typing results in a hospital setting should be achieved within 12-24 h after isolate identification. Now that the turnaround time for WGS testing is decreasing remarkably, this is becoming feasible. For further discussion on outbreak investigation using WGS see Chap. 2.

5.5 Future Perspectives

Typing methods are continuously developing and improving. The methods aim to discriminate different strains beyond the identification level. The first methods used to detect strain-specific differences were based on phenotypic characteristics within specific species e.g. phage typing in *Staphylococcus* and, more general, antibiotic resistance patterns or gel-electrophoresis of total proteins [57–59]. In the molecular era, this changed towards genotypic techniques which were quite laborious (e.g. PFGE) or rapid but non-specific (e.g. RAPD). Nevertheless, these methods provided a new way of identifying specific strains and outbreaks and created an impressive array of new genotypic methods either species-specific or more general. The success of these methods and the applicability was strongly improved by new possibilities in different software packages, making it possible to build databases and compare the outcomes of these techniques with each other taking into account time and location. While some of these genotypic techniques have further improved and reached the level of rapid routine application in the hospital setting, WGS for typing has shown excellent results and is considered superior to all currently used methods owing to its optimal typing results at the highest discriminatory level. This will allow the implementation of these methods in the near future. In order to be able to use these methods for routine typing-guided infection control measures, several issues including storage of typing results, standardisation, cost, speed and ease of use, should be further addressed.

In summary, the field of typing and typing techniques is changing rapidly at the moment. It may well be that typing in the near future will be applied on two different levels: First, a rapid first-line screening with a molecular/proteomic method in health care settings for typing result guided infection control measures, followed by a deeper more thorough second line epidemiological WGS analysis. The following data can be used for further improvement of the first line typing results. In general, it can be expected that WGS based typing will be introduced more and more in health care settings as part of infection control policy to prevent and/or monitor the spread of specific strains and finding the source of these strains. This will play a role at a local, regional, national and international level using different molecular techniques.

References

1. van Belkum A, Tassios PT, Dijkshoorn L, Haeggman S, Cookson B, Fry NK et al (2007) Guidelines for the validation and application of typing methods for use in bacterial epidemiology. Clin Microbiol Infect 13(Suppl 3):1–46
2. Goering RV (2010) Pulsed field gel electrophoresis: a review of application and interpretation in the molecular epidemiology of infectious disease. Infect Genet Evol 10(7):866–875
3. Neoh HM, Tan XE, Sapri HF, Tan TL (2019) Pulsed-field gel electrophoresis (PFGE): a review of the "gold standard" for bacteria typing and current alternatives. Infect Genet Evol 74:103935
4. Lopez-Canovas L, Martinez Benitez MB, Herrera Isidron JA, Flores SE (2019) Pulsed field gel electrophoresis: past, present, and future. Anal Biochem 573:17–29
5. Versalovic J, Koeuth T, Lupski JR (1991) Distribution of repetitive DNA sequences in eubacteria and application to fingerprinting of bacterial genomes. Nucleic Acids Res 19(24):6823–6831
6. Nadon CA, Trees E, Ng LK, Moller Nielsen E, Reimer A, Maxwell N et al (2013) Development and application of MLVA methods as a tool for inter-laboratory surveillance. Euro Surveill 18(35):20565
7. Sabat AJ, Budimir A, Nashev D, Sa-Leao R, van Dijl J, Laurent F et al (2013) Overview of molecular typing methods for outbreak detection and epidemiological surveillance. Euro Surveill 18(4):20380
8. Reuland EA, Al Naiemi N, Kaiser AM, Heck M, Kluytmans JA, Savelkoul PH et al (2016) Prevalence and risk factors for carriage of ESBL-producing Enterobacteriaceae in Amsterdam. J Antimicrob Chemother 71(4):1076–1082
9. Asadollahi P, Farahani NN, Mirzaii M, Khoramrooz SS, van Belkum A, Asadollahi K et al (2018) Distribution of the most prevalent spa types among clinical isolates of methicillin-resistant and -susceptible Staphylococcus aureus around the world: a review. Front Microbiol 9:163
10. Mellmann A, Mosters J, Bartelt E, Roggentin P, Ammon A, Friedrich AW et al (2004) Sequence-based typing of flaB is a more stable screening tool than typing of flaA for monitoring of Campylobacter populations. J Clin Microbiol 42(10):4840–4842
11. Harmsen D, Claus H, Witte W, Rothganger J, Claus H, Turnwald D et al (2003) Typing of methicillin-resistant Staphylococcus aureus in a university hospital setting by using

novel software for spa repeat determination and database management. J Clin Microbiol 41(12):5442–5448
12. Bosch TP, Pluister GN, Luit M, Landman F, van Santen-Verheuvel M, Schot C, Witteveen S, van der Zwaluw K, Heck MEOC, Svhouls LM (2015) Multiple-locus variable number tandem repeat analysis is superior to spa typing and sufficient to characterize MRSA for surveillance purposes. Fut Microbiol 12:1155–1162
13. Freeman J, Bauer MP, Baines SD, Corver J, Fawley WN, Goorhuis B et al (2010) The changing epidemiology of Clostridium difficile infections. Clin Microbiol Rev 23(3):529–549
14. Jolley KAM, M.C.J. (2014) Using multilocus sequence typing to study bacterial variation: prospects in the genomic era. Fut Microbiol 9(5):623–630
15. Pinholt M, Larner-Svensson H, Littauer P, Moser CE, Pedersen M, Lemming LE et al (2015) Multiple hospital outbreaks of vanA Enterococcus faecium in Denmark, 2012-13, investigated by WGS, MLST and PFGE. J Antimicrob Chemother 70(9):2474–2482
16. Bom RJ, Christerson L, van der Loeff MF S, Coutinho RA, Herrmann B, Bruisten SM (2011) Evaluation of high-resolution typing methods for Chlamydia trachomatis in samples from heterosexual couples. J Clin Microbiol 49(8):2844–2853
17. Tankouo-Sandjong B, Sessitsch A, Liebana E, Kornschober C, Allerberger F, Hachler H et al (2007) MLST-v, multilocus sequence typing based on virulence genes, for molecular typing of Salmonella enterica subsp. enterica serovars. J Microbiol Methods 69(1):23–36
18. Tenover FCA, Arbeit RD, Goering RV (1997) How to select and interpret molecular strain typing methods for epidemiological studies of bacterial infections: a review for healthcare epidemiologists. Molecular Typing Working Group of the Society for Healthcare Epidemiology of America. Infect Control Hosp Epidemiol 18(6):429–439
19. Carattoli A (2011) Plasmids in Gram negatives: molecular typing of resistance plasmids. Int J Med Microbiol 301(8):654–658
20. Boot M, Raadsen S, Savelkoul PH et al. (2013) Rapid plasmid replicon typing by real time PCR melting curve analysis. BMC Microbiol 13, 83. https://doi.org/10.1186/1471-2180-13-83
21. Carattoli A (2009) Resistance plasmid families in Enterobacteriaceae. Antimicrob Agents Chemother 53(6):2227–2238
22. Dawson SE, Gibreel T, Nicolaou N, AlRabiah H, Xu Y, Goodacre R et al (2014) Implementation of Fourier transform infrared spectroscopy for the rapid typing of uropathogenic Escherichia coli. Eur J Clin Microbiol Infect Dis 33(6):983–988
23. Dieckmann R, Hammerl JA, Hahmann H, Wicke A, Kleta S, Dabrowski PW et al (2016) Rapid characterisation of Klebsiella oxytoca isolates from contaminated liquid hand soap using mass spectrometry, FTIR and Raman spectroscopy. Faraday Discuss 187:353–375
24. Johler S, Stephan R, Althaus D, Ehling-Schulz M, Grunert T (2016) High-resolution subtyping of Staphylococcus aureus strains by means of Fourier-transform infrared spectroscopy. Syst Appl Microbiol 39(3):189–194
25. Quintelas C, Ferreira EC, Lopes JA, Sousa C (2018) An overview of the evolution of infrared spectroscopy applied to bacterial typing. Biotechnol J 13(1). https://doi.org/10.1002/biot.201700449. Epub 2017 Nov 15. PMID: 29090857
26. Wenning M, Scherer S (2013) Identification of microorganisms by FTIR spectroscopy: perspectives and limitations of the method. Appl Microbiol Biotechnol 97(16):7111–7120
27. Zarnowiec P, Mizera A, Chrapek M, Urbaniak M, Kaca W (2016) Chemometric analysis of attenuated total reflectance infrared spectra of Proteus mirabilis strains with defined structures of LPS. Innate Immun 22(5):325–335
28. Spinali S, van Belkum A, Goering RV, Girard V, Welker M, Van Nuenen M et al (2015) Microbial typing by matrix-assisted laser desorption ionization-time of flight mass spectrometry: do we need guidance for data interpretation? J Clin Microbiol 53(3):760–765
29. Sauget M, Valot B, Bertrand X, Hocquet D (2017) Can MALDI-TOF mass spectrometry reasonably type bacteria? Trends Microbiol 25(6):447–455

30. Gardner SN, Hall BG (2013) When whole-genome alignments just won't work: kSNP v2 software for alignment-free SNP discovery and phylogenetics of hundreds of microbial genomes. PLoS One 8(12):e81760
31. ECDC (2016) European Centre for Disease Prevention and Control: expert opinion on whole genome sequencing for public health surveillance. Stockholm: ECDC. https://www.ecdc.europa.eu/en/publications-data/expert-opinion-whole-genome-sequencing-public-health-surveillance
32. Darmon E, Leach DR (2014) Bacterial genome instability. Microbiol Mol Biol Rev 78(1):1–39
33. Bertelli C, Greub G (2013) Rapid bacterial genome sequencing: methods and applications in clinical microbiology. Clin Microbiol Infect 19(9):803–813
34. de Been M, Pinholt M, Top J, Bletz S, Mellmann A, van Schaik W et al (2015) Core genome multilocus sequence typing scheme for high- resolution typing of Enterococcus faecium. J Clin Microbiol 53(12):3788–3797
35. Carrico JA, Crochemore M, Francisco AP, Pissis SP, Ribeiro-Goncalves B, Vaz C (2018) Fast phylogenetic inference from typing data. Algorithms Mol Biol 13:4
36. Abdelbary MMH, Senn L, Moulin E, Prod'hom G, Croxatto A, Greub G et al (2018) Evaluating the use of whole-genome sequencing for outbreak investigations in the lack of closely related reference genome. Infect Genet Evol 59:1–6
37. Schurch AC, Arredondo-Alonso S, Willems RJL, Goering RV (2018) Whole genome sequencing options for bacterial strain typing and epidemiologic analysis based on single nucleotide polymorphism versus gene-by-gene-based approaches. Clin Microbiol Infect 24(4):350–354
38. Ashton PN, Nair S, Peters T, Tewolde R, Day M, Doumith M, Green J, Jenkins C, Underwood A, Arnold C, de Pinna ED, Dallman T, Grant K (2015) Revolutionising public health reference microbiology using whole genome sequencing: salmonella as an exemplar. bioRxiv preprint
39. Nadon C, Van Walle I, Gerner-Smidt P, Campos J, Chinen I, Concepcion-Acevedo J et al (2017) PulseNet International: Vision for the implementation of whole genome sequencing (WGS) for global food-borne disease surveillance. Euro Surveill 22(23):30544
40. Hammerum AM, Hansen F, Nielsen HL, Jakobsen L, Stegger M, Andersen PS et al (2016) Use of WGS data for investigation of a long-term NDM-1-producing Citrobacter freundii outbreak and secondary in vivo spread of blaNDM-1 to Escherichia coli, Klebsiella pneumoniae and Klebsiella oxytoca. J Antimicrob Chemother 71(11):3117–3124
41. Mathers AJ, Crook D, Vaughan A, Barry KE, Vegesana K, Stoesser N et al (2019) Klebsiella quasipneumoniae provides a window into carbapenemase gene transfer, plasmid rearrangements, and patient interactions with the hospital environment. Antimicrob Agents Chemother 63(6):e02513–18. https://doi.org/10.1128/AAC.02513-18. PMID: 30910889; PMCID: PMC6535554
42. Orlek A, Stoesser N, Anjum MF, Doumith M, Ellington MJ, Peto T et al (2017) Plasmid classification in an era of whole-genome sequencing: application in studies of antibiotic resistance epidemiology. Front Microbiol 8:182
43. Arredondo-Alonso S, Willems RJ, van Schaik W, Schurch AC (2017) On the (im)possibility of reconstructing plasmids from whole-genome short-read sequencing data. Microb Genom 3(10):e000128
44. de Been M, Lanza VF, de Toro M, Scharringa J, Dohmen W, Du Y et al (2014) Dissemination of cephalosporin resistance genes between Escherichia coli strains from farm animals and humans by specific plasmid lineages. PLoS Genet 10(12):e1004776
45. Krawczyk PS, Lipinski L, Dziembowski A (2018) PlasFlow: predicting plasmid sequences in metagenomic data using genome signatures. Nucleic Acids Res 46(6):e35
46. Robertson J, Nash JHE (2018) MOB-suite: software tools for clustering, reconstruction and typing of plasmids from draft assemblies. Microb Genom 4(8):e000206. https://doi.org/10.1099/mgen.0.000206
47. Jesus TF, Ribeiro-Goncalves B, Silva DN, Bortolaia V, Ramirez M, Carrico JA (2019) Plasmid ATLAS: plasmid visual analytics and identification in high-throughput sequencing data. Nucleic Acids Res 47(D1):D188–DD94

48. Koren S, Phillippy AM (2015) One chromosome, one contig: complete microbial genomes from long-read sequencing and assembly. Curr Opin Microbiol 23:110–120
49. Larsen MV, Cosentino S, Rasmussen S, Friis C, Hasman H, Marvig RL et al (2012) Multilocus sequence typing of total-genome-sequenced bacteria. J Clin Microbiol 50(4):1355–1361
50. Bletz S, Mellmann A, Rothganger J, Harmsen D (2015) Ensuring backwards compatibility: traditional genotyping efforts in the era of whole genome sequencing. Clin Microbiol Infect 21(4):347 e1–347 e4
51. Gordon M, Yakunin E, Valinsky L, Chalifa-Caspi V, Moran-Gilad J (2017) Infections ESGfL. A bioinformatics tool for ensuring the backwards compatibility of Legionella pneumophila typing in the genomic era. Clin Microbiol Infect 23(5):306–310
52. Phelan J, O'Sullivan DM, Machado D, Ramos J, Whale AS, O'Grady J et al (2016) The variability and reproducibility of whole genome sequencing technology for detecting resistance to anti-tuberculous drugs. Genome Med 8(1):132
53. Mellmann A, Andersen PS, Bletz S, Friedrich AW, Kohl TA, Lilje B et al (2017) High interlaboratory reproducibility and accuracy of next-generation-sequencing-based bacterial genotyping in a ring trial. J Clin Microbiol 55(3):908–913
54. ECCMID2018 (2018) The importance of transparent boxes: need for reproducible and certifiable solutions for high-throughput sequencing analysis in routine epidemiology [Internet]. https://2018.eccmid.org/scientific_programme/preliminary_programme/educational_workshops/in the session: Common standards and guidelines for whole-genome sequencing (WGS) of microbial pathogens in routine epidemiology
55. Mellmann A, Bletz S, Boking T, Kipp F, Becker K, Schultes A et al (2016) Real-time genome sequencing of resistant bacteria provides precision infection control in an institutional setting. J Clin Microbiol 54(12):2874–2881
56. McGann P, Bunin JL, Snesrud E, Singh S, Maybank R, Ong AC et al (2016) Real time application of whole genome sequencing for outbreak investigation – what is an achievable turnaround time? Diagn Microbiol Infect Dis 85(3):277–282
57. Bouvet PJ, Jeanjean S, Vieu JF, Dijkshoorn L (1990) Species, biotype, and bacteriophage type determinations compared with cell envelope protein profiles for typing Acinetobacter strains. J Clin Microbiol 28(2):170–176
58. Dijkshoorn L, Wubbels JL, Beunders AJ, Degener JE, Boks AL, Michel MF (1989) Use of protein profiles to identify Acinetobacter calcoaceticus in a respiratory care unit. J Clin Pathol 42(8):853–857
59. Wentworth BB (1963) Bacteriophage typing of staphylococci. Bacteriol Rev 27:253–272

Chapter 6
Next-Generation Sequencing in Clinical Virology

Anneloes van Rijn-Klink, Jutte J. C. De Vries, and Eric C. J. Claas

6.1 Metagenomic Sequencing for Pathogen Detection and Discovery

6.1.1 Pathogen Detection

Viruses cause an important part of human infectious diseases. In addition to the already known diseases caused by viruses, they are the suspected, still unidentified, aetiological agent in several other diseases and are the leading cause of newly discovered syndromes [2]. This, in combination with the steady emergence of new (pathogenic) viruses over the past years, suggests new pathogens will continuously be discovered [3]. In order to detect previously unknown viruses, there is a need for advanced detection methods, independent of culture or previous knowledge of nucleotide sequences. Unbiased next-generation sequencing (NGS), namely metagenomic NGS, is such a catch-all method. Metagenomics is the analysis of the complete genomic content in a clinical sample [4, 5]. Next to the detection of viruses, mNGS provides information on viral genotyping, virulence markers, epidemiology, and resistance, as well as the presence of other potentially pathogenic micro-organisms and host information [6–9].

Current routine viral diagnostics are mainly based on nucleic acid amplification tests (NAAT) such as real-time polymerase chain reaction (rtPCR). These techniques however, only target a predefined set of suspected viruses. Other pathogens or genetic diversity of the targeted viruses can lead to false-negative NAAT results. mNGS overcomes these limitations and seems to be a promising tool for routine

A. van Rijn-Klink · J. J. C. De Vries · E. C. J. Claas (✉)
Department of Medical Microbiology, Leiden University Medical Center,
Leiden, The Netherlands
e-mail: e.claas@lumc.nl

J. Moran-Gilad, Y. Yagel (eds.), *Application and Integration of Omics-powered Diagnostics in Clinical and Public Health Microbiology*,
https://doi.org/10.1007/978-3-030-62155-1_6

diagnostics in samples taken from patients with a clinical suspicion of infection but no aetiology detected.

Several proof-of-principle studies applying NGS for the detection of viruses in clinical samples such as cerebrospinal fluid, sera and respiratory samples have shown a good diagnostic yield, with a sensitivity comparable to real-time PCR [10–14].

Several clinical syndromes are presumed to have a viral aetiology, even though one cannot be isolated. These include, among others, meningoencephalitis, myocarditis and acute liver failure. Meningoencephalitis, for instance, is a severe clinical illness, for which in less than half of the patients an aetiologic agent is found. Although in 20–50% of the patients, an infectious cause is likely, in a more substantial portion of patients, infection is suspected. Several studies demonstrated the diagnostic potential of mNGS in patients with meningoencephalitis, and several pathogens previously not associated with this illness such as parvo- and astroviruses, were identified in clinical samples, as well as new pathogens [15–20]. mNGS also appears to have a significant diagnostic value in myocarditis and liver failure [21, 22].

6.1.2 *Virus Discovery*

With its unbiased approach, mNGS appears to be of crucial importance for outbreak management of emerging pathogens. Nearly all relevant outbreaks in recent years are caused by previously unknown viruses [7, 23–27]. To manage and control these outbreaks, it is essential to identify the aetiological agent rapidly. mNGS has been shown to enable this rapid identification of emerging, newly identified, viruses. Two outbreaks of haemorrhagic fever in Africa could be attributed to an arenavirus and a new rhabdovirus, called Bas-Congovirus, whereas a new phlebovirus was the causative agent causing an outbreak of fever and thrombocytopenia in China [28–31].

Examples of other previously unknown viruses discovered using mNGS are a novel astrovirus, VA1-HMO-C, in association with encephalitis [20, 32, 33]; a new Arenavirus, lujovirus, associated with three transplant-related deaths [34]; a new picornavirus, klassevirus, associated with gastroenteritis [35–37]; a novel enterovirus (EV C109) in tropical febrile respiratory illness [38]; and a new phlebovirus, Heartland virus, associated with severe febrile illness [39]. In addition, the discovery of new oncogenic viruses has been described, such as a new polyomavirus associated with Merkel cell skin carcinoma [40]. Proof for an association between the aetiology of cancer and the finding of a virus in cancer tissue is hard to provide, but it is suggested that viruses cause 10–15% of the cancers. mNGS makes it possible to demonstrate viral sequences and viral integration in tumours and discover the role of new oncogenic viruses [40, 41].

Finally, several new polyomaviruses and a novel parvovirus have been described in current studies, but their association with a clinical outcome is less evident [42–46].

These studies clearly demonstrate the usefulness of mNGS in the field of virus discovery, although data should be interpreted with caution. With mNGS all viruses, including non-pathogenic commensal viruses, are detected. Extensive further studies are required to identify the causative role of these newly discovered viruses in diseases. However, the rapid metagenomic identification of the pandemic SARS-CoV2 in Wuhan in 2019 can be seen as a striking example of the added value of mNGS in virus discovery.

Although mNGS shows promising results as described, the application of unbiased NGS as a general routine tool has been implemented in a limited number of laboratories so far. There are still many hurdles that need to be overcome. First of all, the lack of standardised procedures, especially a high-throughput, unbiased pre-treatment protocol for combined DNA and RNA sequencing, but also appropriate (process) controls, and clinically relevant read counts need to be established. Regularly, viral enrichment steps to increase the relatively low amount of viral reads and thus the sensitivity of the assays have been used, but evidently can introduce a detection bias. In addition, standardisation of the bioinformatics analysis, the choice of the pipeline and database used for analysis, quality control requirements, and in particular the interpretation of the data produced, all need optimisation to enable full and appropriate use of mNGS results. Next, the role of the virome needs to be established. Form a practical point of view, the costs and turnaround time of sequencing results are a matter of concern for diagnostic use, but rapid improvements can be observed. Before mNGS can be introduced as a routine diagnostic tool, all these issues need to be addressed and validated carefully in order to fulfil current accreditation requirements.

The use of mNGS as a supplement to current diagnostics has already been established in diagnostic laboratories. For those patients where standard diagnostic methods have been unable to identify a pathogen, mNGS can help detect unexpected or new pathogens. Though the role of mNGS in the clinical diagnosis of viral infections might still be relatively small, its diagnostic potential is huge and its contribution will increase over the next few years.

6.2 Virome Analysis

The human microbiome typically consists of a community of microbes colonising the human body. The part of the microbiome consisting of viruses is the virome. The vast majority of publications on the microbiome of different human body compartments have focused on bacteria. However, a growing number of studies address the virome, with the gut virome being the most prominent. The development of NGS has contributed to the characterisation of the virome with relation to health and disease. Most publications on the virome focus on DNA viruses whereas the RNA virome is understudied. An important explanation is that when focusing on the bacterial part of the microbiome with untargeted next-generation sequencing, double-stranded (ds) DNA viruses can be studied simultaneously. In contrast, in order to study RNA viruses and single-stranded (ss) DNA viruses by untargeted NGS, specific pre-treatment with the conversion of the viral genome to dsDNA is necessary.

It is of importance to realise the bias towards dsDNA viruses in the literature resulting from this technical matter. Pan-virus-specific sequencing remains a challenge since viruses lack a universally conserved gene such as the 16S for prokaryotes that is used for targeted bacterial metagenomics.

Until several years ago, it was assumed that some human body compartments are sterile, based on the lack of detection of micro-organisms with culture and nucleic acid testing (NAT). In contrast, current published sequencing data on the virome suggest that any compartment of the human body is colonised with viruses. The human virome seems to be highly personalised but with dynamic composition depending on several variables (diet, antibiotic usage, age and genetic composition).

6.2.1 The Human Virome in Different Body Compartments

The virome in different body compartments in relation to health and disease will be described in the next section. Figure 6.1 shows an overview of body compartments with their most commonly found prokaryotic (bacteriophages) and eukaryotic viruses thus describing the virome of healthy controls.

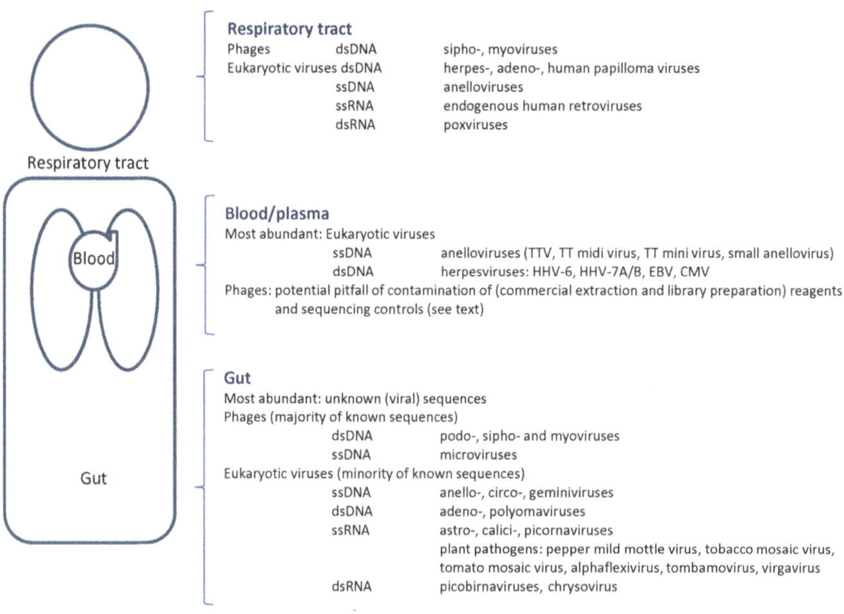

Fig. 6.1 Overview of the virome composition most commonly reported in literature, for different body compartments of *asymptomatic* adults. The focus was on reports with sequencing data on the virome of healthy controls [47, 49, 62–65, 76, 77, 123, 124]

6.2.2 *The Gut Virome*

In 2003, the first metagenomics analysis of viruses present in faeces was published. Faeces of a healthy male were analysed by sequencing of double-stranded DNA viruses. The majority of the sequences detected were not related to sequences in public databases. Of the matched sequences, the majority were bacterial, one fourth were phages and <5% eukaryotic viruses. By modelling, the complete virome was estimated to comprise of approximately 1200 different viral genotypes [47]. Others have confirmed the finding that the human gut virome mainly consists of phages, whereas only a minority consists of viruses infecting human, animals, amoeba and plants [48]. Phages can be either lytic (virulent) or lysogenic (dormant). Lytic (virulent) phages lyse the bacterial host cell with lytic enzymes at the moment of viral release, whereas lysogenic phages integrate their DNA into the bacterial host genome and subsequent replication is dependent on host DNA replication. Lytic phages can have narrow or broad host ranges, respectively infecting one bacterial strain or multiple species of closely related bacteria. The most common phages detected by Breithart et al. [47] were siphoviruses (dsDNA viruses), followed by myoviruses (dsDNA), podoviruses (dsDNA) and microviruses (ssDNA), likely reflecting the presence of enteric bacteria present in the faecal sample. Analysing a larger series of faecal samples, Kim et al. [49] confirmed the predominance of unknown sequences followed by sequences of podoviruses, siphoviruses and myoviruses, present in all samples of healthy individuals. Using a protocol aiming at ss and ds DNA viruses by random amplification with polymerase phi29, microviruses (ssDNA) made up 3–9% of known viruses. To date, this is the only publication focusing on single-stranded DNA viruses in human faeces [49].

The composition of the gut virome changes with age. Shortly after birth, the gut virome has the lowest diversity (corresponding with lower diversity of the bacterial microbiome) and consists mostly of phages that infect Gram-positive bacteria. Rapidly, the virome population becomes more diverse: in a twin study, DNA viruses were detected starting from 3 months after birth, with anelloviruses, circoviruses and geminiviruses being the most abundant (practically all newborns) and nanoviruses, adenoviruses, polyomaviruses and parvoviruses being much less abundant (in the minority of the newborns) [50]. After 3 months of age, eukaryotic DNA virus richness increases further.

In adults, the gut virome is mostly unique to each individual and remains relatively stable over time: 80% of the detected sequences of the gut virome (mainly uncharacterised bacteriophages) persisted over 2–5 years [51]. The composition of the human virome differs more between individuals than within one individual over time, even after a dietary intervention [52]. Thus, the human gut virome is highly personalised and temporally stable.

The gut virome has been studied in relation to several diseases. Changes in the gut virome, viral dysbiosis, have been associated with inflammatory bowel disease (IBD). Decreased amounts of phages were detected in ulcerated mucosal biopsies of patients with Crohn's disease compared with non-ulcerated mucosa and healthy controls [53, 54], though the role in pathogenesis is not clear. A study on faecal microbiota transplantation in *Clostridium difficile* infection found that when more phage taxa were transferred from donor to recipient, a favourable treatment outcome was observed [55]. Adenoviruses have also been detected in faecal samples of IBD patients while absent in controls [56] and an increased number of adenoviruses have been found in HIV patients though without an association with HIV disease status or therapy [57].

Legoff et al. [58] suggested an association of picobirnaviruses with early post-transplant graft-versus-host-disease (GVHD) in adult haematopoietic stem cell transplant patients. Though picobirnaviruses can be detected in a wide range of wild and domestic mammals including humans, pigs, swine, calves, foals and hamsters, a role as the cause of gastroenteritis is unclear [59]. Recently, picobirnavirus infectivity of bacteria has been suggested [60]. The work of Legoff et al. [58] suggests that the presence of picobirnaviruses in faeces may serve as a potential biomarker for identifying the development of GVHD.

Few studies of the RNA gut virome have been published. In healthy donors, shotgun cDNA sequencing of filtrated faeces revealed that the majority of the viruses detected were plant RNA viruses, with pepper mild mottle virus (PMMV, ssRNA) being most abundant [61, 62]. Two third of faecal samples of healthy individuals on two different continents were PMMV positive. Other viruses detected in all individuals were tobacco mosaic virus (ssRNA, pathogenic for tobacco), tomato mosaic virus (ssRNA, pathogenic for tomato) though much lower abundant (0.1–3% of viruses). Plant pathogenic viruses were also detected in pepper-containing food, cereals, fruits, tobacco and other vegetables, suggesting ingestion as a source. Furthermore, picobirnaviruses (dsRNA) were detected at low abundance.

6.2.3 The Blood/Plasma Virome

Several studies detect a variety of viruses present in the blood and plasma of healthy individuals. A recent study examining blood samples from 8000 healthy individuals [63] with a protocol aiming at dsDNA, detected the following viruses as the most abundant: human herpesvirus (HHV)-7 in 20% of individuals, HHV-4 (Epstein-Barr virus) in 15%, anellovirus (torque teno virus, TTV and TTV-like mini virus, TLMV) in 9% and HHV6B in 5%. Only a small minority (1%) of the blood samples contained Merkel cell polyomavirus, HHV-5 (cytomegalovirus), human T-lymphotropic virus)HTLV) and human papillomavirus (HPV). Importantly, blood samples spiked with the positive control phiX174 phage contained extensive phage DNA (not limited to phiX174) whereas samples without phiX174 spike did not, suggesting contamination of commercial phiX174 materials. Contamination of commercial materials will be discussed further below.

Wylie et al. [64] compared the plasma virome of febrile children with afebrile children (up to the age of 3). Anelloviruses were the only viruses detected in afebrile children (three quarters). In contrast, the febrile children carried anelloviruses (all children) and less often roselovirus, enteroviruses, polyomaviruses, astrovirus, and human pegivirus (formerly GB virus C/hepatitis G virus).

In a study on the temporal dynamics of the plasma virome in lung transplant patients, the plasma virome of healthy controls contained herpesviruses (100% of samples), siphoviruses (80%), anelloviruses (70%) and myoviruses (30%) [65]. The anellovirus abundancy was higher in lung transplant recipients than in healthy controls, with up to 48 different anellovirus strains identified within a single lung transplant recipient. Anelloviruses are considered as non-pathogenic, but the results of this study may serve as a potential marker for guiding immunosuppressive drug therapy after transplantation [65]. Higher TTV levels (NAAT) seem to be protective for rejection after kidney transplantation [66].

The proportion of anelloviruses have also been found to be increased in HIV-infected subjects with low CD4+ compared to high CD4+ T cell count, indicating that progression to AIDS is associated with changes in the plasma concentration of commensal viruses [67]. In the plasma virome of healthy controls, anelloviruses were detected, in contrast to HIV/AIDS patients with bacteriophages and human endogenous retrovirus abundance without detection of anelloviruses. These data suggest immune control of the blood virome in healthy individuals.

The finding of viruses in blood samples is of particular interest in the blood transfusion setting, where viral metagenomics through NGS has been proposed as an approach for the identification and surveillance of unknown or unexpected viruses that may be transmitted to recipients by blood products [68]. Plasma-pooled samples from recipients and donors of blood contained mainly anelloviruses and human pegivirus, but also hepatitis B and C viruses. Importantly, contaminants from commercial nucleic acid extraction silica-binding spin columns were detected in several studies on the blood virome: parvovirus-like hybrid genome, PHV/NIH-CQV [68, 69] and circoviruses/densoviruses and iridoviruses [70].

6.2.4 The Oral and Respiratory Tract Virome

Several studies have focused on the virome of the oral cavity and respiratory tract. The saliva virome primarily consists of phages, mainly siphoviruses, myoviruses and podoviruses (detected in healthy individuals at any longitudinal time point) and less frequently papillomaviruses and inoviruses [71]. When compared to their faeces, the salivary virome of healthy individuals more frequently contained papillomaviruses and less frequently microviruses and podoviruses. Principle component analysis of follow-up saliva samples of controls showed that many viruses are stable and specific in an individual [72]. Long-term (6 weeks) intravenous antibiotic therapy has been associated with a high abundancy of papillomaviruses in saliva [73]. The proportion of shared virome sequences in gingival samples was found to be higher in patients with periodontal disease [71].

The virome in nasopharyngeal washings of pediatric children with respiratory symptoms and positive for respiratory pathogen NAAT were studied by Van

Boheemen et al. (manuscript submitted). A protocol with combined detection of both DNA and RNA viruses enabled analysis of both DNA and RNA virome. The pathogens detected by NAAT (mainly influenza virus, rhinovirus, parainfluenzavirus, human metapneumovirus) comprised 65–99% of the virome in the majority of cases, followed by respectively dsDNA bacteriophages (Caudovirales: siphoviruses, myoviruses, podoviruses), ssDNA bacteriophages, betaherpesviruses, human papillomaviruses and human endogenous retroviruses (K113) (Fig. 6.2). Data from other studies in asymptomatic children indicate that the respiratory virome of healthy children is less diverse and mainly dominated by anelloviruses, that were detected in 10–20% of controls with only a small proportion of prevalent epidemic respiratory viruses [64, 74].

Figure 6.2 is an example of a krona plot of the virome detected in one nasopharyngeal washing of a symptomatic pediatric patient infected with respiratory pathogen Adenovirus type non 40–41 (NAAT positive). Spike-in sequence controls: equine arteritis virus, phocine herpes virus and Enterobacteria phage phiX174 (microvirus).

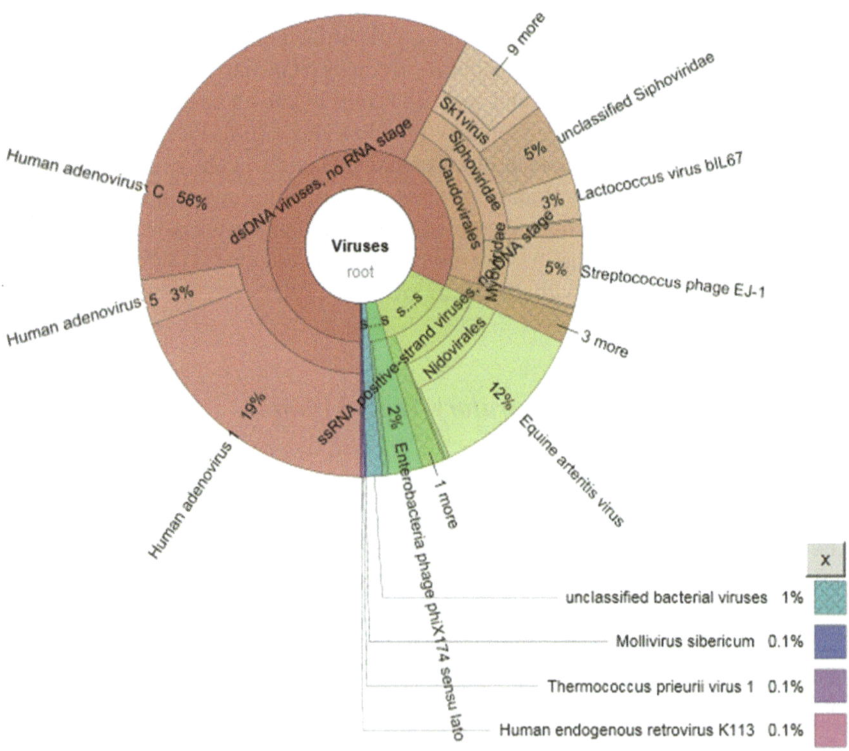

Fig. 6.2 Example of a krona plot of the virome detected in one nasopharyngeal washing of a symptomatic pediatric patient infected with respiratory pathogen Adenovirus type non 40–41 (NAT positive). Spike-in sequence controls: equine arteritis virus, phocine herpes virus and Enterobacteria phage phiX174 (microvirus)

In a study of the virome of bronchoalveolar lavage (BAL) samples from healthy lung transplant recipients, samples were predominated by anelloviruses and herpesviruses (all individuals at all time-points) and in a small minority of the samples coronaviruses, siphoviruses, myoviruses and papillomaviruses [65]. Graft dysfunction was associated with lower anellovirus loads in BAL fluids of lung transplant patients, suggesting that immune activation may restrict TTV abundance in the lung [75].

The virome of the respiratory tract has also been studied in cystic fibrosis (CF) patients. Sputum from non-CF controls was found to have more distinct phage communities whereas phage communities in CF patients were highly similar and potentially with lower richness (viral dysbiosis). Eukaryotic viruses detected in >60% of non-CF controls were retroviruses, adenoviruses, herpesviruses, papillomaviruses, iridoviruses, and poxviruses. Whereas sputum viromes of CF patients were dominated by a few viruses including human herpesviruses and retroviruses, sputum of non-CF controls was higher in the diversity of eukaryotic viruses [76].

6.2.5 The Genitourinary Virome

A limited number of reports have been published on the genitourinary compartment. Phages and low-risk genotypes of human papillomaviruses (95% of the individuals) seem to be the main components of the urinary virome, followed by herpesviruses and polyomaviruses [77, 78]. No association of the urinary virome with urinary tract infections was found in that specific study [78]. In vaginal samples from healthy individuals studied for DNA viruses, the most abundant viruses were human papillomaviruses, followed by cytomegalovirus, anelloviruses, lymphocryptoviruses and roseoloviruses [77].

6.2.6 The Skin Virome

Few studies have focused on DNA in skin swabs and found mainly phages (myoviruses, siphoviruses) and papillomaviruses [79]. Additionally, human polyomaviruses (6, 7 and Merckel cell polyomavirus), adenoviruses, anelloviruses, circoviruses, herpesviruses have been found, interestingly on healthy skin cells [80]. The role of the viral skin virome in health and disease is still unclear.

In summary, the knowledge on the human virome and its role in health and disease is growing since the earliest publication in 2003. More recent findings have resulted in new hypotheses on the potential contribution of the human virome to pathogenicity. Furthermore, a potential role as a useful marker for disease- such as early stages of GVH- has been suggested. In the near future sequencing information will become more widely available and is likely to contribute to the understanding of the role of the human virome and its dysbiosis in health and disease.

6.3 Antiviral Susceptibility and Resistance

Over the last decade, antiviral susceptibility testing has changed from a culture based assay to predominantly molecular testing, either by rapid analysis of SNPs by real-time PCR, hybridisation assays, or nucleotide sequence analysis. This part will discuss the application of NGS in the field of antiviral susceptibility testing.

Using Sanger sequencing, both identification of viruses, their evolution and potential accumulation of SNPs have been extensively studied. The major disadvantage of Sanger sequencing is that only the majority variants of pathogens are identified. If a minor variant represents less than 20–25% of the total genomes, identification becomes problematic. Application of deep sequencing enables identification of these minor variants that comprise up to 1–5% of the population. In this way, more detailed information can be obtained on viral diversity in molecular epidemiology, virus evolution and viral in-host dynamics [81, 82]. Although these studies contribute to a better understanding of viral pathogenesis, spread and evolution, there is no direct impact on patient management. Therefore, the focus of this part is NGS application in antiviral susceptibility and resistance testing. This is especially relevant when treating infections of RNA viruses with intrinsically high mutation frequencies and replication capacity, in which the development of resistance is common. Several studies have shown that by applying deep sequencing, a more accurate overview of the actual presence of these mutations can be obtained, with more resistance-associated mutations (RAM) being detected [83]. The most important question to answer is whether the presence of minor variant RAMs actually correlates with an increased risk for development of antiviral resistance. Initial studies were focusing on the technical possibility to detect minor variants using NGS and those were successful [84]. The long-term clinical effects with relation to the development of resistance in different viral groups needs to be further addressed.

6.3.1 *Herpesviruses*

Herpes virus infections caused by herpes simplex viruses (HSV1 and HSV2), varicella-zoster virus (VZV) and cytomegalovirus (CMV) can be treated successfully with antivirals. The viral thymidine kinase (TK) and polymerase (POL) genes are the most common target for treatment and, as a consequence, mutations in these genes may confer resistance to the antiviral agents. Mutations in the exonuclease domain of the polymerase compromise proofreading capabilities and as such, can contribute to resistance.

In VZV, mainly acyclovir (ACV) resistance has been reported due to mutations in the TK gene (ORF36), although mutations in the POL gene (ORF28) conferring resistance to cidofovir (CDV) or foscarnet (FOS) have also been described [85]. Deep sequencing applications by NGS are rare so far. Mercier-Darty et al. [86] compared Sanger sequencing to NGS in an orthotopic heart transplant patient. The patient received valacyclovir (VACV) for prophylaxis of HSV and VZV for a year. Later, the patient developed herpes zoster (HZ) lesions, which were successfully treated with low dose VACV. A month later, extended HZ lesions appeared. As ACV resistance-associated mutations (RAM) were detected by Sanger sequencing, therapy was switched to intravenous FOS with significant improvement. Subsequently, VACV was reintroduced as prophylaxis and the patient healed completely. Retrospectively, the PCR amplicons obtained for Sanger sequencing were also subjected to deep-sequencing after library preparation using the Nextera® kit on the Miseq® platform (Illumina). After 1.5 months of VACV therapy, Sanger results showed a Q69Stop RAM, which appeared present in 95% of the population by deep-sequencing. In pre-treatment samples, only natural polymorphisms within the TK and POL genes were detected. NGS, however, revealed TK frameshift mutations: T-ins 688–691 (77%) and C-del 493–498 (95%) that were retrospectively detected by Sanger sequencing as well. In addition, minor variants, not reported by Sanger, were detected: 13% C-del 664, 4% C-del 493–498. No POL RAMS were identified under FOS therapy. In conclusion, deep sequencing by NGS provided improved detection of minor variants and showed the presence of mixed populations.

HSV-1 is another clinically significant alpha herpesvirus, being a major cause of morbidity and mortality in haematopoietic stem cell transplant (HSCT) patients, usually mandating ACV prophylaxis. Especially in immunocompromised patients, prolonged treatment may result in ACV resistant isolates in up to 5–10% of HSCT patients [87]. Fujii et al. [88], have compared Sanger sequencing to NGS for detection of acyclovir-resistant HSV-1 variants in this patient group. In four patients analysed, all mutations detected by Sanger sequencing were also detected by NGS. However, in two patients, additional RAMs were detected. In one patient 5% variants carried a G-del 615–619 in the TK gene, and in another patient 11% C-del 620–622 was detected, illustrating that mutations were detected earlier by NGS and low abundance variants were identified that were missed by Sanger sequencing. An essential drawback of this study is that sequencing was performed on cultured isolates, as direct amplification of the amplicons from clinical materials was unsuccessful. Because of this and the higher costs, the authors were uncertain whether this method was suitable for use in a clinical setting.

Most studies in the field of antiviral resistance of herpesvirus have been performed on CMV as it is one of the most significant pathogens after solid organ transplantation with a highly diverse range of clinical complications [85]. Here, ganciclovir (GCV) and oral valganciclovir (VGCV) are used for treatment and prophylaxis. Detection of minor CMV resistant variants has already been achieved several years ago by pyrosequencing [89–92].

Chou et al. [93] describe a fatal case of CMV infection in a chronic lymphocytic leukaemia (CLL) patient, where a CLL relapse was accompanied by CMV reactivation with increasing viral load despite valganciclovir (vGCV) treatment. Suspected pneumonia resulted in a switch to intravenous FOS, which resulted in clinical improvement. Later the patient developed diffuse CMV pneumonia, which stabilised using high dose vGCV and intravenous CDV. Upon de-escalation of the treatment regimen, a new sharp rise in viral load was observed in the presence of CLL relapse, and at this point, treatment was stopped and eventually the patient died. Five samples obtained during 19 weeks of treatment were analysed by Sanger sequencing. Retrospectively, deep sequencing using PCR amplification and Roche 454 GS sequencing was performed. Unfortunately, no pre-treatment samples were available. The deep sequencing clearly illustrated rapid evolution of mutations conferring multidrug resistance, of which resistant subpopulations were detected weeks earlier.

Garrigue et al. [94] quantified the difference between Sanger and deep sequencing by mixing wild type and mutant plasmids and observed reliable detection of 2% variants by NGS and 20% by Sanger sequencing. This was also shown in clinical practice; with a 9.6% variant threshold detected using NGS, while Sanger reported a wild type virus. In addition, viral diversity of the TK gene (UL97) was studied in 5–8 samples per patient from four patients with and without antiviral treatment, and RAMs in the UL97 gene were detected as well.

6.3.2 Influenza Viruses

For influenza, antiviral treatment dates back to 1966, when Symmetrel® (Amantadine) was introduced. Amantadine blocks the influenza M2 ion channel, a product of the matrix gene segment of influenza A (but not influenza B) viruses. Resistance is induced by mutations in the M2 gene coding for the ion channel pore, most frequently the S31N mutation. Resistance testing is possible by testing for this mutation by allele-specific PCR. However, currently, over 99% of all circulating human influenza virus A strains, the subtypes A(H1N1)pdm09 as well as A(H3N2), are resistant to Amantadine, and therefore this antiviral is no longer of use in clinical practice.

Neuraminidase inhibitors (NAI) are the most effective antivirals for treatment of influenza infections, and despite alternatives, oseltamivir has been the most widely prescribed NAI worldwide. The emergence of resistant viruses is infrequent since 2009 and can be induced by prolonged treatment with NAI [95]. Rapid analysis of potential resistance is readily done by molecular assays detecting several mutations in both the viral neuraminidase (NA) and the hemagglutinin (HA). In influenza A virus (H1N1)pdm09, the H275Y mutation has been most frequently found, whereas for influenza A virus (H3N2)- E119V, R292K, and N294S are the primary mutations conferring antiviral resistance. In influenza B viruses, changes at amino acids E117, D197, H273, and R374 are associated with resistance, but other RAMs and combinations of RAMs have been identified as well. SNP analysis by real-time PCR has been used as a tool for immediate screening of antiviral resistance in only a few hours [85].

Prior culture of viruses may select for variants not representing the virus population in clinical specimens, and therefore a direct application to clinical samples should be preferably performed [96].

NGS has been applied in high throughput diversity and evolution studies [97], but not frequently for resistance testing, mainly because of the limited number of

SNPs playing a role in resistance and the relatively long turnaround time. Therefore NGS does not seem to be particularly beneficial for optimal patient management, although earlier detection of resistance markers has been shown. For detection of minor variants, Pichon et al. [98] used digital droplet PCR (ddPCR) and were able to detect minor (less than 1%) variants of oseltamivir-resistant viruses in a wild type (WT) population, enabling earlier detection of resistant viruses. Zhou et al. [99] generated amplicons of the NA, HA and matrix genes of influenza A and B in a multiplex PCR. Application of these amplicons to NGS resulted in an in-depth analysis of antigenic evolution and antiviral resistance. WGS of influenza viruses has also been used as a tool for influenza surveillance [100].

Although at the moment the role for NGS in testing for antiviral resistance is limited, this clearly may change in the future. If NGS is applied as a catch-all method in the microbiology laboratory for diagnosis of acute infections (see the first part of this chapter), all additional information on potential resistance markers and even compensatory mutations [101] will be available to the clinician as well.

6.3.3 Hepatitis Viruses

6.3.3.1 Hepatitis B Virus (HBV)

Approximately 257 million people are chronically infected with HBV, and in 2015, an estimated 887,000 patients died (WHO factsheet hepatitis B, 18th July 2018). An interesting application of NGS in the field of HBV was the identification of risk factors for developing viral variants involved in hepatocellular carcinoma (HCC) as reviewed by Wu et al. [102] and Liu [103]. Antiviral treatment consisted of nucleoside/nucleotide analogues as lamivudine and tenofovir, which readily resulted in resistance development. Deep sequencing, mainly using the Roche/454 GS pyrosequencing platform, has been applied for early detection of RAMs and the presence of minor variants [104, 105]. Minor variants carrying RAMs have also been detected in patients not receiving treatment, 1.2% by Sanger sequencing and 15% by NGS [106]. However, Jones et al. [107] showed that additional resistant variants may emerge as well once antiviral therapy had been initiated. Nowadays, tenofovir and entecavir are predominantly used for the treatment of HBV as they currently are the most potent antiviral agents. Important additional benefit of these drugs is the high genetic barrier for the development of resistance [108]. An important role for NGS in the analysis of antiviral resistance in HBV is therefore unlikely.

6.3.3.2 Hepatitis C Virus (HCV)

An estimated 71 million people are living with chronic HCV infection of which approximately 399.000 per year die, mainly because of liver cirrhosis or HCC (WHO factsheet hepatitis,18th July 2018). Over the last decade, a huge advancement in treating patients with hepatitis C was made direct-acting antiviral agents (DAA). Different classes of DAA are available- protease inhibitors, polymerase inhibitors, and the NS5A replication complex inhibitors. Combination therapy with NS5B polymerase inhibitor sofosbuvir (SOF) in combination with NS5A phosphoprotein inhibitors daclatasvir or ledipasvir (Harvoni®) has been shown very successful in treating a wide range of HCV genotypes, with a sustained viral response of 95% after 12 weeks of therapy. This already indicates that the development of resistance is not a significant problem. Despite an extremely high replication rate of a trillion genomes per day and a high mutation rate of the polymerase, the genetic barrier for developing resistance to this drug is high. *In vitro* studies have been able to generate mutations that could be detected in resistant isolates, but clinically this does not seem to be of importance [85]. In addition, resistance mutations in HCV are not archived in chronic HCV patients and therefore it is unlikely that minor variants will play a role in the risk of treatment failure, especially with the currently highly effective DAA treatment combinations [109].

6.3.4 Human Immunodeficiency Virus (HIV)

By the end of 2017, 36.9 million people were living with HIV, and still, yearly 1.8 million new infections were detected. Despite significant advances in antiretroviral treatment, 940,000 people died of AIDS in 2017, the majority in Africa. Availability of appropriate combined antiretroviral therapy (cART) is still limited in those parts of the world with the largest infected population. In developed countries, HIV infection can be considered a chronic, incurable, infection and monitoring of viral loads and analysis of antiretroviral drug resistance is part of the standard care for HIV-infected patients. Amplification and sequencing of (parts of) the *pol* gene, coding for the protease, reverse transcriptase, and integrase enzymes provides detailed information of RAMs that may result in decreased susceptibility to the major classes of antiviral drugs for HIV, the protease inhibitors (PI), nucleoside reverse-transcriptase inhibitors (NRTI), non-nucleoside reverse-transcriptase inhibitors (NNRTI) and the integrase strand transfer inhibitors (INSTI). Additional sequence analysis of the *env* gene can determine the tropism i.e. co-receptor usage of the virus for which another limited class of co-receptor inhibitors is available. Sanger sequencing, either as part of commercial kits or in house assays [85] has been used for almost two decades to determine antiretroviral treatment failure in patients with increased viral loads. Web-based interpretation tools are available to aid in decision making [110].

With the high replication rate and high mutation frequency, variant viruses are likely to be part of the HIV quasi-species and HIV cART management may be improved by including NGS analysis to identify minor variants of RAMs. Primarily pyrosequencing was used to generate HIV deep sequencing results but by now other, improved platforms are available. Opportunities, limitations and potential for

HIV-1 clinical management have been recently reviewed [111].To aid analysis, a pipeline for minor variant analysis is also available [112].

The first application of deep sequencing for HIV was described over 10 years ago [113]. Since then, numerous studies have been carried out that support the concept of improved and earlier detection of minor variants [114, 115]. More importantly, an increased risk for virological failure to first-line NNRTI based ART with the presence of minor variant RAMs at baseline has been described [116, 117]. Moreover, detection of minor variant co-receptor tropism sequences in sequences of the V3 region of the *env* gene by NGS has been shown to be predictive of the failure of treatment with maraviroc, a CCR5 receptor blocking antiviral [118].

For protease inhibitors, the presence of minor variant RAMs before treatment may be less problematic because of the high genetic barrier for resistance development. This means that a combination of mutations is required to get a measurable effect on the antiviral susceptibility and these mutations will generally harm the viral fitness [119].

Nowadays, integrase inhibitors are being used more frequently as first-line antiviral agents. Few clinical studies are available on the use of deep-sequencing and the risk of virological failure. Pou et al. [120] estimated that a quarter of patients that experienced virological failure would benefit from more detailed information on RAMs deep sequencing resulting in more reliable salvage therapy.

6.4 Future Perspective

Although NGS technology is improving, questions concerning reliable application remain, especially in viruses with a high mutation frequency in combination with a high replication rate. First of all, there are practical issues such as sampling bias and unequal amplification of viral quasi-species that may result in over- or under-representation of variants [109]. This is especially true when using NGS technology generating short fragments of sequence data (all platforms except Pacific Biosciences and Oxford Nanopore) where it is difficult to establish the relevance of the sequences harbouring minor variant RAMs as the genetic context is unclear: is the identified RAM part of a viable virus genome?

Although NGS shows the potential for detection of minor variants at an earlier time-point, the most important question is whether this will actually lead to increased risk of treatment failure. Currently, there is no definite answer to this question. As shown in the overview provided by Casadella and Rogers [111], these clinical consequences have been studied in only a few NGS studies and these do not provide a clear answer. Other reports speculate that the presence of minor variant RAMs does not determine the risk on development of resistance over time, but rather that a high diversity and diversifying selection of the virus population under treatment is the most important factor [121, 122]. For analysing this hypothesis, the broad application of NGS remains an appropriate tool. In summary, more clinical studies are required to show the clinical benefit of deep-sequencing using NGS platforms in optimising antiviral treatment.

References

1. Cong J, Zhang X (2018) How human microbiome talks to health and disease. Eur J Clin Microbiol Infect Dis 37(9):1595–1601
2. Woolhouse M, Gaunt E (2007) Ecological origins of novel human pathogens. Crit Rev Microbiol 33(4):231–242
3. Woolhouse ME, Howey R, Gaunt E, Reilly L, Chase-Topping M, Savill N (2008) Temporal trends in the discovery of human viruses. Proc Biol Sci 275(1647):2111–2115
4. Thomas T, Gilbert J, Meyer F (2012) Metagenomics - a guide from sampling to data analysis. Microb Inform Exp 2(1):3
5. Wooley JC, Godzik A, Friedberg I (2010) A primer on metagenomics. PLoS Comput Biol 6(2):e1000667
6. van Boheemen S, de Graaf M, Lauber C, Bestebroer TM, Raj VS, Zaki AM et al (2012) Genomic characterization of a newly discovered coronavirus associated with acute respiratory distress syndrome in humans. mBio 3(6) e00473-12
7. Zaki AM, van Boheemen S, Bestebroer TM, Osterhaus AD, Fouchier RA (2012) Isolation of a novel coronavirus from a man with pneumonia in Saudi Arabia. N Engl J Med 367(19):1814–1820
8. Hoffmann B, Scheuch M, Hoper D, Jungblut R, Holsteg M, Schirrmeier H et al (2012) Novel orthobunyavirus in Cattle, Europe, 2011. Emerg Infect Dis 18(3):469–472
9. Mongkolrattanothai K, Naccache SN, Bender JM, Samayoa E, Pham E, Yu G et al (2017) Neurobrucellosis: unexpected answer from metagenomic next-generation sequencing. J Pediatr Infect Dis Soc 6(4):393–398
10. Petty TJ, Cordey S, Padioleau I, Docquier M, Turin L, Preynat-Seauve O et al (2014) Comprehensive human virus screening using high-throughput sequencing with a user-friendly representation of bioinformatics analysis: a pilot study. J Clin Microbiol 52(9):3351–3361
11. Yang J, Yang F, Ren L, Xiong Z, Wu Z, Dong J et al (2011) Unbiased parallel detection of viral pathogens in clinical samples by use of a metagenomic approach. J Clin Microbiol 49(10):3463–3469
12. Parker J, Chen J (2017) Application of next generation sequencing for the detection of human viral pathogens in clinical specimens. J Clin Virol 86:20–26
13. Prachayangprecha S, Schapendonk CM, Koopmans MP, Osterhaus AD, Schurch AC, Pas SD et al (2014) Exploring the potential of next-generation sequencing in detection of respiratory viruses. J Clin Microbiol 52(10):3722–3730
14. Hasan MR, Rawat A, Tang P, Jithesh PV, Thomas E, Tan R et al (2016) Depletion of human DNA in spiked clinical specimens for improvement of sensitivity of pathogen detection by next-generation sequencing. J Clin Microbiol 54(4):919–927
15. Kawada J, Okuno Y, Torii Y, Okada R, Hayano S, Ando S et al (2016) Identification of viruses in cases of pediatric acute encephalitis and encephalopathy using next-generation sequencing. Sci Rep 6:33452
16. Guan H, Shen A, Lv X, Yang X, Ren H, Zhao Y et al (2016) Detection of virus in CSF from the cases with meningoencephalitis by next-generation sequencing. J Neurovirol 22(2):240–245
17. Brown JR, Bharucha T, Breuer J (2018) Encephalitis diagnosis using metagenomics: application of next generation sequencing for undiagnosed cases. J Infect 76(3):225–240
18. Chan BK, Wilson T, Fischer KF, Kriesel JD (2014) Deep sequencing to identify the causes of viral encephalitis. PLoS One 9(4):e93993
19. Benjamin LA, Lewthwaite P, Vasanthapuram R, Zhao G, Sharp C, Simmonds P et al (2011) Human parvovirus 4 as potential cause of encephalitis in children, India. Emerg Infect Dis 17(8):1484–1487
20. Quan PL, Wagner TA, Briese T, Torgerson TR, Hornig M, Tashmukhamedova A et al (2010) Astrovirus encephalitis in boy with X-linked agammaglobulinemia. Emerg Infect Dis 16(6):918–925

21. Takeuchi S, Kawada JI, Okuno Y, Horiba K, Suzuki T, Torii Y et al (2018) Identification of potential pathogenic viruses in patients with acute myocarditis using next-generation sequencing. J Med Virol 90(12) 1814–1821
22. Suzuki T, Kawada JI, Okuno Y, Hayano S, Horiba K, Torii Y et al (2017) Comprehensive detection of viruses in pediatric patients with acute liver failure using next-generation sequencing. J Clin Virol 96:67–72
23. Nichol ST, Spiropoulou CF, Morzunov S, Rollin PE, Ksiazek TG, Feldmann H et al (1993) Genetic identification of a hantavirus associated with an outbreak of acute respiratory illness. Science 262(5135):914–917
24. Rota PA, Oberste MS, Monroe SS, Nix WA, Campagnoli R, Icenogle JP et al (2003) Characterization of a novel coronavirus associated with severe acute respiratory syndrome. Science 300(5624):1394–1399
25. Dawood FS, Jain S, Finelli L, Shaw MW, Lindstrom S, Garten RJ et al (2009) Emergence of a novel swine-origin influenza A (H1N1) virus in humans. N Engl J Med 360(25):2605–2615
26. Shinde V, Bridges CB, Uyeki TM, Shu B, Balish A, Xu X et al (2009) Triple-reassortant swine influenza A (H1) in humans in the United States, 2005-2009. N Engl J Med 360(25):2616–2625
27. Gao R, Cao B, Hu Y, Feng Z, Wang D, Hu W et al (2013) Human infection with a novel avian-origin influenza A (H7N9) virus. N Engl J Med 368(20):1888–1897
28. Briese T, Paweska JT, McMullan LK, Hutchison SK, Street C, Palacios G et al (2009) Genetic detection and characterization of Lujo virus, a new hemorrhagic fever-associated arenavirus from southern Africa. PLoS Pathog 5(5):e1000455
29. Grard G, Fair JN, Lee D, Slikas E, Steffen I, Muyembe JJ et al (2012) A novel rhabdovirus associated with acute hemorrhagic fever in central Africa. PLoS Pathog 8(9):e1002924
30. Xu B, Liu L, Huang X, Ma H, Zhang Y, Du Y et al (2011) Metagenomic analysis of fever, thrombocytopenia and leukopenia syndrome (FTLS) in Henan Province, China: discovery of a new bunyavirus. PLoS Pathog 7(11):e1002369
31. Yu XJ, Liang MF, Zhang SY, Liu Y, Li JD, Sun YL et al (2011) Fever with thrombocytopenia associated with a novel bunyavirus in China. N Engl J Med 364(16):1523–1532
32. Brown JR, Morfopoulou S, Hubb J, Emmett WA, Ip W, Shah D et al (2015) Astrovirus VA1/HMO-C: an increasingly recognized neurotropic pathogen in immunocompromised patients. Clin Infect Dis 60(6):881–888
33. Fremond ML, Perot P, Muth E, Cros G, Dumarest M, Mahlaoui N et al (2015) Next-generation sequencing for diagnosis and tailored therapy: a case report of astrovirus-associated progressive encephalitis. J Pediatr Infect Dis Soc 4(3):e53–e57
34. Palacios G, Druce J, Du L, Tran T, Birch C, Briese T et al (2008) A new arenavirus in a cluster of fatal transplant-associated diseases. N Engl J Med 358(10):991–998
35. Holtz LR, Finkbeiner SR, Zhao G, Kirkwood CD, Girones R, Pipas JM et al (2009) Klassevirus 1, a previously undescribed member of the family Picornaviridae, is globally widespread. Virol J 6:86
36. Greninger AL, Runckel C, Chiu CY, Haggerty T, Parsonnet J, Ganem D et al (2009) The complete genome of klassevirus – a novel picornavirus in pediatric stool. Virol J 6:82
37. Li L, Victoria J, Kapoor A, Blinkova O, Wang C, Babrzadeh F et al (2009) A novel picornavirus associated with gastroenteritis. J Virol 83(22):12002–12006
38. Yozwiak NL, Skewes-Cox P, Gordon A, Saborio S, Kuan G, Balmaseda A et al (2010) Human enterovirus 109: a novel interspecies recombinant enterovirus isolated from a case of acute pediatric respiratory illness in Nicaragua. J Virol 84(18):9047–9058
39. McMullan LK, Folk SM, Kelly AJ, MacNeil A, Goldsmith CS, Metcalfe MG et al (2012) A new phlebovirus associated with severe febrile illness in Missouri. N Engl J Med 367(9):834–841
40. Feng H, Shuda M, Chang Y, Moore PS (2008) Clonal integration of a polyomavirus in human Merkel cell carcinoma. Science 319(5866):1096–1100
41. Flippot R, Malouf GG, Su X, Khayat D, Spano JP (2016) Oncogenic viruses: lessons learned using next-generation sequencing technologies. Eur J Cancer 61:61–68

42. Siebrasse EA, Reyes A, Lim ES, Zhao G, Mkakosya RS, Manary MJ et al (2012) Identification of MW polyomavirus, a novel polyomavirus in human stool. J Virol 86(19):10321–10326
43. Yu G, Greninger AL, Isa P, Phan TG, Martinez MA, de la Luz SM et al (2012) Discovery of a novel polyomavirus in acute diarrheal samples from children. PLoS One 7(11):e49449
44. Buck CB, Phan GQ, Raiji MT, Murphy PM, McDermott DH, McBride AA (2012) Complete genome sequence of a tenth human polyomavirus. J Virol 86(19):10887
45. Sauvage V, Foulongne V, Cheval J, Ar Gouilh M, Pariente K, Dereure O et al (2011) Human polyomavirus related to African green monkey lymphotropic polyomavirus. Emerg Infect Dis 17(8):1364–1370
46. Phan TG, Vo NP, Bonkoungou IJ, Kapoor A, Barro N, O'Ryan M et al (2012) Acute diarrhea in West African children: diverse enteric viruses and a novel parvovirus genus. J Virol 86(20):11024–11030
47. Breitbart M, Hewson I, Felts B, Mahaffy JM, Nulton J, Salamon P et al (2003) Metagenomic analyses of an uncultured viral community from human feces. J Bacteriol 185(20):6220–6223
48. Carding SR, Davis N, Hoyles L (2017) Review article: the human intestinal virome in health and disease. Aliment Pharmacol Ther 46(9):800–815
49. Kim MS, Park EJ, Roh SW, Bae JW (2011) Diversity and abundance of single-stranded DNA viruses in human feces. Appl Environ Microbiol 77(22):8062–8070
50. Lim ES, Zhou Y, Zhao G, Bauer IK, Droit L, Ndao IM et al (2015) Early life dynamics of the human gut virome and bacterial microbiome in infants. Nat Med 21(10):1228–1234
51. Minot S, Bryson A, Chehoud C, Wu GD, Lewis JD, Bushman FD (2013) Rapid evolution of the human gut virome. Proc Natl Acad Sci U S A 110(30):12450–12455
52. Minot S, Sinha R, Chen J, Li H, Keilbaugh SA, Wu GD et al (2011) The human gut virome: inter-individual variation and dynamic response to diet. Genome Res 21(10):1616–1625
53. Lepage P, Colombet J, Marteau P, Sime-Ngando T, Dore J, Leclerc M (2008) Dysbiosis in inflammatory bowel disease: a role for bacteriophages? Gut 57(3):424–425
54. Norman JM, Handley SA, Baldridge MT, Droit L, Liu CY, Keller BC et al (2015) Disease-specific alterations in the enteric virome in inflammatory bowel disease. Cell 160(3):447–460
55. Zuo T, Wong SH, Lam LYK, Lui R, Cheung K, Tang W et al (2017) Bacteriophage transfer during fecal microbiota transplantation is associated with treatment response in *Clostridium Difficile* infection. Gastroenterology 152(5):S140–S1S1
56. Wang W, Jovel J, Halloran B, Wine E, Patterson J, Ford G et al (2015) Metagenomic analysis of microbiome in colon tissue from subjects with inflammatory bowel diseases reveals interplay of viruses and bacteria. Inflamm Bowel Dis 21(6):1419–1427
57. Monaco CL, Gootenberg DB, Zhao G, Handley SA, Ghebremichael MS, Lim ES et al (2016) Altered virome and bacterial microbiome in human immunodeficiency virus-associated acquired immunodeficiency syndrome. Cell Host Microbe 19(3):311–322
58. Legoff J, Resche-Rigon M, Bouquet J, Robin M, Naccache SN, Mercier-Delarue S et al (2017) The eukaryotic gut virome in hematopoietic stem cell transplantation: new clues in enteric graft-versus-host disease. Nat Med 23(9):1080–1085
59. Rosen BI, Fang ZY, Glass RI, Monroe SS (2000) Cloning of human picobirnavirus genomic segments and development of an RT-PCR detection assay. Virology 277(2):316–329
60. Krishnamurthy SR, Wang D (2018) Extensive conservation of prokaryotic ribosomal binding sites in known and novel picobirnaviruses. Virology 516:108–114
61. Zhang T, Breitbart M, Lee WH, Run JQ, Wei CL, Soh SW et al (2006) RNA viral community in human feces: prevalence of plant pathogenic viruses. PLoS Biol 4(1):e3
62. Nakamura S, Yang CS, Sakon N, Ueda M, Tougan T, Yamashita A et al (2009) Direct metagenomic detection of viral pathogens in nasal and fecal specimens using an unbiased high-throughput sequencing approach. PLoS One 4(1):e4219
63. Moustafa A, Xie C, Kirkness E, Biggs W, Wong E, Turpaz Y et al (2017) The blood DNA virome in 8,000 humans. PLoS Pathog 13(3):e1006292
64. Wylie KM, Mihindukulasuriya KA, Sodergren E, Weinstock GM, Storch GA (2012) Sequence analysis of the human virome in febrile and afebrile children. PLoS One 7(6):e27735

65. Segura-Wang M, Gorzer I, Jaksch P, Puchhammer-Stockl E (2018) Temporal dynamics of the lung and plasma viromes in lung transplant recipients. PLoS One 13(7):e0200428
66. Schiemann M, Puchhammer-Stockl E, Eskandary F, Kohlbeck P, Rasoul-Rockenschaub S, Heilos A et al (2017) Torque Teno virus load-inverse association with antibody-mediated rejection after kidney transplantation. Transplantation 101(2):360–367
67. Li L, Deng X, Linsuwanon P, Bangsberg D, Bwana MB, Hunt P et al (2013) AIDS alters the commensal plasma virome. J Virol 87(19):10912–10915
68. Sauvage V, Eloit M (2016) Viral metagenomics and blood safety. Transfus Clin Biol 23(1):28–38
69. Naccache SN, Greninger AL, Lee D, Coffey LL, Phan T, Rein-Weston A et al (2013) The perils of pathogen discovery: origin of a novel parvovirus-like hybrid genome traced to nucleic acid extraction spin columns. J Virol 87(22):11966–11977
70. Lysholm F, Wetterbom A, Lindau C, Darban H, Bjerkner A, Fahlander K et al (2012) Characterization of the viral microbiome in patients with severe lower respiratory tract infections, using metagenomic sequencing. PLoS One 7(2):e30875
71. Ly M, Abeles SR, Boehm TK, Robles-Sikisaka R, Naidu M, Santiago-Rodriguez T et al (2014) Altered oral viral ecology in association with periodontal disease. mBio 5(3):e01133–e01114
72. Abeles SR, Robles-Sikisaka R, Ly M, Lum AG, Salzman J, Boehm TK et al (2014) Human oral viruses are personal, persistent and gender-consistent. ISME J 8(9):1753–1767
73. Abeles SR, Ly M, Santiago-Rodriguez TM, Pride DT (2015) Effects of long term antibiotic therapy on human Oral and fecal viromes. PLoS One 10(8):e0134941
74. Wang Y, Zhu N, Li Y, Lu R, Wang H, Liu G et al (2016) Metagenomic analysis of viral genetic diversity in respiratory samples from children with severe acute respiratory infection in China. Clin Microbiol Infect 22(5):458.e1–458.e9
75. Abbas AA, Diamond JM, Chehoud C, Chang B, Kotzin JJ, Young JC et al (2017) The perioperative lung transplant virome: torque Teno viruses are elevated in donor lungs and show divergent dynamics in primary graft dysfunction. Am J Transplant Off J Am Soc Transplant Am Soc Transplant Surg 17(5):1313–1324
76. Willner D, Furlan M, Haynes M, Schmieder R, Angly FE, Silva J et al (2009) Metagenomic analysis of respiratory tract DNA viral communities in cystic fibrosis and non-cystic fibrosis individuals. PLoS One 4(10):e7370
77. Wylie KM, Mihindukulasuriya KA, Zhou Y, Sodergren E, Storch GA, Weinstock GM (2014) Metagenomic analysis of double-stranded DNA viruses in healthy adults. BMC Biol 12:71
78. Santiago-Rodriguez TM, Ly M, Bonilla N, Pride DT (2015) The human urine virome in association with urinary tract infections. Front Microbiol 6:14
79. Hannigan GD, Meisel JS, Tyldsley AS, Zheng Q, Hodkinson BP, SanMiguel AJ et al (2015) The human skin double-stranded DNA virome: topographical and temporal diversity, genetic enrichment, and dynamic associations with the host microbiome. mBio 6(5):e01578–e01515
80. Zarate S, Taboada B, Yocupicio-Monroy M, Arias CF (2017) Human Virome. Arch Med Res 48(8):701–716
81. Leung P, Eltahla AA, Lloyd AR, Bull RA, Luciani F (2017) Understanding the complex evolution of rapidly mutating viruses with deep sequencing: beyond the analysis of viral diversity. Virus Res 239:43–54
82. Wohl S, Schaffner SF, Sabeti PC (2016) Genomic analysis of viral outbreaks. Ann Rev Virol 3(1):173–195
83. Tzou PL, Ariyaratne P, Varghese V, Lee C, Rakhmanaliev E, Villy C et al (2018) Comparison of an in vitro diagnostic next-generation sequencing assay with sanger sequencing for HIV-1 genotypic resistance testing. J Clin Microbiol 56(6) e00105-18
84. Fernandez-Caballero JA, Chueca N, Poveda E, Garcia F (2017) Minimizing next-generation sequencing errors for HIV drug resistance testing. AIDS Rev 19(4):231–238
85. van der Beek MT, Claas ECJ (2016) Phenotypic and genotypic antiviral susceptibility testing. p.201-228. Clinical virology manual, 5th edn. Eds. Loeffelholz, Hodinka, Young, Pinsky. American Society of Microbiology, Wiley, Hoboken, NJ.

86. Mercier-Darty M, Boutolleau D, Lepeule R, Rodriguez C, Burrel S (2018) Utility of ultra-deep sequencing for detection of varicella-zoster virus antiviral resistance mutations. Antivir Res 151:20–23
87. Chen Y, Scieux C, Garrait V, Socie G, Rocha V, Molina JM et al (2000) Resistant herpes simplex virus type 1 infection: an emerging concern after allogeneic stem cell transplantation. Clin Infect Dis 31(4):927–935
88. Fujii H, Kakiuchi S, Tsuji M, Nishimura H, Yoshikawa T, Yamada S et al (2018) Application of next-generation sequencing to detect acyclovir-resistant herpes simplex virus type 1 variants at low frequency in thymidine kinase gene of the isolates recovered from patients with hematopoietic stem cell transplantation. J Virol Methods 251:123–128
89. Sahoo MK, Lefterova MI, Yamamoto F, Waggoner JJ, Chou S, Holmes SP et al (2013) Detection of cytomegalovirus drug resistance mutations by next-generation sequencing. J Clin Microbiol 51(11):3700–3710
90. Kampmann SE, Schindele B, Apelt L, Buhrer C, Garten L, Weizsaecker K et al (2011) Pyrosequencing allows the detection of emergent ganciclovir resistance mutations after HCMV infection. Med Microbiol Immunol 200(2):109–113
91. Schnepf N, Dhedin N, Mercier-Delarue S, Andreoli A, Mamez AC, Ferry C et al (2013) Dynamics of cytomegalovirus populations harbouring mutations in genes UL54 and UL97 in a haematopoietic stem cell transplant recipient. J Clin Virol 58(4):733–736
92. Benzi F, Vanni I, Cassina G, Ugolotti E, Di Marco E, Cirillo C et al (2012) Detection of ganciclovir resistance mutations by pyrosequencing in HCMV-infected pediatric patients. J Clin Virol 54(1):48–55
93. Chou S, Ercolani RJ, Sahoo MK, Lefterova MI, Strasfeld LM, Pinsky BA (2014) Improved detection of emerging drug-resistant mutant cytomegalovirus subpopulations by deep sequencing. Antimicrob Agents Chemother 58(8):4697–4702
94. Garrigue I, Moulinas R, Recordon-Pinson P, Delacour ML, Essig M, Kaminski H et al (2016) Contribution of next generation sequencing to early detection of cytomegalovirus UL97 emerging mutants and viral subpopulations analysis in kidney transplant recipients. J Clin Virol 80:74–81
95. Gooskens J, Jonges M, Claas EC, Meijer A, Kroes AC (2009) Prolonged influenza virus infection during lymphocytopenia and frequent detection of drug-resistant viruses. J Infect Dis 199(10):1435–1441
96. Qi Y, Fan H, Qi X, Zhu Z, Guo X, Chen Y et al (2014) A novel pyrosequencing assay for the detection of neuraminidase inhibitor resistance-conferring mutations among clinical isolates of avian H7N9 influenza virus. Virus Res 179:119–124
97. Roosenhoff R, van der Linden A, Schutten M, Fouchier RAM (2017) A9 Deep sequencing analysis to investigate the importance of within host genetic diversity and evolution of influenza A viruses for the development of resistance against neuraminidase inhibitors. Virus Evol 3(Suppl 1) vew036.008
98. Pichon M, Gaymard A, Josset L, Valette M, Millat G, Lina B et al (2017) Characterization of oseltamivir-resistant influenza virus populations in immunosuppressed patients using digital-droplet PCR: comparison with qPCR and next generation sequencing analysis. Antivir Res 145:160–167
99. Zou X, Guo Q, Zhang W, Chen H, Bai W, Lu B et al (2018) Dynamic variation and reversion in the signature amino acids of H7N9 virus during human infection. J Infect Dis 218(4):586–594
100. McGinnis J, Laplante J, Shudt M, George KS (2016) Next generation sequencing for whole genome analysis and surveillance of influenza A viruses. J Clin Virol 79:44–50
101. Wu NC, Young AP, Dandekar S, Wijersuriya H, Al-Mawsawi LQ, Wu TT et al (2013) Systematic identification of H274Y compensatory mutations in influenza A virus neuraminidase by high-throughput screening. J Virol 87(2):1193–1199
102. Wu IC, Liu WC, Chang TT (2018) Applications of next-generation sequencing analysis for the detection of hepatocellular carcinoma-associated hepatitis B virus mutations. J Biomed Sci 25(1):51

103. Liu WC, Wu IC, Lee YC, Lin CP, Cheng JH, Lin YJ et al (2017) Hepatocellular carcinoma-associated single-nucleotide variants and deletions identified by the use of genome-wide high-throughput analysis of hepatitis B virus. J Pathol 243(2):176–192
104. Chen S, Wu J, Gu E, Shen Y, Wang F, Zhang W (2016) Evaluation of the dynamic pattern of viral evolution in patients with virological breakthrough during treatment with nucleoside/nucleotide analogs by ultradeep pyrosequencing. Mol Med Rep 13(1):651–660
105. Lowe CF, Merrick L, Harrigan PR, Mazzulli T, Sherlock CH, Ritchie G (2016) Implementation of next-generation sequencing for hepatitis B virus resistance testing and genotyping in a clinical microbiology laboratory. J Clin Microbiol 54(1):127–133
106. Lok AS, Ganova-Raeva L, Cloonan Y, Punkova L, Lin HS, Lee WM et al (2017) Prevalence of hepatitis B antiviral drug resistance variants in North American patients with chronic hepatitis B not receiving antiviral treatment. J Viral Hepat 24(11):1032–1042
107. Jones LR, Sede M, Manrique JM, Quarleri J (2016) Hepatitis B virus resistance substitutions: long-term analysis by next-generation sequencing. Arch Virol 161(10):2885–2891
108. Cho H, Ahn H, Lee DH, Lee JH, Jung YJ, Chang Y et al (2018) Entecavir and tenofovir reduce hepatitis B virus-related hepatocellular carcinoma recurrence more effectively than other antivirals. J Viral Hepat 25(6):707–717
109. Smit E (2014) Antiviral resistance testing. Curr Opin Infect Dis 27(6):566–572
110. Liu TF, Shafer RW (2006) Web resources for HIV type 1 genotypic-resistance test interpretation. Clin Infect Dis 42(11):1608–1618
111. Casadella M, Paredes R (2017) Deep sequencing for HIV-1 clinical management. Virus Res 239:69–81
112. Huber M, Metzner KJ, Geissberger FD, Shah C, Leemann C, Klimkait T et al (2017) MinVar: a rapid and versatile tool for HIV-1 drug resistance genotyping by deep sequencing. J Virol Methods 240:7–13
113. Wang C, Mitsuya Y, Gharizadeh B, Ronaghi M, Shafer RW (2007) Characterization of mutation spectra with ultra-deep pyrosequencing: application to HIV-1 drug resistance. Genome Res 17(8):1195–1201
114. Stelzl E, Proll J, Bizon B, Niklas N, Danzer M, Hackl C et al (2011) Human immunodeficiency virus type 1 drug resistance testing: evaluation of a new ultra-deep sequencing-based protocol and comparison with the TRUGENE HIV-1 Genotyping Kit. J Virol Methods 178(1–2):94–97
115. Avidor B, Girshengorn S, Matus N, Talio H, Achsanov S, Zeldis I et al (2013) Evaluation of a benchtop HIV ultradeep pyrosequencing drug resistance assay in the clinical laboratory. J Clin Microbiol 51(3):880–886
116. Cozzi-Lepri A, Noguera-Julian M, Di Giallonardo F, Schuurman R, Daumer M, Aitken S et al (2015) Low-frequency drug-resistant HIV-1 and risk of virological failure to first-line NNRTI-based ART: a multicohort European case-control study using centralized ultrasensitive 454 pyrosequencing. J Antimicrob Chemother 70(3):930–940
117. Li JZ, Paredes R, Ribaudo HJ, Svarovskaia ES, Metzner KJ, Kozal MJ et al (2011) Low-frequency HIV-1 drug resistance mutations and risk of NNRTI-based antiretroviral treatment failure: a systematic review and pooled analysis. JAMA 305(13):1327–1335
118. Swenson LC, Chui CK, Brumme CJ, Chan D, Woods CK, Mo T et al (2013) Genotypic analysis of the V3 region of HIV from virologic nonresponders to maraviroc-containing regimens reveals distinct patterns of failure. Antimicrob Agents Chemother 57(12):6122–6130
119. Wensing AM, Calvez V, Gunthard HF, Johnson VA, Paredes R, Pillay D et al (2017) 2017 update of the drug resistance mutations in HIV-1. Top Antivir Med 24(4):132–133
120. Pou C, Noguera-Julian M, Perez-Alvarez S, Garcia F, Delgado R, Dalmau D et al (2014) Improved prediction of salvage antiretroviral therapy outcomes using ultrasensitive HIV-1 drug resistance testing. Clin Infect Dis 59(4):578–588
121. Fun A, Leitner T, Vandekerckhove L, Daumer M, Thielen A, Buchholz B et al (2018) Impact of the HIV-1 genetic background and HIV-1 population size on the evolution of raltegravir resistance. Retrovirology 15(1):1

122. Kearney MF, Spindler J, Wiegand A, Shao W, Haubrich R, Riddler S et al (2018) Lower pre-ART intra-participant HIV-1 pol diversity may not be associated with virologic failure in adults. PLoS One 13(1):e0190438
123. McElvania TeKippe E, Wylie KM, Deych E, Sodergren E, Weinstock G, Storch GA (2012) Increased prevalence of anellovirus in pediatric patients with fever. PLoS One 7(11):e50937
124. Young JC, Chehoud C, Bittinger K, Bailey A, Diamond JM, Cantu E, Haas AR, Abbas A, Frye L, Christie JD, Bushman FD, Collman RG (2015) Viral metagenomics reveal blooms of anelloviruses in the respiratory tract of lung transplant recipients. Am J Transplant 15(1):200–209

Chapter 7
Metagenomic Applications for Infectious Disease Testing in Clinical Laboratories

Laura Filkins and Robert Schlaberg

7.1 Introduction

An explosion of technological advancements in clinical microbiology over the past two decades is rapidly transforming the laboratory diagnosis of infectious disease. Some of the most influential advancements include introduction of rapid organism identification by matrix-assisted laser desorption-ionisation time-of-flight (MALDI-TOF) mass spectrometry and DNA sequencing of marker genes, increased availability of direct-from-specimen nucleic acid detection tests (NAAT, including syndromic panels), targeted detection of genetic markers to rapidly predict antimicrobial resistance [1, 2]. These methods decrease time-to-results, provide accurate identification and improved sensitivity compared to classic methods, enable clinicians to select optimal antimicrobial therapy sooner, and reduce overuse of antibiot ics [3, 4].

While the current clinical microbiology methods have greatly improved routine diagnostics, these approaches have limitations. Both culture-dependent and independent methods are only able to detect a limited repertoire of organisms. Utilising these methods, only targeted (pre-selected), viable, and/or culturable microorganisms will be detected. Additionally, strains exhibiting non-standard phenotypes (biochemical identification), altered protein expression profiles (MALDI-TOF), or genetic variation (NAAT) within the targeted micro-organism groups may lead to incorrect or false-negative results. For NAAT, frequent test redesign may be

L. Filkins
Departement of Pathology, University of Texas Southwestern Medical Center, Dallas, TX, USA

R. Schlaberg (✉)
Departement of Pathology, University of Utah, Salt Lake City, UT, USA
e-mail: robert.schlaberg@path.utah.edu

J. Moran-Gilad, Y. Yagel (eds.), *Application and Integration of Omics-powered Diagnostics in Clinical and Public Health Microbiology*,
https://doi.org/10.1007/978-3-030-62155-1_7

necessary, especially when new pathogens emerge as has recently been highlighted by the need to design, manufacture, validate, and distribute new NAAT to detect the emerging SARS-CoV-2. Further, differentiating strains of the same species (strain typing) for diagnostic, surveillance, and infection prevention purposes usually requires additional testing, which limits availability and timeliness of results. Metagenomic next-generation sequencing (NGS) directly from patient specimens in clinical laboratories (clinical metagenomics) helps overcome these challenges as it provides a hypothesis-free, genome-based, high-resolution alternative to conventional testing. Clinical metagenomics enables detection of organisms that are difficult to culture, slow growing, genetically divergent, while also providing genotypic information for the purpose of strain-typing or prediction of antimicrobial resistance.

As clinical metagenomic testing is adopted by a rapidly growing number of laboratories the need for standardised, streamlined, high quality, and compliant workflows increases. In this chapter, we present an overview of current technologies, remaining challenges, and approaches to overcome them. We define metagenomic sequencing as the process of sequencing nucleic acid (RNA and/or DNA) directly from clinical specimens, including the use of workflows that apply target enrichment, host depletion or other pre-sequencing steps.

7.2 Clinical Need for Advanced Testing

The efficacy of conventional diagnostics varies based on the clinical syndrome, patient population, and breadth of available diagnostic resources. The most challenging clinical syndromes to diagnose are those that present with non-specific symptoms, have a broad differential, and are unresponsive to empiric therapy. Strong interest is placed on the application of metagenomic testing for the diagnosis of meningitis/encephalitis, pneumonia, fever of unknown origin (FUO), bone and joint infections, intraocular infections, and others. Glaser and colleagues reported that a likely aetiologic agent of encephalitis was identified in less than 40% of patients enrolled in the California Encephalitis Project [5]. Similarly, diagnosis of community acquired and healthcare associated pneumonias is challenging with current testing approaches returning negative results in 20–60% of cases [6–8]. Further, determining the true aetiologic agent of pneumonia when one or more potential pathogens are detected often requires additional scrutiny and clinical interpretation, especially with pathogens that are highly prevalent, can also be commensals, persist after an acute infection, or causes varying disease severity [9]. In prosthetic joint infections, conventional culture methods fail to identify the causative micro-organism in about 40–50% of cases. Broad-range PCR or NGS can increase the diagnostic yield by 25% or more compared to culture [10, 11]. Sequencing of cell-free DNA (cfDNA) from plasma has recently been applied for the detection of micro-organisms associated with numerous clinical indications including sepsis, FUO, pneumonia, deep-seated infections, and others [12–15]. Finally, clinical metagenomics is a promising approach for the diagnosis of intraocular infections.

The very small specimen quantity that is obtainable from intraocular sources limits the number of NAAT and culture testing that can be performed. Using current molecular methods, fungi and viruses can be detected with >90% sensitivity and 75–90% sensitivity for bacterial detection from ocular sources, but achieving these sensitivities requires multiple assays and relatively large specimen volume [16]. Metagenomic sequencing provides a unified testing alternative that requires less specimen volume than a combination of bacterial, mycobacterial, fungal and viral culture, and multiple pathogen-specific NAAT [17].

Metagenomics can provide a diagnosis in many challenging diagnostic scenarios when conventional methods may be unsuccessful, as discussed below (Fig. 7.1). Additional applications of NGS in clinical microbiology include antimicrobial resistance (AMR) prediction, molecular epidemiology, and microbiome community profiling which are not covered here [18–21] but in other chapters of this book.

Clinical metagenomics can decrease time-to-results for slow growing or hard to diagnose micro-organisms, provide rapid, high-resolution micro-organism identification, resistance prediction to support optimal treatment choices, and reduce costs

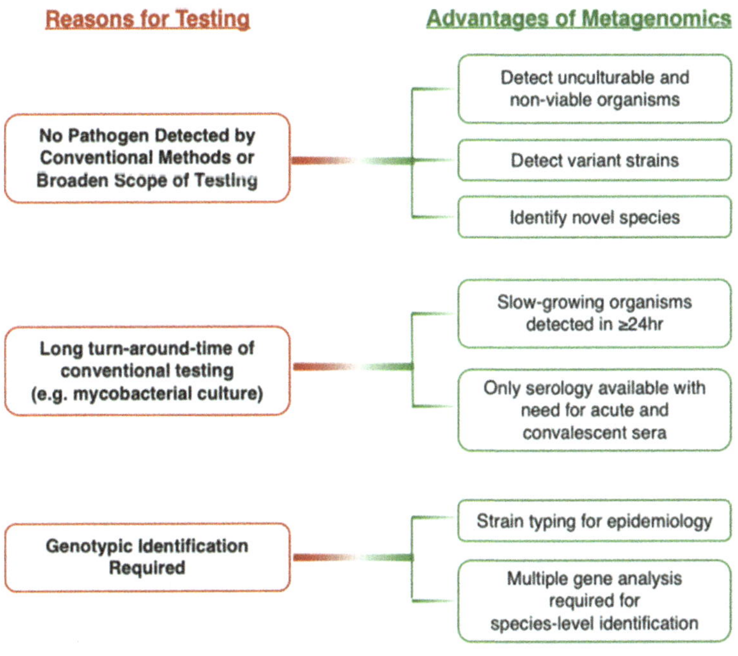

Fig. 7.1 Applications for metagenomic pathogen detection. Untargeted metagenomic next-generation sequencing is a culture-independent method that identifies pathogens by microbial nucleic acid detection directly from the patient specimen. This method detects nucleic acid from viable or non-viable cells and extracellular nucleic acids. Detection of variant strains and novel organisms is possible. Turnaround time for metagenomics test results is variable, typically ranging from 24 hours to 2 weeks, depending on the test design. Finally, whole genome or multiple gene sequencing enables specific classification of micro-organisms, even those that are phylogenetically closely related

by providing a comprehensive approach to answering multiple diagnostic questions [22]. However, for patients and clinicians to benefit from these advantages, significant changes in diagnostic algorithms and laboratory workflows will be required. While case reports have been published in diverse disease areas highlighting the power of clinical metagenomics, few clinical trials systematically comparing diagnostic yield and clinical benefit to standard of care have been conducted. Some of the most notable case reports and case series are based on cerebrospinal fluid (CSF) testing yielding unexpected aetiologies, including neuroleptospirosis in a pediatric patient with severe combined immunodeficiency, chronic meningoencephalitis due to Cache Valley Virus in a patient with X-linked agammaglobulinaemia, and neurobrucellosis in a female paediatric patient [23–25]. In a multicentre clinical trial for diagnosis of meningitis and encephalitis, 27.9% of enrolled patients were ultimately diagnosed with an infectious pathogen; of which, 33% of pathogens were detected by both conventional testing and metagenomic NGS on CSF, 45% by conventional testing only, and 22% by metagenomic NGS only [26]. The SEP-SEQ study employed metagenomic pathogen detection from plasma cfDNA and demonstrated detection of probable infectious causes of sepsis in an additional 15% of patients undiagnosed by conventional testing [15]. Similarly, Long et al. showed increased bacterial detection using plasma cfDNA compared to standard blood culture alone in patients with suspected sepsis in addition to detection of viral pathogens in 18% of the patients [27]. Testing of nasopharyngeal/oropharyngeal swabs from children hospitalised for community-acquired pneumonia by next-generation RNA sequencing identified previously missed putative pathogens in approximately 30% of patients [28].

Metagenomics is also a powerful tool to discover novel or emerging pathogens that escape detection by conventional methods. An early example was the identification of a novel rhabdovirus in serum from a patient with haemorrhagic fever [29]. Since then, pathogen (especially viral) discovery has been accelerated by the use of metagenomic sequencing and led to the detection of Henan Fever Virus, a novel bunyavirus in patients with fever, thrombocytopenia, and leukaemia syndrome; a novel arenavirus related to the lymphocytic choriomeningitis viruses in a cluster of fatal organ transplants; Lujo virus, an arenavirus first discovered from an outbreak of five cases of hemorrhagic fever in South Africa [30–32]; the recently discovered SARS-CoV-2; and many others.

These and other success stories were among the first evidence demonstrating the power of metagenomics-based pathogen detection for clinical diagnosis. However, until recently testing workflows and equipment were too slow, too expensive, required too much expert knowledge, and bioinformatics skills to be implemented in clinical laboratories. As these barriers are being removed, clinical metagenomics is increasingly being implemented in routine diagnostic algorithms. Further optimising, streamlining, and accelerating specimen preparation and sequencing technologies, standardising micro-organism identification, result interpretation, and quality control methods will facilitate adoption by clinical laboratories.

7.3 Test Design and Development

The potential benefits of utilising NGS technologies in clinical microbiology has been strongly demonstrated with case reports and initial clinical studies, including those described above. While the gains are substantial, as with any new technology, performance must be characterised for each clinical application and testing approach so that risks can be mitigated. Here, we describe technical and clinical challenges of metagenomic analyses for infectious disease diagnosis and suggest approaches to improve test characteristics while minimising sources of potential error.

7.3.1 Pre-Analytic Factors

As with any laboratory test, pre-analytic factors affect performance. Relevant factors include appropriate patient selection, defining relevant specimen types, specimen collection, preservation, transport, and storage conditions, and determining specimen stability. Pre-analytical steps need to be controlled and specimen rejection criteria need to be defined [33]. These factors are not unique to metagenomic testing but can affect the testing outcome differently than other microbiologic methods. For example, leaving a sputum specimen at room temperature for extended periods may result in reduced viability of fastidious pathogens, which can affect their recovery by culture. Results for that same specimen could be affected by over-growth of normal flora or degradation of pathogen nucleic acids limiting the sensitivity of metagenomics. Either scenario could cause decreased test sensitivity and would likely change the interpretation of results.

7.3.2 Specimen Preparation

At minimum, wet bench processes include nucleic acid extraction, library preparation, and sequencing. Additional wet bench procedures can enhance detection of pathogen-derived nucleic acids during sequencing, such as pathogen enrichment and removal of host nucleic acid or cells.

7.3.2.1 Nucleic Acid Extraction

Efficient, nucleic acid isolation is essential for producing high-quality sequencing libraries and reliable pathogen identification results. The target nucleic acid of interest (RNA versus DNA, or both) must be determined during test design to best address the needs of the test. DNA is useful when evaluating bacterial, fungal, and eukaryotic targets and for detection of DNA viruses. However, RNA extraction sequencing is required for detecting RNA viruses. DNA enables whole-genome sequencing, whereas RNA sequencing is limited to those genes that are actively expressed. RNA can be advantageous for detection of pathogens that have high levels of gene expression, as the number of nucleic acids in the sequencing dataset is amplified compared to the amount of nucleic acid that would be present from a genome. Conversely, latent infections with quiescent micro-organisms may be more difficult to detect using RNA compared to DNA.

Specimen type and relevant pathogens guide selection of extraction methods [34]. Tissue and stool typically require more aggressive methods, such as mechanical lysis or bead beating, to release nucleic acids due to the physical composition of the specimen, whereas for other specimens, such as CSF or synovial fluid, chemical lysis is sufficient. Target micro-organisms with thick cell walls, including many fungi, usually require a mechanical, bead beating lysis method. Finally, the use of high purity plasticware and reagents (i.e. tubes, columns, buffers) with low levels of contaminating nucleic acids reduces detection of background organisms.

7.3.2.2 Pathogen Enrichment

A common challenge of untargeted metagenomic analysis for pathogen detection is the significant proportion of sequencing reads that are derived from host nucleic acid. Host cells or free nucleic acids compete with pathogen nucleic acid during sequencing and can reduce the analytical strength. Methods to enrich pathogen-derived and/or reduce host-derived nucleic acids can improve analytical sensitivity while reducing sequencing costs by reducing the depth of sequencing required to detect low abundance organisms. Target enrichment can be achieved by capture of pathogen nucleic acid or PCR-based enrichment. A common approach in microbiome studies focused specifically on bacterial or fungal communities is PCR amplification of marker genes followed by next-generation sequencing of PCR amplicons. Broad-range primers usual target conserved regions of the 16S rRNA gene (bacteria) or ITS region (fungi), or other highly conserved genes and are applied for amplification from total DNA (or less commonly cDNA) [35]. The resulting bacterial or fungal-enriched nucleic acid pool is then used for library preparation. This broadly targeted approach is also utilised for analysis of clinical specimens when suspicion for bacterial or fungal aetiology is strong, however detection of a causative micro-organism is limited to the selected category. Multiple primer enrichment can similarly be used to increase nucleic acid quantities for viral detection [36]. Capture-based enrichment methods have also been employed to select for sequences from organisms of interest [37]. However, bias is introduced by both broad targeting amplification methods, random amplification methods, and sequence capture [38–40]. Therefore, bias should be closely evaluated and characterised during clinical test development when these enrichment methods are used.

For untargeted metagenomic approaches, host-depletion is an important consideration and can increase detection of pathogens [41]. A variety of host-depletion methods exist and are applied at different steps within the sequencing workflow.

One approach is to deplete host nucleic acids before extraction. Allander et al. demonstrated improved detection of enveloped DNA viruses after treatment of serum with DNase to reduce extracellular host DNA [42]. For RNA sequencing, the removal of highly abundant host RNA includes ribosomal RNAs (most specimen types) or globin transcripts (whole blood specimens). Common methods of targeted RNA depletion include probe-based removal and target cleavage after nucleic acid extraction. See further discussion in Chap. 8.

7.3.2.3 Library Preparation

Sequencing libraries preparation methods have improved rapidly involving fewer and fewer steps, becoming faster to perform (often within a few hours or less), and can be automated on routine liquid handling instruments. Workflows are further streamlined by methods that limit the need for quality control and quantification of sequencing libraries. For optimal efficiency and to reduce costs, laboratories usually pool multiple barcoded libraries for sequencing on one sequencing run. Barcodes should be selected and demultiplexing parameters should be defined to limit misassociation of sequencing reads ("index cross-talk", "barcode hopping") as this can cause false-positive results [43, 44]. Strategies include dual indexing and design or selection of barcodes with maximal edit distance. Ideal sequencing datasets are diverse, containing numerous different reads mapping to each target microorganism. Therefore, library preparation methods that produce libraries with minimal duplication and increased diversity of reads typically yields higher-quality sequencing datasets and higher-confidence pathogen detection.

7.3.2.4 Sequencing

The selection of a sequencing platform is a critical step in the design phase of test development. Considerations should include resources already available at the institution, capital expense requirements, complexity of specimen preparation, reagent and sequencing run cost, desired read length, total read number per run, sequencing run time, sequencing error profile, and bioinformatics/analysis support. Prioritisation of these variables for individual applications in clinical microbiology will vary making one sequencing platform preferable over another given the precise needs of the test. Sequencing platform characteristics have been summarised elsewhere [35, 45, 46]. A comparison of Illumina sequencing platform (short read) versus Oxford Nanopore MinION platform (long read) of stool from pre-term infants demonstrated that long reads improved species-level detection for some bacteria, while the high error rate of the MinION prevented species-level identification for other bacterial genera that were successfully identified by Illumina [47]. This example highlights the challenges of either approach (long versus short read length) and the importance of tailoring the test design for the goal of an individual test.

The sensitivity of detection and specificity of micro-organism identification in specimens with low pathogen load is improved with increased sequencing depth, especially when de novo genome assembly is required for identification [48]. Unfortunately, increasing sequencing depth comes with increased cost and often longer run times. Speeding up the sequencing step of metagenomics workflows is a high priority for clinical applications as infectious disease testing requires a more rapid turn-around time than other genomics applications. In contrast to short read sequencing platforms, some long read technologies allow real time analysis which can be used to terminate sequencing when sufficient data has been generated. Using the Oxford Nanopore MinION, Greninger et al. demonstrated that sufficient sequencing data could be achieved to identify viral pathogens in high load serum specimens with <10 minutes of sequencing time, whereas moderate load specimens required 30–40 minutes [49].

7.3.3 Sequence Analysis

Clinical metagenomics presents unique challenges when compared to academic discovery applications. In research settings, the focus is often on comprehensive analyses (e.g. whole genome sequences) and increased time for computation and manual analysis by experts are acceptable. Additionally, multiple different analysis approaches are frequently used, often in a batched mode for all specimens that are part of a given study, to extract all pertinent genetic information and/or enable quantification of gene expression or organism abundance. In contrast, clinical testing requires testing and interpretation of results on a daily basis by a number of operators, strict adherence to pre-determined and validated procedures and interpretative criteria. Software used for data analysis needs to be diagnostic grade, version controlled, regularly updated, and meet data protection and privacy requirements. All procedures must be thoroughly vetted and turnaround time (TAT) for computational analysis steps are essential to the clinical utility of metagenomic tests. The selection of all analysis steps, including run quality pass/fail, read quality filtering, read classification (for organism detection) and/or contig assembly (for strain typing and de novo discovery), micro-organism determination, and reporting needs to be carefully determined based on the clinical application.

7.3.3.1 Sequence Analysis Tools

Preferred sequence data analysis methods may depend on the intended use of the test and the type of results that need to be generated [50–56]. Numerous bioinformatics tools have been published for research applications and vary in their approach to analysing sequences, accuracy and sensitivity of read classification, run time, and other characteristics [51, 52, 54, 56]. Requirements for data analysis software used in diagnostic workflows and need for bioinformatics support have to be taken into consideration when determining sequence analysis strategies. General approaches for sequence data analysis and read classification include alignment-based and alignment-free methods (k-*mer* based), use of whole genome or marker gene-based approaches (e.g. *rRNA*s, other conserved genes) [57–66].

Analysis time is a critical characteristic for clinical NGS-based tests, as extended TAT limits clinical utility. General approaches to faster read classification include reducing the number of sequence comparisons by limiting the number of query (i.e. reads per specimen) or reference sequences (i.e. database size) and utilising faster sequence comparisons tools (i.e. faster alignment or alignment-free methods) [57]. Reducing the number of query sequences is most commonly achieved by removal of duplicate reads, binning or clustering of sequences before querying and subsequent querying of a single representative sequence for each cluster and assembly of sequences into longer contigs [67]. Database sizes can be reduced by limiting redundancy while representing as much sequence diversity as possible [68]. However, for clinical diagnostics reducing reference database sizes carries substantial risk for loss of performance via higher rates of false negative (pathogen-derived reads do not match the representative sequence closely enough to be identified) and false positive results (mis-assignment of reads to the next-closest reference sequence if a better, correct match is missing). Thus, database design is a critical component of clinical metagenomics tests. Many open source sequence analysis tools (e.g. Kraken) allow users to provide their own reference sequences, allowing customisation to specific requirements and applications [52]. However, extreme bias and limited quality of public reference databases pose substantial challenges when broad pathogen detection requirements necessitate comprehensive databases [69, 70]. In recent years, rapid read classification tools have been developed that reduce the need to limit the size of reference sequence databases. Analyses that took days or longer can now be performed within an hour or less [51, 60, 63]. In addition, to speed the ease of use, reliability and accuracy, independence of expertise of the user, and version control are other important features for data analysis tools to be used as part of clinical metagenomics workflows.

7.3.3.2 Organism Classification and Result Interpretation

Independent of the selected bioinformatic analysis tool, criteria for micro-organism classification and result interpretation must be defined. Important considerations include relative importance of sensitivity vs. specificity of pathogen detection, relevant micro-organism abundance (e.g. are low-positive results relevant?), composition and abundance of normal microbiota (e.g. do pathogens need to be differentiated from closely related commensals?, which ones?), expected biologic sequence diversity for relevant pathogens (e.g. RNA viruses), and prioritisation and interpretation of results (e.g. do certain commensals need to be excluded or high-impact pathogens be prioritised for reporting purposes?). In general, if the focus is on sensitivity, less stringent classification and interpretation criteria may be appropriate whereas applications that require high specificity will need to employ more stringent classification and interpretation criteria. In addition, stringency may have to vary substantially between different taxa and require adjustment for given sequencing read lengths and sequencing error profiles. For example, classification of pathogens with divergent genomes (e.g. RNA viruses) may require laxer sequence comparison conditions (smaller k in k-*mer* based approaches, shorter seed length and higher tolerance for gaps and mismatches) or protein-level analyses (i.e. comparison of translated nucleotide query sequences against a protein or translated nucleotide database) to maximise sensitivity. While traditionally slow, these searches can now be performed at rapid speed [51, 58, 60, 63]. Final classification and interpretation criteria for a test as a whole or given micro-organism will impact test performance and should be acknowledged in the test information provided to clinicians.

7.3.3.3 Identifying Contamination

Sequencing artifacts (e.g. low-quality reads) and sequencing data representing contamination introduced during specimen collection or processing (e.g. reagent contamination) need to be anticipated, identified as such, and differentiated from relevant, specimen-derived sequences. Contamination may arise from reagents containing microbial DNA (e.g. due to environmental contamination, as part of recombinant enzymes, etc.), may be introduced during specimen collection, storage, or processing, mis-inoculation or impurities of barcode sequences, carry-over of within sequencing instruments, index hopping, and other mechanisms [44, 71, 72]. The use of ultra-pure reagents in well controlled molecular laboratory settings reduces but often cannot completely eliminate the risk of contamination. Therefore, carefully selected external (positive and negative) controls and internal (spike-in) controls are needed throughout the entire workflow to identify sequences not derived from the clinical specimen [73].

7.3.3.4 Result Interpretation

Some of the consideration for determining which detected micro-organisms should be included in a diagnostic report may include: (1) comparison of micro-organisms detected in patient specimens with those identified in external controls; (2) in shotgun metagenomic workflows, the detection level for a given micro-organism depends on the presence and abundance of other organisms and host nucleic acid; because those may differ between patient specimens and external controls, simply excluding micro-organisms found in external controls may not yield the optimal results; approaches that take the biomass and composition of the specimen into consideration have been developed [74, 75]; (3) as discussed above, *a priori* defining those organisms that are relevant for a given test and prioritising those for reporting may be beneficial; (4) adjusting confidence thresholds for reporting of organisms based on the intended use of the test, impact of a given detection, completeness of reference databases and/or genetic variability of relevant micro-organisms; and (5) reporting only of those organisms that meet a validated minimum reporting detection thresholds. Thresholds may be based on a number of individual metrics or combinations of criteria including minimum total number of reads assigned to a given organism, establishing a minimum proportion of genes or genome that needs to be identified, minimum depth of coverage over a pre-determined region of the genome, and others. Thresholds may need to be customised for specific micro-organisms. In particular, taxa from dense parts of the phylogenetic tree (i.e. with genetically similar neighbours) may require particular attention. By tailoring detection and reporting criteria to individual micro-organisms, sensitivity and specificity can be maximised.

7.3.3.5 Approach to Test Validation

Ideally, validations would include clinical specimens with known results based on high-quality predicate tests, with known quantities, covering all relevant micro-organisms detectable by the sequencing test, in all relevant specimen matrices, combined with clinical specimens that contain micro-organisms that need to be differentiated from relevant organisms to avoid false-positive results. However, due to limited availability of well-characterised specimens, lack of a universal reference method, and the sheer scope of clinical metagenomics tests, this is generally not realistic. There is currently no consensus on how laboratories should strike a balance between sufficiently characterising test performance while using limited resources judiciously. Approaches often include a combination of positive and negative patient specimens (based on conventional tests), spiked patient specimens, reference materials (as individual positives or mock communities, with or without matrix), and *in silico* generated mock specimens (based on simulated micro-organism sequences with or without real or mock matrix sequences) [15, 28, 76]. Usually, positives at least for the most common pathogens and commensals can be sourced for the relevant specimen types. Mock specimens (laboratory spiked or *in silico* generated) can help assess performance for detection of clinically important but less widely available micro-organisms. Serial dilution studies (again, laboratory spiked or *in silico* generated) can be used to assess sensitivity while specificity can also be tested using negative patient specimens, blanks, and *in silico* generated specimens. As with other diagnostic tests, routine performance characteristics (accuracy, reproducibility, sensitivity, specificity, stability, etc.) should be considered.

Testing of *in silico* generated specimens enables assessment of a much larger number of relevant pathogens and commensals at low cost and with complete knowledge of the expected results. Sequencing data of the same size, read length, and error profile can be constructed computationally (*in silico*) and analysed with the diagnostic pipeline. As discussed above, sequencing data from real patient specimens often contains sequence artefacts and sequences that did not originate from the specimen. If using *in silico* generated specimens, this should be taken into consideration. Relevant sample composition can be recapitulated by generating sequencing data from the host (human DNA sequences), common contaminants, and commensals [28]. A large number of metagenomics datasets are also available from public databases and may help avoid over-training when *in silico* data are generated from sequences contained in classification databases (i.e. perfect matches exist for simulated specimens) [77, 78]. *In silico* testing is especially important for validation of rare but clinically important pathogens, including emerging pathogens and biosafety level (BSL)-3 or BSL-4 agents that may not be practical to handle for spiking experiments. This approach can also be useful for studying closely related taxa that may be common but difficult to differentiate (e.g. *Streptococcus pneumoniae* and *S. mitis*) as specimen composition can be fully controlled, including their relative abundance.

7.3.4 *Quality Management*

Quality control and quality assurance must be implemented throughout the metagenomic testing process. All steps of testing, including pre-analytic, analytic, and post-analytic should be assessed through the laboratory's quality procedures [79]. There is no consensus yet on the specifics and extent of quality control measures. Some approaches are listed below.

7.3.4.1 Quality Control and Assessment

Analysis of specimen-level and run-level quality metrics is recommended throughout the specimen processing and data analysis workflow, including pre-analytic specimen checks, nucleic acid yield, assessment of library quantity and quality, evaluation of sequencing data quality and quantity for the entire run (including results for external controls) and for each specimen [28]. Sequencing error rate and base call quality are among the commonly used metrics to assess run performance. Pass/fail criteria should be defined to ensure high quality results without being overly stringent, resulting in unnecessary costs and delays. For positive control specimens, the expected identity and relative abundance of detectable organisms is known, and expected results need to be obtained. Negative controls can consist of matrix-matched or blank specimens and help identify contamination (see above). Matrix-matched controls can also identify problems that are dependent on specimen characteristics (e.g. viscosity, presence of inhibitors). Internal controls (e.g. whole micro-organisms also controlling for extraction, or synthetic nucleic acid) should be selected so that they can be readily differentiated from micro-organisms of interest and can be spiked into a master mix that is used for all specimens (e.g. lysis buffer) or be used as specimen-specific spike-in control with a unique sequence [28]. Depending on the specific strategy, internal controls can be used as processing controls, to monitor specimen composition, and identify specimen-to-specimen contamination. The number of sequencing reads and/or sequence coverage of spike-in controls can also be used to assess specimen adequacy.

7.3.4.2 TAT

To be clinically actionable, results need to be reported in a timely manner. Longer TAT tests may have clinical utility for chronic infections. At least for short read platforms, sequencing library preparation and NGS contribute the most to the overall TAT. Often, host depletion or target enrichment steps can further increase processing times. When determining the need for automation, the rate of errors during sample processing, repeat rates, reproducibility, as well as impact on TAT should be taken into consideration. Time to result can also be highly variable for different sequencing platforms and throughput needs, ranging from less than an hour to multiple days [45]. Data analysis (even for diagnostic applications) can now be performed in well below an hour [51, 60, 63] and data analysis steps often do not significantly contribute to the overall TAT any more. Workflow management further impact TAT. To minimize TAT, organising workflows in at least two shifts may be required. Implementation of clear protocols including repeat algorithms and multiple pass/fail check points throughout testing and special considerations for specimens with short storage stability that may not support repeat testing is especially important for minimizing TAT during non-ideal testing situations.

7.4 Remaining Challenges for NGS in Clinical Diagnostics

Breakthroughs in specimen preparation, sequencing technology, and computational biology enabled introduction of the first clinical metagenomics tests at select reference and public health laboratories. Protocols and technologies evolve rapidly and implementing clinical metagenomics tests is becoming feasible for a growing number of laboratories. To further increase access, future workflow improvements will likely increase analytical sensitivity, reduce TAT and costs (both per sample costs and capital expense requirements), streamline test development. Clinical outcome and test utilisation studies are needed to establish guidelines for best ordering practices.

7.4.1 Sample Processing

An ongoing challenge for metagenomics-based testing is the fact that host nucleic acid and pathogen nucleic acid compete during library preparation and sequencing. Numerous methods for both pathogen target enrichment and host (nucleic acids or cell) depletion exist aiming at increasing sensitivity and decreasing the required sequencing depth, and therefore cost [41, 42, 80]. However, most available methods have considerable limitations, requiring fresh specimens, high molecular weight nucleic acid, long incubation times, or off-target effects. For RNA-sequencing-based workflows, ribosomal RNA (rRNA) and globin depletion (for bloody specimens) are commercially available. In addition, greater ease-of-use and lower costs of customized depletion probes makes it feasible to also consider removal of other highly abundant transcripts. Similar technology has also made it possible to design target enrichment workflows that allow for broad pathogen detection [37, 81–84]. Potential cross-reactivity between host and pathogen sequences - that may be difficult to exclude or quantify - remains a challenge for hybridisation-based depletion methods. Another concern is loss of specimen nucleic acid and pathogen yield in additional processing steps. Commercially available depletion or enrichment methods are needed that reduce cost and workflow barriers for diagnostic laboratories and maximise analytical sensitivity of broad NGS-based pathogen detection tests.

Clean reagents that are free of contaminating nucleic acids and workflows that reduce the risk for environmental contamination are essential for molecular testing in general but problems are amplified for clinical metagenomics tests due to their broad scope [85, 86]. The impact of any improvements will be greatest on low biomass specimens that are most vulnerable to artifacts introduced by reagent and environmental contamination.

Complexity of metagenomics specimen preparation workflows provide a barrier for laboratories. Resources including laboratory space for unidirectional workflow, personnel training, and expertise for data analysis and interpretation have to anticipate and accommodate workflow complexities [87]. Future development will have to focus on simplifying workflows, minimising hands-on time, reducing expertise needed for post-sequencing steps, including quality control/quality assurance of metagenomics workflows. Many of these problems have been addressed in other areas of NGS testing already and lessons can be applied to clinical metagenomics, and the next years are likely to bring substantial improvements in ease-of-use and performance of metagenomics tests.

7.4.2 Sequencing and Data Analysis

In addition to user-friendly data analysis and reporting tools designed for use by clinical laboratories, the combination of fast (within approximately 4 h), reliable, and economical sequencing platforms will be essential for broad adoption of clinical metagenomics in clinical and public health laboratories. Decreased costs could also open NGS technology to a number of additional microbiology applications. For example, laboratories might consider more general use of whole genome sequencing for identification of clinical isolates.

Sequence data analysis, organism identification, and reporting will need to be further standardised [88]. Currently, most laboratories use customised analysis tools and criteria limiting reproducibility of results and external validity of published studies [89]. Standardised data analysis will also reduce the effort needed for laboratories to develop metagenomics tests. NGS data analysis software should include user interfaces designed for laboratory staff (i.e. not requiring bioinformatics skills), reporting functionality, including interfacing of results with laboratory information systems, and support routine result review and release workflows [90].

7.4.3 *Test Utilization*

As with any new technology, optimal applications for clinical metagenomics need to be established. More clinical utility studies need to be performed with specific application, patient enrollment criteria, comprehensive predicate testing, defined specimen collection, preservation, and processing protocols, and clinical outcome data. Currently, the most common scenario for ordering clinical metagenomics tests is in critically ill patients in addition to standard diagnostic workup or after standard testing is unsuccessful. The use as a test of last resort has the disadvantages of further prolonging the time to diagnosis and limiting testing to patients with low pre-test probability. In addition, current testing approaches often provide an incomplete picture of the potential pathogens detected. At least on some specimen types (e.g. respiratory specimens) identification of one potential pathogen does not exclude the possibility that additional, possibly more relevant pathogens may have gone undetected. Incorporating metagenomics tests earlier may benefit patients and reduce unnecessary testing but adequate patient selection criteria need to be defined. For example, in patients with risk factors or clinical presentations that lead to a long list of differential diagnoses, broad pathogen detection with a single test early on could shorten the time to diagnosis and reduce costs for unnecessary testing and inadequate treatment. Another application is specimens that usually have very limited volume available but require testing for a number of organisms (e.g. vitreous or intraocular fluid). Limited specimen volume may allow clinicians and laboratories to perform only a few pathogen-specific tests. Being able to test for a much larger number of potential pathogens with a single test provides an advantage to metagenomics tests [91]. Further clinical studies are required to identify high yield testing situations with positive clinical impact.

7.4.4 *Incidental Findings*

One potential consequence of untargeted testing is the inadvertent detection of host genomic variants, unexpected pathogens (e.g. sexually transmitted infections), or non-validated micro-organisms with confident detection and clear clinical significance. Thus, the question "should the additional information be disclosed to the patient?" becomes relevant. To avoid incidental detection of host genomic variants, human sequence data can be removed or not analysed further [92] and patient privacy considerations or requirements may dictate methods for storing and processing data [93]. The possibility of generating incidental findings requires balancing best clinical care with patient privacy [94–96]. The American College of Medical Genetics and Genomics (ACMG) has published recommendations for reporting of specific conditions, genes, or variants when discovered incidentally [97]. Similar guidelines have not been published yet for incidental results generated by metagenomics tests.

7.5 Conclusions

The development of metagenomic tests for pathogen detection has the potential to change the face of laboratory testing for infectious diseases. Published cases and early clinical studies demonstrate the promise of detecting unexpected, uncommon, slow growing, co-infecting pathogens in difficult-to-diagnose patients [12–15]. This technology can be particularly useful for diagnosis of rare micro-organisms for which there is a lack of available clinical tests and detection of uncommon variants of common pathogens [26, 98]. The untargeted nature of testing enables broad pathogen detection from a single, low-volume specimen, which is especially important for testing in children, precious specimens (e.g. intraocular fluid, CSF), or those that are difficult to recollect (e.g. specimens collected before initiation of antimicrobial therapy). In addition to clinical diagnoses, metagenomics also has many important applications in public health testing and infection control (e.g. strain typing, profiling for molecular resistance determinants, or surveillance).

Understanding and defining appropriate clinical indications for metagenomics testing remains a challenge and clinical utility studies will be needed. Conducting those studies and continuously improving metagenomics tests will require a multidisciplinary approach, involving clinical, laboratory, computational biology, and data science teams. Because of the heavy dependence on sequencing and data analysis technologies, collaborations between laboratory experts and test developers will also be required. Analytic phase improvements include optimisation of wet-bench methods, sequencing technology, and data analysis procedures. Result analysis and reporting can be improved to better assist clinicians in interpretation of results. Test development and validation will likely continue to provide challenges to laboratories until methods are more standardised and guidance documents become available. In their absence, laboratories will have to use judgment, a risk-based approach, and consider a combination of the different validation strategies outlined above. Microbiology test results are generally reported as "detected" or "not detected". Given the vast quantity and resolution of data acquired by metagenomic approaches, the laboratory has the opportunity to provide additional, clinically relevant information to assist result interpretation. Reporting may include not only an micro-organism name, but the quantity at which it was detected, genotypic information, genetic markers of drug resistance, and even gene expression activities of detected pathogens. By their sheer breadth, metagenomics tests also require a paradigm that relies less on extensive expertise in a certain class of pathogens as the same workflow will produce results across all categories of pathogens.

References

1. Muldrew K (2009) Molecular diagnostics of infectious diseases. Curr Opin Pediatr 21(1):102–111
2. Tan KE, Ellis BC, Lee R, Stamper PD, Zhang SX, Carroll KC (2012) Prospective evaluation of a matrix-assisted laser desorption ionization-time of flight mass spectrometry system in a hospital clinical microbiology laboratory for identification of bacteria and yeasts: a bench-by-bench study for assessing the impact on time to identification and cost-effectiveness. J Clin Microbiol 50(10):3301–3308
3. Banerjee R, Teng CB, Cunningham SA, Ihde SM, Steckelberg JM, Moriarty JP et al (2015) Randomized trial of rapid multiplex polymerase chain reaction-based blood culture identification and susceptibility testing. Clin Infect Dis 61(7):1071–1080
4. Pavia AT (2011) Viral infections of the lower respiratory tract: old viruses, new viruses, and the role of diagnosis. Clin Infect Dis 52(Suppl 4):S284–S289
5. Glaser CA, Honarmand S, Anderson LJ, Schnurr DP, Forghani B, Cossen CK et al (2006) Beyond viruses: clinical profiles and etiologies associated with encephalitis. Clin Infect Dis 43(12):1565–1577
6. Choi SH, Hong SB, Ko GB, Lee Y, Park HJ, Park SY et al (2012) Viral infection in patients with severe pneumonia requiring intensive care unit admission. Am J Respir Crit Care Med 186(4):325–332
7. Jain S, Self WH, Wunderink RG, Fakhran S, Balk R, Bramley AM et al (2015) Community-acquired pneumonia requiring hospitalization among U.S. adults. N Engl J Med 373(5):415–427
8. Jain S, Williams DJ, Arnold SR, Ampofo K, Bramley AM, Reed C et al (2015) Community-acquired pneumonia requiring hospitalization among U.S. children. N Engl J Med 372(9):835–845
9. Self WH, Williams DJ, Zhu Y, Ampofo K, Pavia AT, Chappell JD et al (2016) Respiratory viral detection in children and adults: comparing asymptomatic controls and patients with community-acquired pneumonia. J Infect Dis 213(4):584–591
10. Tarabichi M, Shohat N, Goswami K, Alvand A, Silibovsky R, Belden K et al (2018) Diagnosis of Periprosthetic joint infection: the potential of next-generation sequencing. J Bone Joint Surg Am 100(2):147–154
11. Gallo J, Kolar M, Dendis M, Loveckova Y, Sauer P, Zapletalova J et al (2008) Culture and PCR analysis of joint fluid in the diagnosis of prosthetic joint infection. New Microbiol 31(1):97–104
12. Hong DK, Blauwkamp TA, Kertesz M, Bercovici S, Truong C, Banaei N (2018) Liquid biopsy for infectious diseases: sequencing of cell-free plasma to detect pathogen DNA in patients with invasive fungal disease. Diagn Microbiol Infect Dis 92(3):210–213
13. Farnaes L, Wilke J, Ryan Loker K, Bradley JS, Cannavino CR, Hong DK et al (2019) Community-acquired pneumonia in children: cell-free plasma sequencing for diagnosis and management. Diagn Microbiol Infect Dis 94(2):188–191
14. Hogan CA, Yang S, Garner OB, Green DA, Gomez CA, Dien Bard J et al (2020) Clinical impact of metagenomic next-generation sequencing of plasma cell-free DNA for the diagnosis of infectious diseases: a multicenter retrospective cohort study. Clin Infect Dis. https://doi.org/10.1093/cid/ciaa035
15. Blauwkamp TA, Thair S, Rosen MJ, Blair L, Lindner MS, Vilfan ID et al (2019) Analytical and clinical validation of a microbial cell-free DNA sequencing test for infectious disease. Nat Microbiol 4(4):663–674

16. Taravati P, Lam D, Van Gelder RN (2013) Role of molecular diagnostics in ocular microbiology. Curr Ophthalmol Rep 1(4). https://doi.org/10.1007/s40135-013-0025-1
17. Doan T, Pinsky BA (2016) Current and future molecular diagnostics for ocular infectious diseases. Curr Opin Ophthalmol 27(6):561–567
18. Deurenberg RH, Bathoorn E, Chlebowicz MA, Couto N, Ferdous M, Garcia-Cobos S et al (2017) Application of next generation sequencing in clinical microbiology and infection prevention. J Biotechnol 243:16–24
19. Dunne WM Jr, Jaillard M, Rochas O, Van Belkum A (2017) Microbial genomics and antimicrobial susceptibility testing. Expert Rev Mol Diagn 17(3):257–269
20. Kashyap PC, Chia N, Nelson H, Segal E, Elinav E (2017) Microbiome at the frontier of personalized medicine. Mayo Clin Proc 92(12):1855–1864
21. Koser CU, Ellington MJ, Cartwright EJ, Gillespie SH, Brown NM, Farrington M et al (2012) Routine use of microbial whole genome sequencing in diagnostic and public health microbiology. PLoS Pathog 8(8):e1002824
22. Motro Y, Moran-Gilad J (2017) Next-generation sequencing applications in clinical bacteriology. Biomol Detect Quantif 14:1–6
23. Mongkolrattanothai K, Naccache SN, Bender JM, Samayoa E, Pham E, Yu G et al (2017) Neurobrucellosis: unexpected answer from metagenomic next-generation sequencing. J Pediatric Infect Dis Soc 6(4):393–398
24. Wilson MR, Naccache SN, Samayoa E, Biagtan M, Bashir H, Yu G et al (2014) Actionable diagnosis of neuroleptospirosis by next-generation sequencing. N Engl J Med 370(25):2408–2417
25. Wilson MR, Suan D, Duggins A, Schubert RD, Khan LM, Sample HA et al (2017) A novel cause of chronic viral meningoencephalitis: Cache Valley virus. Ann Neurol 82(1):105–114
26. Wilson MR, Sample HA, Zorn KC, Arevalo S, Yu G, Neuhaus J et al (2019) Clinical metagenomic sequencing for diagnosis of meningitis and encephalitis. N Engl J Med 380(24):2327–2340
27. Long Y, Zhang Y, Gong Y, Sun R, Su L, Lin X et al (2016) Diagnosis of Sepsis with cell-free DNA by next-generation sequencing technology in ICU patients. Arch Med Res 47(5):365–371
28. Schlaberg R, Chiu CY, Miller S, Procop GW, Weinstock G, Professional Practice C et al (2017) Validation of metagenomic next-generation sequencing tests for universal pathogen detection. Arch Pathol Lab Med 141(6):776–786
29. Grard G, Fair JN, Lee D, Slikas E, Steffen I, Muyembe JJ et al (2012) A novel rhabdovirus associated with acute hemorrhagic fever in Central Africa. PLoS Pathog 8(9):e1002924
30. Palacios G, Druce J, Du L, Tran T, Birch C, Briese T et al (2008) A new arenavirus in a cluster of fatal transplant-associated diseases. N Engl J Med 358(10):991–998
31. Xu B, Liu L, Huang X, Ma H, Zhang Y, Du Y et al (2011) Metagenomic analysis of fever, thrombocytopenia and leukopenia syndrome (FTLS) in Henan Province, China: discovery of a new bunyavirus. PLoS Pathog 7(11):e1002369
32. Briese T, Paweska JT, McMullan LK, Hutchison SK, Street C, Palacios G et al (2009) Genetic detection and characterization of Lujo virus, a new hemorrhagic fever-associated arenavirus from southern Africa. PLoS Pathog 5(5):e1000455
33. Rehm HL, Bale SJ, Bayrak-Toydemir P, Berg JS, Brown KK, Deignan JL et al (2013) ACMG clinical laboratory standards for next-generation sequencing. Genet Med 15(9):733–747
34. Ali N, Rampazzo RCP, Costa ADT, Krieger MA (2017) Current nucleic acid extraction methods and their implications to point-of-care diagnostics. Biomed Res Int 2017:9306564
35. Lefterova MI, Suarez CJ, Banaei N, Pinsky BA (2015) Next-generation sequencing for infectious disease diagnosis and management: a report of the Association for Molecular Pathology. J Mol Diagn 17(6):623–634
36. Deng X, Achari A, Federman S, Yu G, Somasekar S, Bartolo I et al (2020) Metagenomic sequencing with spiked primer enrichment for viral diagnostics and genomic surveillance. Nat Microbiol 5(3):443–454
37. Wylie TN, Wylie KM, Herter BN, Storch GA (2015) Enhanced virome sequencing using targeted sequence capture. Genome Res 25(12):1910–1920

38. Klindworth A, Pruesse E, Schweer T, Peplies J, Quast C, Horn M et al (2013) Evaluation of general 16S ribosomal RNA gene PCR primers for classical and next-generation sequencing-based diversity studies. Nucleic Acids Res 41(1):e1
39. Kim KH, Bae JW (2011) Amplification methods bias metagenomic libraries of uncultured single-stranded and double-stranded DNA viruses. Appl Environ Microbiol 77(21):7663–7668
40. Rosseel T, Van Borm S, Vandenbussche F, Hoffmann B, van den Berg T, Beer M et al (2013) The origin of biased sequence depth in sequence-independent nucleic acid amplification and optimization for efficient massive parallel sequencing. PLoS One 8(9):e76144
41. Hasan MR, Rawat A, Tang P, Jithesh PV, Thomas E, Tan R et al (2016) Depletion of human DNA in spiked clinical specimens for improvement of sensitivity of pathogen detection by next-generation sequencing. J Clin Microbiol 54(4):919–927
42. Allander T, Emerson SU, Engle RE, Purcell RH, Bukh J (2001) A virus discovery method incorporating DNase treatment and its application to the identification of two bovine parvovirus species. Proc Natl Acad Sci U S A 98(20):11609–11614
43. Costello M, Fleharty M, Abreu J, Farjoun Y, Ferriera S, Holmes L et al (2018) Characterization and remediation of sample index swaps by non-redundant dual indexing on massively parallel sequencing platforms. BMC Genomics 19(1):332
44. MacConaill LE, Burns RT, Nag A, Coleman HA, Slevin MK, Giorda K et al (2018) Unique, dual-indexed sequencing adapters with UMIs effectively eliminate index cross-talk and significantly improve sensitivity of massively parallel sequencing. BMC Genomics 19(1):30
45. Goldberg B, Sichtig H, Geyer C, Ledeboer N, Weinstock GM (2015) Making the leap from research laboratory to clinic: challenges and opportunities for next-generation sequencing in infectious disease diagnostics. MBio 6(6):e01888–e01815
46. Goodwin S, McPherson JD, McCombie WR (2016) Coming of age: ten years of next-generation sequencing technologies. Nat Rev Genet 17(6):333–351
47. Leggett RM, Alcon-Giner C, Heavens D, Caim S, Brook T, Kujawska M et al (2017) Rapid MinION metagenomic profiling of the preterm infant gut microbiota to aid in pathogen diagnostics. bioRxiv
48. Cheval J, Sauvage V, Frangeul L, Dacheux L, Guigon G, Dumey N et al (2011) Evaluation of high-throughput sequencing for identifying known and unknown viruses in biological samples. J Clin Microbiol 49(9):3268–3275
49. Greninger AL, Naccache SN, Federman S, Yu G, Mbala P, Bres V et al (2015) Rapid metagenomic identification of viral pathogens in clinical samples by real-time nanopore sequencing analysis. Genome Med 7:99
50. Breitwieser FP, Lu J, Salzberg SL (2019) A review of methods and databases for metagenomic classification and assembly. Brief Bioinform 20(4):1125–1136
51. Lindgreen S, Adair KL, Gardner PP (2016) An evaluation of the accuracy and speed of metagenome analysis tools. Sci Rep 6:19233
52. McIntyre ABR, Ounit R, Afshinnekoo E, Prill RJ, Henaff E, Alexander N et al (2017) Comprehensive benchmarking and ensemble approaches for metagenomic classifiers. Genome Biol 18(1):182
53. Nooij S, Schmitz D, Vennema H, Kroneman A, Koopmans MPG (2018) Overview of virus metagenomic classification methods and their biological applications. Front Microbiol 9:749
54. Peabody MA, Van Rossum T, Lo R, Brinkman FS (2015) Evaluation of shotgun metagenomics sequence classification methods using in silico and in vitro simulated communities. BMC Bioinf 16:363
55. Quince C, Walker AW, Simpson JT, Loman NJ, Segata N (2017) Shotgun metagenomics, from sampling to analysis. Nat Biotechnol 35(9):833–844
56. Sczyrba A, Hofmann P, Belmann P, Koslicki D, Janssen S, Droge J et al (2017) Critical assessment of metagenome interpretation-a benchmark of metagenomics software. Nat Methods 14(11):1063–1071
57. Ames SK, Hysom DA, Gardner SN, Lloyd GS, Gokhale MB, Allen JE (2013) Scalable metagenomic taxonomy classification using a reference genome database. Bioinformatics 29(18):2253–2260

58. Buchfink B, Xie C, Huson DH (2015) Fast and sensitive protein alignment using DIAMOND. Nat Methods 12(1):59–60
59. Caporaso JG, Kuczynski J, Stombaugh J, Bittinger K, Bushman FD, Costello EK et al (2010) QIIME allows analysis of high-throughput community sequencing data. Nat Methods 7(5):335–336
60. Flygare S, Simmon K, Miller C, Qiao Y, Kennedy B, Di Sera T et al (2016) Taxonomer: an interactive metagenomics analysis portal for universal pathogen detection and host mRNA expression profiling. Genome Biol 17(1):111
61. Huson DH, Auch AF, Qi J, Schuster SC (2007) MEGAN analysis of metagenomic data. Genome Res 17(3):377–386
62. Huson DH, Beier S, Flade I, Gorska A, El-Hadidi M, Mitra S et al (2016) MEGAN Community edition - interactive exploration and analysis of large-scale microbiome sequencing data. PLoS Comput Biol 12(6):e1004957
63. Naccache SN, Federman S, Veeraraghavan N, Zaharia M, Lee D, Samayoa E et al (2014) A cloud-compatible bioinformatics pipeline for ultrarapid pathogen identification from next-generation sequencing of clinical samples. Genome Res 24(7):1180–1192
64. Ounit R, Wanamaker S, Close TJ, Lonardi S (2015) CLARK: fast and accurate classification of metagenomic and genomic sequences using discriminative k-mers. BMC Genomics 16:236
65. Rosen G, Garbarine E, Caseiro D, Polikar R, Sokhansanj B (2008) Metagenome fragment classification using N-mer frequency profiles. Adv Bioinforma 2008:205969
66. Sunagawa S, Mende DR, Zeller G, Izquierdo-Carrasco F, Berger SA, Kultima JR et al (2013) Metagenomic species profiling using universal phylogenetic marker genes. Nat Methods 10(12):1196–1199
67. Leung HC, Yiu SM, Yang B, Peng Y, Wang Y, Liu Z et al (2011) A robust and accurate binning algorithm for metagenomic sequences with arbitrary species abundance ratio. Bioinformatics 27(11):1489–1495
68. Segata N, Waldron L, Ballarini A, Narasimhan V, Jousson O, Huttenhower C (2012) Metagenomic microbial community profiling using unique clade-specific marker genes. Nat Methods 9(8):811–814
69. Breitwieser FP, Pertea M, Zimin AV, Salzberg SL (2019) Human contamination in bacterial genomes has created thousands of spurious proteins. Genome Res 29(6):954–960
70. Edgar R (2018) Taxonomy annotation and guide tree errors in 16S rRNA databases. PeerJ 6:e5030
71. D'Amore R, Ijaz UZ, Schirmer M, Kenny JG, Gregory R, Darby AC et al (2016) A comprehensive benchmarking study of protocols and sequencing platforms for 16S rRNA community profiling. BMC Genomics 17:55
72. Nelson MC, Morrison HG, Benjamino J, Grim SL, Graf J (2014) Analysis, optimization and verification of Illumina-generated 16S rRNA gene amplicon surveys. PLoS One 9(4):e94249
73. Lusk RW (2014) Diverse and widespread contamination evident in the unmapped depths of high throughput sequencing data. PLoS One 9(10):e110808
74. Zinter MS, Mayday MY, Ryckman KK, Jelliffe-Pawlowski LL, DeRisi JL (2019) Towards precision quantification of contamination in metagenomic sequencing experiments. Microbiome 7(1):62
75. Eisenhofer R, Minich JJ, Marotz C, Cooper A, Knight R, Weyrich LS (2019) Contamination in low microbial biomass microbiome studies: issues and recommendations. Trends Microbiol 27(2):105–117
76. Miller S, Naccache SN, Samayoa E, Messacar K, Arevalo S, Federman S et al (2019) Laboratory validation of a clinical metagenomic sequencing assay for pathogen detection in cerebrospinal fluid. Genome Res 29(5):831–842
77. Bokulich NA, Rideout JR, Mercurio WG, Shiffer A, Wolfe B, Maurice CF et al (2016) Mockrobiota: a public resource for microbiome bioinformatics benchmarking. mSystems 1(5):e00062–16

78. Sichtig H, Minogue T, Yan Y, Stefan C, Hall A, Tallon L et al (2019) FDA-ARGOS is a database with public quality-controlled reference genomes for diagnostic use and regulatory science. Nat Commun 10(1):3313
79. Aziz N, Zhao Q, Bry L, Driscoll DK, Funke B, Gibson JS et al (2015) College of American Pathologists' laboratory standards for next-generation sequencing clinical tests. Arch Pathol Lab Med 139(4):481–493
80. Gu W, Crawford ED, O'Donovan BD, Wilson MR, Chow ED, Retallack H et al (2016) Depletion of abundant sequences by hybridization (DASH): using Cas9 to remove unwanted high-abundance species in sequencing libraries and molecular counting applications. Genome Biol 17:41
81. Paskey AC, Frey KG, Schroth G, Gross S, Hamilton T, Bishop-Lilly KA (2019) Enrichment post-library preparation enhances the sensitivity of high-throughput sequencing-based detection and characterization of viruses from complex samples. BMC Genomics 20(1):155
82. Wylie KM, Wylie TN, Buller R, Herter B, Cannella MT, Storch GA (2018) Detection of viruses in clinical samples by use of metagenomic sequencing and targeted sequence capture. J Clin Microbiol 56(12): e01123–18
83. O'Flaherty BM, Li Y, Tao Y, Paden CR, Queen K, Zhang J et al (2018) Comprehensive viral enrichment enables sensitive respiratory virus genomic identification and analysis by next generation sequencing. Genome Res 28(6):869–877
84. Briese T, Kapoor A, Mishra N, Jain K, Kumar A, Jabado OJ et al (2015) Virome capture sequencing enables sensitive viral diagnosis and comprehensive virome analysis. mBio 6(5):e01491–e01415
85. Chiu CY, Miller SA (2019) Clinical metagenomics. Nat Rev Genet 20(6):341–355
86. Strong MJ, Xu G, Morici L, Splinter Bon-Durant S, Baddoo M, Lin Z et al (2014) Microbial contamination in next generation sequencing: implications for sequence-based analysis of clinical samples. PLoS Pathog 10(11):e1004437
87. Miller S, Chiu C, Rodino K, Miller M. (2020) Point-Counterpoint: Should We be Performing Metagenomic Next-Generation Sequencing for Infectious Disease Diagnosis in the Clinical Laboratory? J Clin Microbiol 58(3):e01739–19
88. Mitchell SL, Simner PJ (2019) Next-generation sequencing in clinical microbiology: are we there yet? Clin Lab Med 39(3):405–418
89. White DJ, Wang J, Hall RJ (2017) Assessing the impact of assemblers on virus detection in a De Novo metagenomic analysis pipeline. J Comput Biol 24(9):874–881
90. Allcock RJN, Jennison AV, Warrilow D (2017) Towards a universal molecular microbiological test. J Clin Microbiol 55(11):3175–3182
91. Doan T, Acharya NR, Pinsky BA, Sahoo MK, Chow ED, Banaei N et al (2017) Metagenomic DNA sequencing for the diagnosis of intraocular infections. Ophthalmology 124(8):1247–1248
92. Hall RJ, Draper JL, Nielsen FGG, Dutilh BE (2015) Beyond research: a primer for considerations on using viral metagenomics in the field and clinic. Front Microbiol 6:224
93. Chiu C, Miller S (2016) Molecular microbiology: diagnostic principles and Practice. Persing DH, editor. ASM Press, Washington, DC
94. Rahimzadeh V, Avard D, Senecal K, Knoppers BM, Sinnett D (2015) To disclose, or not to disclose? Context matters. Eur J Hum Genet 23:279–284
95. Clarke AJ (2014) Managing the ethical challenges of next-generation sequencing in genomic medicine. Br Med Bull 111(1):17–30
96. Davey S (2014) Next generation sequencing: considering the ethics. Int J Immunogenet 41(6):457–462
97. Green RC, Berg JS, Grody WW, Kalia SS, Korf BR, Martin CL et al (2013) ACMG recommendations for reporting of incidental findings in clinical exome and genome sequencing. Genet Med 15(7):565–574
98. Graf EH, Simmon KE, Tardif KD, Hymas W, Flygare S, Eilbeck K et al (2016) Unbiased detection of respiratory viruses by use of RNA sequencing-based metagenomics: a systematic comparison to a commercial PCR panel. J Clin Microbiol 54(4):1000–1007

Chapter 8
Integrating Metagenomics in the Routine Lab

Etienne Ruppé, Yannick Charretier, Vladimir Lazarevic, and Jacques Schrenzel

8.1 Introduction

The broad availability of next-generation sequencing (NGS) tools as well as the avalanche of papers, reports [1–4] and scientific conferences [5–7] on the applicability of such methods in routine clinical practice cast little doubts about their potential. Yet, the timing for their implementation in routine remains actively debated. Metagenomics, in general, describes the sequencing of a diverse set of micro-organisms from a sample. This can be further divided into amplicon-based – in which certain elements are amplified before sequencing (usually 16S rRNA bacterial genes) thus identifying only the amplicon, and shotgun metagenomics where the entire genomic content of a sample is sequenced in an unbiased manner. The methods described in this chapter will refer to the latter.

Clinical metagenomics (CMg), based on whole genome sequencing (WGS) of clinical samples, offers the potential to directly detect all micro-organisms present in a sample or even detect RNA viruses if coupled to a reverse-transcription step.

E. Ruppé
Laboratoire de Bactériologie, Université de Paris, Hôpital Bichat, Paris, France

Y. Charretier · V. Lazarevic
Genomic Research Laboratory, Service of Infectious Diseases, Geneva University Hospitals and Faculty of Medicine, Geneva, Switzerland

J. Schrenzel (✉)
Genomic Research Laboratory, Service of Infectious Diseases, Geneva University Hospitals and Faculty of Medicine, Geneva, Switzerland

Bacteriology Laboratory, Service of Laboratory Medicine, Geneva University Hospitals, Geneva, Switzerland
e-mail: jacques.schrenzel@genomic.ch

J. Moran-Gilad, Y. Yagel (eds.), *Application and Integration of Omics-powered Diagnostics in Clinical and Public Health Microbiology*,
https://doi.org/10.1007/978-3-030-62155-1_8

This approach could therefore provide the detection of all micro-organisms present in the sample, including those organisms that are fastidious or even cannot be cultivated. By skipping the cultivation step, CMg could constitute a rapid and generic approach providing all medically-actionable information: the presence or the absence of micro-organisms (detection), their identification to the species level or even beyond (speciation, and potentially some genotyping capabilities), the detection of antimicrobial resistance determinants (i.e. antimicrobial resistance-conferring genes and mutational events associated with resistance) with the potential to guide antibiotic therapy and virulence-associated genes. In theory, sequencing-based diagnostics could compete and maybe replace conventional methods.

This is, of course, an overly optimistic description, and current results do only address some of those expectations. Remarkably, the potential of CMg to replace conventional methods requires to benchmark it against a tremendous number of parameters which have been developed and validated over decades, in order for it to run smoothly in our routine laboratories. These new methods will have to show their superiority (or lack thereof) against all such parameters and in the meantime, they cannot be considered as substitution methods, but should instead be seen as complementary tools. The massive routine implementation of CMg might happen in the next years, after having addressed these numerous elements. In the meantime, academic labs have rapidly adopted these tools as a growing research field and as advanced methods for trying to elucidate complex cases.

What is then required to use and offer CMg assays? Moreover, what is to be further demonstrated for routine implementation of such methods to take place?

This chapter aims at describing current common hurdles encountered when attempting to utilise CMg to clinical samples and discuss ways to overcome them. These include sample preparation and wet lab issues such as managing human DNA, analysis and bioinformatic challenges, medical value and reimbursement, certification and documentation ethics and logistics aspects.

8.2 Sample Preparation

Several problems arise from the fact that CMg involves direct sequencing from clinical samples (WGS metagenomics) without prior amplification of the desired target (amplicon-based metagenomics). The major ones being (i) the presence of human DNA in high abundance (often more than 90% of the entire genomic content), and (ii) the low copy number of the micro-organisms present. The next paragraphs will address sample preparation procedures that have been developed to increase the yield of desired nucleic acids. The choice will depend on the specimen sampled and the clinical question to be answered. However, in all cases, the user will have to make sure (and document) that desired nucleic acids are correctly preserved, extracted and sufficiently purified to be sequenced.

8.2.1 *Human DNA Removal*

A fundamental dilemma -not yet solved, is whether to remove human DNA or not during the nucleic acid extraction step. Human DNA removal allows a more in-depth exploration of the sample by saving DNA sequencing capabilities, but at the expense of non-specifically reducing the abundance of microbial DNA. Removal of human DNA might also display very different effects across all microbial species. For example, and according to our experience, the method used to "selectively" lyse human cells and then remove human DNA might affect the recovery of Gram-negative more than that of Gram-positive bacteria.

Human DNA removal methods should, therefore, be developed and calibrated to minimise the unwanted effects on the most sensitive bacterial cells. Do we need to provide a guaranteed minimal analytical sensitivity, i.e. a kind of guaranteed detection threshold? Or should we focus only on the most relevant bacterial targets expected in that type of sample? One might argue that all sample types are not equal and that such a generic statement is impossible, as in some cases, human DNA removal is not needed.

The following paragraphs will elaborate on the different methods for human DNA removal using different techniques at different steps of the sample preparation.

8.2.1.1 Pre-extraction Depletion

To decrease the proportion of sequencing reads derived from human cells, selective lysis of human cells by chaotropic agents, detergents or osmotic shock have been used, followed by free DNA degradation (enzymatic or chemical).

First, the Ultra-Deep Microbiome Prep kit from Molzym (Bremen, Germany) uses guanidine hydrochloride and detergents to lyse blood cells and a nuclease able to degrade human DNA in a chaotropic/detergent environment before extraction of DNA from micro-organisms [9]. As an example, the above-described protocol was successfully put in practice with a bronco-alveolar lavage sample from an immunocompromised patient [10]. When the Ultra-Deep Microbiome Prep protocol was compared to mechanical disruption, a 310-fold increase in the bacterial/human DNA ratio was observed. The enrichment for solid tissues was not as efficient as for a liquid sample, but it may not necessarily be needed, as shown in a case of brucellosis identified without an enrichment process in liver tissue biopsy [11].

Second, the QIAamp DNA Microbiome kit from Qiagen (Hilden, Germany) uses a differential lysis of the human host cells with the AHL buffer (unknown composition) followed by a benzonase (a genetically engineered endonuclease from *Serratia marcescens* able to degrade all forms of DNA and RNA) digestion of exposed nucleic acids before microbial DNA extraction [12, 13]. Different surfactants have been tested for their ability to lyse human cells selectively [14]. For nasopharyngeal aspirate and cerebrospinal fluid specimens spiked with bacterial and viral control strains, the best results were obtained with saponin and Triton-X-100.

Third, Marotz et al. described a procedure of chemical degradation of human DNA that consisted of selective osmotic lysis of mammalian cells followed by a propidium monoazide (PMA) treatment [15]. PMA is a photoreactive DNA intercalator used to detect viable micro-organisms. Upon exposure to visible light, the azide group of PMA is cleaved and a covalent bond is formed between PMA and DNA. This chemical modification is thought to fragment DNA. By applying this method to saliva samples, exposed DNA of human host cells were efficiently depleted (90%).

The choice of the method for host cell depletion will depend on the sample type. Methods designed to be used for the pre-treatment of whole blood would require adaptations when they are used for other sample types (notably solid tissue).

8.2.1.2 Post-extraction Enrichment/Depletion

(i) *Selective enrichment based on DNA methylation density*

DNA methylation is essential for many important biological functions, including protection of self DNA in both prokaryotes and eukaryotes. The three methylated bases N^6-methyladenine (m6A), C^5-methyl-cytosine (m5C) and N^4-methyl-cytosine (m4C) are differentially present in prokaryotes and eukaryotes [16]. Bacterial DNA enrichment procedures based on the specific binding of methylated or non-methylated cytidylate-phosphate-deoxyguanylate (CpG) motifs have been used for samples from different body sites such as blood, saliva, synovial fluid or sonicate fluid of the explanted orthopaedic device and showed various efficacy [13–18].

(ii) *Depletion of abundant sequences by hybridisation (DASH)*

Bacterial adaptive immune systems (CRISPR/Cas systems) that rely on binding and cutting programmed target sequences, were exploited to perform selective depletion of human mitochondrial rRNA sequences by DASH. The DASH method could be employed during the preparation of sequencing libraries, after the transposon-mediated fragmentation but prior to the amplification step. For example, using DASH, the sequencing depth required to detect the fungus, amoeba and the tapeworm reads from the cerebrospinal fluid RNA samples of infected patients, was efficiently lowered [19].

(iii) *Depletion of rRNA and mRNA*

Shotgun sequencing of enriched rRNA or total RNA, which mainly consists of rRNA molecules, offers an alternative to commonly used meta-taxonomic approaches based on amplification and sequencing of 16S rRNA gene fragments [20]. The advantage of this method is that it does not introduce amplification bias. In addition, it is expected to be less affected by contaminant DNA present in reagents.

The Ribo-Zero rRNA enrichment/depletion method is based on the complementarity between affinity-tagged antisense rRNA and rRNA molecules [20]. Both eukaryotic (cytoplasmic and mitochondrial) and prokaryotic rRNA molecules can be either depleted or isolated using this approach. The Ribo-Zero rRNA depletion method enriched up to 40-fold for non-rRNA transcripts and resulted in profiles similar to those obtained without rRNA depletion for both bacterial culture and stool samples [21]. The rRNA removal worked equally well for intact and fragmented total RNA, a feature important for clinical samples that could contain partly degraded RNA. Avraham et al. took advantage of this method to develop a simultaneous analysis of host and pathogen transcriptomes [22]. Their protocol, including efficient lysis of host (bead-beating) and bacterial (enzymatic) cells, is applicable to different host cell types and it is well adapted for intracellular organisms such as *Salmonella enterica*.

The eukaryotic mRNA is polyadenylated at 3′ in contrast to most prokaryotic mRNA. This feature has been used for the enrichment of prokaryotic mRNA based on oligo-(dT) cellulose-binding such as the poly(A)Purist™ MAG kit (Thermo Fischer Scientific, Waltham, MA) [23]. The MICROB*Enrich*™ kit (Thermo Fischer Scientific) was designed to specifically deplete mammalian 18S rRNA, 28S rRNA and mRNA while the MICROB*Express*™ kit (Thermo Fischer Scientific) was designed to specifically deplete bacterial 16S rRNA and 23S rRNA [24]. These kits can be combined to enrich bacterial mRNA from the samples containing both eukaryotic and bacterial cells.

In the same manner, the Ribo-Zero Gold rRNA removal kit (Illumina Inc., San Diego, CA) can be combined with the Poly(A) Purist™ MAG kit to enrich bacterial mRNA. The Ribo-Zero based depletion performed better than MICROB*Enrich*™/MICROB*Express*™ approach in the study of the metatranscriptome of the termite gut microbiota [25].

8.2.2 Direct Sequencing Approaches

8.2.2.1 Use of Whole Genome Amplification for Low-Input Samples, Rare or Single Cells

Whole genome amplification (WGA) is a useful tool to increase the amount of input DNA of low-biomass samples. The most used WGA technique is based on multiple displacement amplification (MDA) [26]. The high yield of amplified DNA and the low error rate of the Phi29 polymerase used for MDA are significant advantages of this technique. The amplification bias is the first limitation of WGA that could be lowered by emulsion-based partitioning of WGA reactions, allowing reproducible metagenomic studies of low biomass samples [27]. A second limitation of WGA (but also of other NGS techniques for microbial community assessment [28]) is the presence of contaminating DNA in reagents [29, 30]. Different visual, statistical, methodological and ecological approaches have been described to recognise reagent contamination, especially for low-biomass samples [31]. Hansen et al. reported the combination of MDA and Nanopore (Oxford Nanopore Technologies, Oxford, United Kingdom) sequencing for a portable identification of the causative agent of an outbreak [32]. They established the protocol for the rapid identification of RNA viruses, using Zika virus as a model, in a suitcase laboratory in less than 7 hours after the samples were taken.

8.2.2.2 Direct Sequencing of Circulating Cell-Free DNA

Cell-free circulating nucleic acids present in biological fluids such as serum, plasma or urine are usually bound to proteins or enveloped in vesicles. Grumaz et al. described a method for diagnosis of septicaemia based on WGS of cell-free circulating DNA (cfDNA) in plasma samples from septic patients and uninfected controls [33] with enhanced performances compared to standard diagnostic [34]. Their procedure included removal of cells by two centrifugation steps, followed by recovery of cfDNA from its complexes in highly denaturing conditions at high temperature using a Circulating Nucleic Acid kit (Qiagen). A microbial cfDNA test developed by Karius Inc. was analytically validated using sheared genomic DNA from a panel of 13 micro-organisms [35]. Their study established the performances both in 358 contrived plasma samples and in 580 clinical samples as well as using *in silico* simulated infections.

8.2.3 Targeted Sequencing or Selective Sequencing

8.2.3.1 On-Target Sequencing (Bait-Capture Approach)

A method initially developed for human exome enrichment [36], based on hybridisation of biotinylated RNA baits, was expanded to selectively capture target DNA (or RNA) and enrich pathogen transcripts, virome, resistome and virulome [37–40]. Before sequencing, the captured DNA (or RNA) are eluted and amplified by PCR with universal primers, which induce a bias in the representation of nucleic acid fragments.

Loss of molecules and amplification artefacts could cause underestimation and overestimation of the molecule count, respectively. The use of unique molecular identifiers (UMIs), alone or in combination with algorithms that take into account stochastic properties of PCR efficacy and sequencing depth, increases the accuracy of the counting molecules in NGS-based methods [41, 42]. The UMI method can reproducibly count the molecules after PCR amplification bypassing the need for a normalisation step.

8.2.3.2 Selective Sequencing: The 'Read Until' Approach

Nanopore sequencing enables real-time data analysis [43]. As DNA molecules pass through the nanopore, an electrical signal is produced that depends on the specific bases in contact with the pore. The channels can be controlled independently in real-time by reversing the voltage across the pore, rejecting undesired DNA molecules and enabling selective sequencing of fragments of interest. This conceptual approach, named 'Read Until' by Oxford Nanopore Technologies, was shown to improve genome coverage of specific regions of Lambda and Ebola viruses [44].

8.3 Bioinformatic Challenges

Clinical metagenomics (CMg) raises several challenges in terms of bioinformatic analyses, the main one being the quality of the databases used for input. As CMg aims at being an exhaustive method, it is expected to detect a broad range of possible micro-organisms that could be present in a clinical sample: bacteria, archaea, viruses (RNA and DNA), fungi and parasites. Moreover, beyond the identification of micro-organisms, CMg is expected to provide further information such as antimicrobial resistance determinants, which requires other specific databases.

Most CMg studies collected publicly available genomes from the NCBI RefSeq or nt databases and applied filters to select for high-quality genomes, such as SURPI [45]. Other studies have used pre-formatted databases such as Kraken miniDB [46] or the specific bacterial marker database MetaPhlAn2 [47]. Building a database for clinical metagenomics requires both exhaustivity and accuracy. Genomes from RefSeq may not span all possible micro-organisms expected in a clinical sample, and conversely genomes from the nt database may lack curation or harbor contaminating DNA. Hence, a curation step of the genome database is a crucial one, which should involve bacteriologists, virologists, parasitologists, mycologists and bioinformaticians.

The same applies to specific databases such as antibiotic resistance databases. Several of them are now available [48], the most popular being ResFinder [49] and CARD [50]. None of them is exhaustive in that they do not totally overlap. Indeed, CARD harbours genes associated with antibiotic resistance (such as expression regulators of resistance genes) that do not fit the operational definition of antibiotic resistance genes proposed by Martinez et al. [51] and are not present in ResFinder. In addition, CARD includes some mutational patterns in bacterial intrinsic genes (such as mutations in topoisomerases associated with fluoroquinolone resistance) that ResFinder does not include.

Furthermore, one should keep in mind that the available databases mostly include antibiotic resistance determinants identified and characterised from cultivable bacteria, while CMg has the capacity of identifying uncultivable bacteria. Hence, detecting antibiotic resistance determinants in those bacteria would be challenging. The databases of antibiotic resistance genes identified in functional metagenomic studies such as FARME DB [52] or ResFinderFG (https://cge.cbs.dtu.dk/services/ResFinderFG/) could be used, but they do not meet the exhaustiveness required for CMg. Accordingly, one should be cautious when inferring a resistance phenotype from metagenomic data when uncultivable bacteria are identified.

Although the data generated by CMg are highly diverse and complex, the bioinformatics pipeline is expected to be fast. Managing giga bases of sequence data is hardly possible on standard laboratory computers, and either a local calculation cluster or a cloud-based one is necessary. Nonetheless, bioinformatics tools based on *k*-merised databases (that is reducing the complexity of the genome reference database by using a *k*-mer profile of the database instead of the sequences themselves) are considerably faster than solutions based on the direct mapping of reads onto reference databases. By considering species-specific marker genes instead of whole genomes, MetaPhlAn2 is also faster since the database is substantially reduced. Another computational resource-consuming step is the metagenomic assembly that is necessary to identify specific genes or small nucleotide variants. Still, the assembly can in some places be skipped by direct mapping of the reads onto a specific database, such as done by the software ARIBA [53] for the identification of antibiotic resistance genes.

Differentiating the micro-organisms that were indeed present in the clinical sample from contaminants introduced during the entire process is another challenge. Indeed, metagenomic sequencing always reports the presence of micro-organisms, even on negative controls [28]. Given that some of those micro-organisms can be involved in infections, defining them as contaminants is problematic, and carries a significant clinical dilemma. For instance, *Cutibacterium* (formerly *Propionibacterium*) *acnes* is a frequent contaminant in CMg output [28, 54], but can also be involved in bone and joint infections [55] or nosocomial meningitis [56]. The identification of contaminants seems to be more frequent with low-biomass samples as observed in a study from our group (see Extended Data Fig. 6 from [54]). Hence, the bioinformatics pipeline should take into account the results from negative controls in order to assess the likelihood of the presence of micro-organisms in clinical samples. Grumaz et al. proposed a sepsis indicating quantifier (SIQ) score based on the metagenomic sequencing of blood samples collected from non-septic patients, in order to assess whether the relative abundances of micro-organisms found in clinical samples is significantly different from that observed in negative controls. The pipeline developed by PathoQuest uses an *in-house* score based on various criteria such as the genome coverage, alignment distribution metric p-value, and the number of segments for segmented viruses. A score below 100 suggests that the micro-organism is a contaminant while a score > 1000 suggests that it is truly present. Wilson et al. sequenced the DNA from 94 cerebrospinal fluid (CSF) samples obtained from patients with non-infectious inflammatory disorders and 24 negative controls (water and reagents) [57]. From that, they developed a weighted *z* score–based scoring algorithm aiming at reducing the taxonomic noise made by contaminants. They analysed 7 CSF samples with this algorithm and found the causative agent to be among the top 2 micro-organisms reported by metagenomic sequencing.

Linking antibiotic resistance determinants (ARD) with their host is highly challenging. Most clinically-relevant antibiotic resistance genes are located on mobile genetic elements (MGEs, e.g. plasmids, phages or transposons) that spread between species. A possible way of linking antibiotic resistance genes to their host is to use the normalised depth of coverage of the antibiotic resistance gene and the median depth of sequencing of the bacteria found in the sample, with the assumption that the host harbours at least one copy of its ARD-encoding gene. Accordingly, the normalised depth of sequencing of the ARD ($nDOS_{ARD}$) should not theoretically be smaller than the median normalised depth of sequencing $nDOS_{HOST}$ of the chromosomal genes of the host, assuming that the extraction process does not alter the proportion between chromosome and MGEs. Thus, if the ARD and the host are linked, the ratio $\frac{nDOS_{ARD}}{nDOS_{HOST}}$ should be greater or equal to 1. Nevertheless, such an association might be hazardous when several hosts are present. We tested this hypothesis on a subset of bone and joint infections samples in which only a single bacterial species had been found both in culture and metagenomic sequencing. Unexpectedly, we found such ratio to be <1 for *Staphylococcus aureus* and its penicillinase-encoding gene *blaZ* (see the Extended Data Fig. 9 from [54]), suggesting that a subset of the *S. aureus* population does not carry *blaZ*. Subsequently, to the best of our knowledge, no bioinformatic solution linking ARD-encoding genes and their host is available. Some authors have proposed to physically connect the DNA (chromosome and MGEs) within the cell before the extraction so that during the bioinformatics process one could identify which reads derive from the same cell [58], but the protocol is still too complex to consider for routine implementation.

Another challenge is the output delivered by the bioinformatics pipeline. This output should be clear to clinical microbiologists and clinicians in order to make clinical decisions (e.g. adapt the antimicrobial regimen). The CMg report should mention the micro-organisms found with a level of confidence (see the previous paragraph about contaminants), their quantification (relative abundance) and the information about antimicrobial resistance (presence of antibiotic resistance genes and/or mutational events associated with resistance). In addition, other bioinformatics parameters allowing a better interpretation of the results should be provided such as the estimated genome coverage. For example, finding an *Escherichia coli* in a clinical sample with no antibiotic resistance gene but a genome coverage of <10% indicates that some antibiotic resistance genes may be present but that the depth of sequencing of the sample was not sufficient to detect them. Conversely, a genome coverage >90% will provide confidence in assessing the absence of antibiotic resistance genes.

8.4 Examples of CMg Applications (Table 8.1)

8.4.1 Bone and Joint Infections (BJI)

BJI are severe infections (most of them involving bacteria) that affect a growing number of patients [59]. Along with surgical intervention, the microbiological diagnosis is a keystone to the management of BJI in (i) identifying the bacteria causing the infection and (ii) assessing their susceptibility to antibiotics. Currently, this is achieved by culturing surgical samples on various media and conditions, together with an extended incubation time to recover fastidiously-growing bacteria that can be involved in BJI. Still, some bacteria will not grow under these conditions because of extreme

Table 8.1 Examples of CMg applications

Type of sample	Study	Number of patients	Population	Bacterial DNA enrichmement	Chemistry
Blood	Grumaz et al. [33]	25 (62 samples)	Healthy controls, septic shock and post-operative abdominal surgery.	No	Illumina
	Gyarmati et al. [61]	9 (27 samples)	Neutropenic patients.	No	Illumina
	Parize et al. [62]	101	Immunosuppressed patients.	No	Illumina
Bone and joint infections	Ruppé et al. [54]	24	Patients with bone and joint infections.	MolYsis	Illumina
Bronchoalveolar lavage	Langelier et al. [65]	22	Haematopoetic cells transplant recipients.	No	Illumina
	Leo et al. [10]	1 (case report)	One immunosuppressed patient.	Yes (MolYsis) and none	Illumina
Cerebrospinal fluid	Greninger et al. [75]	1 (case report)	15-year-old girl with primary amoebic meningoencephalitis.	No (RNAseq)	Illumina
	Ortiz-Alcántara et al. [72]	1 (case report)	13-year-old boy with meningitis.	No (RNAseq)	454
	Perlejewski et al. [69]	12	Patients with multiple sclerosis.	No (RNAseq)	Illumina
	Simner et al. [68]	10	Patients with meningitis.	Yes (saponin and DNAse) and no	Illumina
	Wilson et al. [67]	1 (case report)	14-year-old, immunosuppressed boy with meningitis.	No	Illumina
	Wilson et al. [57]	101	7 patients with chronic meningitis and 94 patients with noninfectious neuroinflammatory disorders.	No (RNAseq)	Illumina
	Wilson et al. [71]	1 (case report)	74-year-old woman with amoebic endophtalmitis and meningoencephalitis.	No (RNAseq)	Illumina
	Wylie et al. [56]	8 (case report)	Patient with chronic meningits (3 samples) and 5 controls.	No	Illumina

Knee synovial fluid	Ivy et al. [60]	168	107 patients with infection and 61 without infection.	MolYsis	Illumina
Mini bronchoalveolar lavage	Pendleton et al. [63]	2 (case reports)	Patients with ventilatory associated pneumonia.	No	Nanopore
Nasopharyngeal/oropharyngeal aspirations	Schlaberg et al. [64]	160	5-year-old children with pneumonia ($n = 70$) or without respiratory symptoms ($n = 90$)	No (RNAseq)	Illumina
Respiratory samples	Charalampous et al. [76]	41	Patients with suspected pneumonia.	Yes (saponine and DNAse)	Nanopore
Sonicate fluid samples from resected hip and knee arthroplasties	Thoendel et al. [55]	408	213 patietns with infection and 195 patients without infection.	MolYsis	Illumina
Sonication fluid samples from prosthetic joint and other orthopaedic device infections	Street et al. [18]	97	Patients with bone and joint infections.	5 µm filter, and a subset of samples with NebNext microbiome DNA enrichment kit	Illumina
Urine	Hasman et al. [73]	19	Patients with urine samples positive in culture.	No	IonTorrent
	Schmidt et al. [74]	15	Patients with urine samples positive in culture ($n = 10$) and 5 spiked samples.	No	Nanopore

oxygen sensitivity, prior antibiotic treatment or metabolic issues (e.g. quiescent bacteria in chronic infections). Consequently, the antibiotic treatment may not span all the bacteria involved in the infection, which can favour relapse and the need for repeated surgical procedures or prolonged wide-spectrum antibiotic treatments. For all those reasons, CMg appears to be a promising technology for the diagnosis of BJI.

At the time of writing, three studies have applied CMg on BJI samples [18, 54, 55]. The study from our group sequenced 24 BJI samples which were reported as positive by conventional culturing [54]. For polymicrobial samples (n = 16), 32/55 bacteria (58.2%) were detected at the species level and 41/55 [74.5%] at the genus level. Conversely, 273 bacteria not found in culture were identified, 182 being possible pathogens and 91 contaminants. Street et al. compared metagenomic sequencing with standard aerobic and anaerobic culture in 97 sonication fluid samples from prosthetic joint and other orthopaedic device infections [18]. Compared to sonication fluid culture, the sensitivity of metagenomic sequencing was 61/69 (88%) at the species level and 64/69 (93%) at the genus level, while specificity was 85/97 (88%) at the species level. Thoendel et al. sequenced 408 fluid samples obtained from patients who underwent hip resection or knee arthroplasties, including 213 with infections and 195 without infection (aseptic failure) [55]. When compared to conventional methods (culture), metagenomics was able to identify known pathogens in 95% (109/115) of culture-positive samples, with additional potential pathogens detected in 9.6% (11/115). New potential pathogens were detected in 44% (43/98) of culture-negative samples. Conversely in samples from uninfected patients, the detection of micro-organisms was rare (7/195, 3.6%). The same group sequenced 168 synovial fluids obtained by synovial puncture in patients with failed knee arthroplasty including 107 with infection and 61 without [60]. Metagenomic sequencing yielded the same micro-organism as the one found in culture in 74 (90%, genus level) and 68 (83%, species level) of the 82 culture-positive samples. For the 25 culture-negative samples from infected patients, metagenomic sequencing identified 4 (16%) samples with potential pathogens detected at the species level. As for the 60 culture-negative aseptic failure cases, metagenomic sequencing identified potential clinically-significant organisms in 11 (18.3%) samples (11.7%, genus level and 4 (6.7%) species level), and 1 probable contaminant. Hence, these studies show the advantage of combining standard methods (i.e. culture) and CMg when dealing with serious infection as neither approach alone has a 100% sensitivity.

8.4.2 Blood Samples

The area of sequencing-based microbiological diagnosis of blood samples is still evolving but has the potential to make a significant clinical impact, as standard culturing has a relatively low yield for growing potential pathogens.

Gyarmati et al. sequenced 27 blood samples from 9 patients with acute leukaemia and suspected bloodstream infections (BSI) at different time points [61]. They identified diverse bacteria (*C. acnes*, *Staphylococcus* spp., *Corynebacterium* spp), viruses (Torque Teno virus, *Cutibacterium* phages) and fungi (*Fusarium oxysporum*, *Botryotinia fuckeliana*, *Aspergillus* spp., *Malassezia globosa*) that were not identified in the negative controls. Of note, the reads assigned to bacteria were detected mostly at fever onset and in persistent fever (69% in both) but in less than 1% in samples after antibiotic therapy. Likewise, Grumaz et al. found that the cfDNA tended to be the most abundant in the context of septic shock and post-operative abdominal surgery [33].

Parize et al. sequenced 101 blood samples from immunosuppressed patients in France [62]. An expert panel analysed the sequencing results. Metagenomic sequencing identified more micro-organisms than conventional methods (36/101 (36%) vs 11/101 (11%), respectively). In 27 patients, a micro-organism was found by metagenomic sequencing but not using conventional methods, mainly *Pseudomonas* spp. ($n = 17$) and *Streptococcus* spp. ($n = 6$). Besides, two false negatives were observed: 1 CMV and 1 *E. coli* (not considered as a true positive by the bioinformatic algorithm due to the *E. coli* DNA contamination of reagents).

8.4.3 Respiratory Samples

Analysing respiratory samples poses a challenge, as these are taken from a non-sterile site, and differentiating between colonisation and infection is complicated.

Our group reported the application of CMg on bronchoalveolar lavage (BAL) in an immunosuppressed patient [12]. Metagenomic sequencing found *Mycobacterium abscessus* and *Corynebacterium jeikeium* that were obtained in culture, as well as other anaerobic bacteria from the oropharyngeal microbiota.

Pendleton et al. reported two cases of hospital-acquired pneumonia caused by *P. aeruginosa* and *S. aureus*, respectively [63]. They sequenced mini BAL samples with the MinION (Oxford Nanopore Technologies) sequencers and in the first case identified one 3217 bp read assigned to *P. aeruginosa* and in the second case six high-quality reads assigned to *S. aureus*. Of note, the read assigned to *P. aeruginosa* was obtained 9 hours after the BAL was performed.

Schlaberg et al. sequenced the retrotranscribed RNA of nasopharyngeal/oropharyngeal swabs from <5 year-old children (70 with community-associated pneumonia without known aetiology and 90 with no respiratory symptoms) [64]. In children with pneumonia, RNA-seq detected 90% of pathogens detected by conventional methods, but also viruses that were not reported, such as anelloviruses, astroviruses, human herpes virus (HHV)-6 and HHV-7 in both children with and without pneumonia. Of note, adenoviruses, which have a DNA genome, were not detected using RNA-seq.

Langelier et al. sequenced the DNA and cDNA of BAL samples obtained from 22 haematopoietic stem cell transplants (HSCT) recipients [65]. Conventional clinical diagnostics identified micro-organisms in seven (32%) patients, all of which were detected by CMg, and six of which were considered pathogens. In six other patients, CMg identified micro-organisms not detected by conventional methods but considered to be pathogens by the authors: two human coronavirus (HCoV) 229E, two human rhinovirus (HRV)-A, one *Corynebacterium propinquum* and one *Streptococcus mitis*. Besides, the authors measured the expression level of some of the host's genes related to innate and adaptive immune responses. They found that the level of expression of these genes was higher in patients with confirmed pneumonia compared to others without pneumonia.

Lastly, Charalampous et al. sequenced 41 respiratory samples (37 sputa, three endotracheal aspirations and one BAL sample) from patients with suspected pneumonia, using Nanopore chemistry, after applying an optimised human DNA depletion based on saponin [66]. Compared to culture results, they found a 96.6% sensitivity and 41.7% specificity. The turn-around time was notably fast, with 6 hours from DNA extraction to results. The limit of detection of bacterial microorganisms was 10^4 colony-forming units per mL. In eight cases, CMg detected an additional bacterium not found in culture but potentially pathogenic: *Moraxella catarrhalis* ($n = 2$), *Klebsiella pneumoniae* ($n = 1$), *P. aeruginosa* ($n = 1$), *S. aureus* ($n = 1$) and *Streptococcus pneumoniae* ($n = 3$). The majority of bacteria found in conventional cultures were susceptible to the tested antibiotics. Out of 33 observed resistance to antibiotics, 14 could be explained by the antibiotic resistance genes found in CMg leaving 19 instances where no definitive explanation could be given, One of them being that, several antibiotic resistance genes such as *tet(M)* and *mefA* were likely borne by commensal bacteria.

8.4.4 Central Nervous System Infections

Since the first cerebrospinal fluid (CSF) sequencing by Wilson et al. and the finding of *Leptospira santarosai* that was undetected by other clinical diagnostic tests [67], several studies have used metagenomic sequencing for the detection of microorganisms from CSF samples in the context of meningitis and meningoencephalitis. Indeed, these infections may be caused by a wide variety of bacteria, viruses and fingi, which raise challenges in terms of bacterial recovery (as previous antibiotic treatment may lead to a negative culture) and identification of the aetiological agent.

Recently, Simner et al. evaluated the performances of nine different protocols to detect pathogens in CSF using CMg [68]. They tested eight CSF samples which , in standard tests, were positive for viruses ($n = 3$), bacteria ($n = 3$) or *Cryptococcus* ($n = 2$), or were negative. Overall, the neat CSF performed better than pelleted or supernatant samples. In one sample, the identification of *Cryptococcus* was, however, unsuccessful by all protocols.

Perljewski et al. sequenced the cDNA from 12 CSF samples obtained from patients with multiple sclerosis [69]. They identified the varicella-zoster virus (VZV) in 11 out of 12 samples, raising questions whether VZV has a possible role in multiple sclerosis, or merely represents chronic reactivation due to immunomodulation. They also found several micro-organisms which were considered as contaminants.

Finally, the application of CMg on CSF allowed the identification of rare pathogens such as *Balamuthia*, involved in amoebic meningoencephalitis [70, 71], *Psychrobacter* sp. [72] and *Cutibacterium* [56].

8.4.5 Urinary Tract Infections

Urinary-tract infections (UTIs) are common infections that are mostly caused by Enterobacterales and *Enterococcus* sp. The microbiological diagnosis of UTI relies on semi-quantitative cultures of urine samples and concentrations thresholds.

Hasman et al. sequenced the total DNA recovered from 23 urine samples, 19 of which had high bacterial load in culture analysis [73]. Twelve urine samples did not yield enough DNA to be sequenced. When the urine culture yielded one bacterial species (n = 17), it was also identified by metagenomic sequencing. Besides the usual uropathgens, other bacteria were identified such as *Aerococcus urinae*.

In another study, Schmidt et al. sequenced the DNA extracted from 10 urine samples obtained from patients and five spiked samples, using the Nanopore MinION [74]. In the 4-h time frame from sampling to results, metagenomic sequencing identified the main bacteria obtained by culture. In both studies, however, linking the relative abundances of sequencing reads to the bacterial load in urine was not attempted.

8.5 Challenges in CMg Implementation

Overall, the routine clinical usage of CMg remains limited by a series of practical hurdles. We do firmly believe that the technical issues described above, related to the wet-lab or bioinformatics, will be progressively solved. Nevertheless, this will not suffice to ensure a straightforward transfer of CMg into medical laboratories. Indeed, implementing these new methods will face challenges both from the clinician and the administration (payers, diagnostic companies, hospitals and issues regarding reimbursement) standpoints.

Clinicians may raise the following questions: does CMg provide similar and reliable information as traditional lab-work for making medical decisions? Does the medical literature support this statement? Will the analysis report be understandable by every clinician? When will these new methods be routinely taught to young physicians, to help them integrate CMg analyses in their future practice?

Payers will also require documentation that CMg, implying new wet-lab and IT tools, but also new human skills, is indeed "cost-efficient". This concept has to be calculated for each of the potential payers. For hospital administrators, CMg implementation might be balanced against savings in other lab/radiological testings or by earlier discharge. For antimicrobial stewardship teams, CMg as a companion diagnostics might help advance using targeted therapies more rapidly. For public health authorities, improved traceability will help address the ever-growing request of the lay public for transparency and more importantly, it will also likely provide more accurate outbreak analyses by retrieving relevant genomic data for the micro-organisms of interest.

Finally, these assays will need certification and documentation thereof, including participation in external quality control programs for proficiency testing. This also implies considering ethical aspects of collected samples and data, for building bio-banks and providing a guarantee that the derived data will be used according to the law, trying as much as possible not to impair but to promote research using these data.

8.6 Future Perspectives

CMg is one of the most promising methods for the diagnosis of infectious diseases in the short/mid-term horizon. Indeed, CMg bears the capacity to identify a wide array of micro-organisms in an unbiased fashion, provided that the micro-oranisms are already known and sequenced. Moreover, CMg can provide some actionable information about the antimicrobial resistance determinants of the detected micro-organisms, the genotype and the virulence genes content faster than current methods. Many technical hurdles remain to be addressed, but we assume that they should be overcome in the coming years. An important issue related to the routine implementation of CMg may indeed be the clinicians themselves. For decades, the diagnosis of infectious diseases relied on cultivation methods and, more recently, on the detection of nucleic acids of selected pathogens (targeted approach). Infectious diseases specialists still follow the paradigm of the detection of a pathogen in a potentially infected site to support the diagnosis of an infection. CMg will change this paradigm by bringing much more information than current methods, mainly about the presence of bacteria not recognised as pathogens to date, sometimes considered to be "unwanted" information or data of uncertain clinical significance by clinicians. Furthermore, CMg could also provide some information related to the host's response, challenging the current diagnostic paradigm. CMg will, therefore, probably reflect a new way to perceive and diagnose infectious diseases.

References

1. Westblade LF, van Belkum A, Grundhoff A, Weinstock GM, Pamer EG, Pallen MJ et al (2016) Role of clinicogenomics in infectious disease diagnostics and public health microbiology. J Clin Microbiol 54(7):1686–1693
2. Rossen JWA, Friedrich AW, Moran-Gilad J (2018) Practical issues in implementing whole-genome-sequencing in routine diagnostic microbiology. Clin Microbiol Infect 24(4):355–360
3. Greninger AL (2018) The challenge of diagnostic metagenomics. Expert Rev Mol Diagn 18(7):605–615

4. Forbes JD, Knox NC, Peterson CL, Reimer AR (2018) Highlighting clinical metagenomics for enhanced diagnostic decision-making: a step towards wider implementation. Comput Struct Biotechnol J 16:108–120
5. Ruppe E, Greub G, Schrenzel J (2017) Messages from the first international conference on clinical metagenomics (ICCMg). Microbes Infect 19(4–5):223–228
6. Ruppe E, Schrenzel J (2018) Messages from the second international conference on clinical metagenomics (ICCMg2). Microbes Infect 20(4):222–227
7. Ruppe E, Schrenzel J (2019) Messages from the third international conference on clinical metagenomics (ICCMg3). Microbes Infect 21(7):273–277
8. Charretier Y, Lazarevic V, Schrenzel J, Ruppé E (2020) Messages from the Fourth International Conference on Clinical Metagenomics. Microbes Infect 22:635–641. https://doi.org/10.1016/j.micinf.2020.07.007
9. Lorenz M, inventor; Molzym Gmbh & Co. Kg, assignee (2007) Use of nucleases for degrading nucleic acids in the presence of chaotropic agents and/or surfactants patent EP1861495 (A1) Abstract of corresponding document: DE102005009479 (A1). 2007-12-05
10. Leo S, Gaia N, Ruppe E, Emonet S, Girard M, Lazarevic V et al (2017) Detection of bacterial pathogens from broncho-alveolar lavage by next-generation sequencing. Int J Mol Sci 18(9)
11. Lazarevic V, Gaia N, Girard M, Leo S, Cherkaoui A, Renzi G et al (2018) When bacterial culture fails, metagenomics can help: a case of chronic hepatic brucelloma assessed by next-generation sequencing. Frontiers Microbiol 9:1566
12. QIAamp DNA Microbiome Handbook - QIAGEN
13. Hohnadel M, Jouette S, inventors; Merck Patent GmbH, assignee. Method for isolating micro organisms from a complex sample patent US20160257987A1. 2016 2016-09-08
14. Hasan MR, Rawat A, Tang P, Jithesh PV, Thomas E, Tan R et al (2016) Depletion of human DNA in spiked clinical specimens for improvement of sensitivity of pathogen detection by next-generation sequencing. J Clin Microbiol 54(4):919–927
15. Marotz CA, Sanders JG, Zuniga C, Zaramela LS, Knight R, Zengler K (2018) Improving saliva shotgun metagenomics by chemical host DNA depletion. Microbiome 6(1):42
16. Sanchez-Romero MA, Cota I, Casadesus J (2015) DNA methylation in bacteria: from the methyl group to the methylome. Curr Opin Microbiol 25:9–16
17. Thoendel M, Jeraldo PR, Greenwood-Quaintance KE, Yao JZ, Chia N, Hanssen AD et al (2016) Comparison of microbial DNA enrichment tools for metagenomic whole genome sequencing. J Microbiol Methods 127:141–145
18. Street TL, Sanderson ND, Atkins BL, Brent AJ, Cole K, Foster D et al (2017) Molecular diagnosis of orthopedic-device-related infection directly from sonication fluid by metagenomic sequencing. J Clin Microbiol 55(8):2334–2347
19. Gu W, Crawford ED, O'Donovan BD, Wilson MR, Chow ED, Retallack H et al (2016) Depletion of abundant sequences by hybridization (DASH): using Cas9 to remove unwanted high-abundance species in sequencing libraries and molecular counting applications. Genome Biol 17:41
20. Sooknanan RR, inventor; Epicentre Technologies Corporation, assignee (2018) Methods, compositions, and kits for generating rRNA-depleted samples or isolating rRNA from samples patent US2018044660 (A1). 2018-02-15
21. Giannoukos G, Ciulla DM, Huang K, Haas BJ, Izard J, Levin JZ et al (2012) Efficient and robust RNA-seq process for cultured bacteria and complex community transcriptomes. Genome Biol 13(3):R23
22. Avraham R, Haseley N, Fan A, Bloom-Ackermann Z, Livny J, Hung DT (2016) A highly multiplexed and sensitive RNA-seq protocol for simultaneous analysis of host and pathogen transcriptomes. Nat Protoc 11(8):1477–1491
23. Conrad RC, inventor; Ambion, Inc, assignee (2004) High efficiency mrna isolation methods and compositions patent US2004230048 (A1). 2004-11-18
24. Mendoza L, Moturi S, Setterquist R, Whitley J, inventors; Ambion, Inc; Mendoza, Leopoldo, G; Moturi, Sharmili; Setterquist, Robert; Whitley, John, Penn, assignee (2006) Methods

and compositions for depleting abundant RNA transcripts patent WO2006110314 (A2). 2006-10-19
25. Marynowska M, Goux X, Sillam-Dusses D, Rouland-Lefevre C, Roisin Y, Delfosse P et al (2017) Optimization of a metatranscriptomic approach to study the lignocellulolytic potential of the higher termite gut microbiome. BMC Genomics 18(1):681
26. Dean FB, Hosono S, Fang L, Wu X, Faruqi AF, Bray-Ward P et al (2002) Comprehensive human genome amplification using multiple displacement amplification. Proc Natl Acad Sci U S A 99(8):5261–5266
27. Hammond M, Homa F, Andersson-Svahn H, Ettema TJ, Joensson HN (2016) Picodroplet partitioned whole genome amplification of low biomass samples preserves genomic diversity for metagenomic analysis. Microbiome 4(1):52
28. Salter SJ, Cox MJ, Turek EM, Calus ST, Cookson WO, Moffatt MF et al (2014) Reagent and laboratory contamination can critically impact sequence-based microbiome analyses. BMC Biol 12:87
29. Blainey PC, Quake SR (2011) Digital MDA for enumeration of total nucleic acid contamination. Nucleic Acids Res 39(4):e19
30. Thoendel M, Jeraldo P, Greenwood-Quaintance KE, Yao J, Chia N, Hanssen AD et al (2017) Impact of contaminating DNA in whole-genome amplification kits used for metagenomic shotgun sequencing for infection diagnosis. J Clin Microbiol 55(6):1789–1801
31. de Goffau MC, Lager S, Salter SJ, Wagner J, Kronbichler A, Charnock-Jones DS et al (2018) Recognizing the reagent microbiome. Nat Microbiol 3(8):851–853
32. Hansen S, Faye O, Sanabani SS, Faye M, Bohlken-Fascher S, Faye O et al (2018) Combination random isothermal amplification and nanopore sequencing for rapid identification of the causative agent of an outbreak. J Clin Virol 106:23–27
33. Grumaz S, Stevens P, Grumaz C, Decker SO, Weigand MA, Hofer S et al (2016) Next-generation sequencing diagnostics of bacteremia in septic patients. Genome Med 8(1):73
34. Grumaz S, Grumaz C, Vainshtein Y, Stevens P, Glanz K, Decker SO et al (2019) Enhanced performance of next-generation sequencing diagnostics compared with standard of care microbiological diagnostics in patients suffering from septic shock. Crit Care Med 47(5):e394–e402
35. Blauwkamp TA, Thair S, Rosen MJ, Blair L, Lindner MS, Vilfan ID et al (2019) Analytical and clinical validation of a microbial cell-free DNA sequencing test for infectious disease. Nat Microbiol 4(4):663–674
36. Gnirke A, Melnikov A, Maguire J, Rogov P, LeProust EM, Brockman W et al (2009) Solution hybrid selection with ultra-long oligonucleotides for massively parallel targeted sequencing. Nat Biotechnol 27(2):182–189
37. Bent ZW, Tran-Gyamfi MB, Langevin SA, Brazel DM, Hamblin RY, Branda SS et al (2013) Enriching pathogen transcripts from infected samples: a capture-based approach to enhanced host-pathogen RNA sequencing. Anal Biochem 438(1):90–96
38. Briese T, Kapoor A, Mishra N, Jain K, Kumar A, Jabado OJ et al (2015) Virome capture sequencing enables sensitive viral diagnosis and comprehensive virome analysis. mBio 6(5):e01491–e01415
39. Wylie TN, Wylie KM, Herter BN, Storch GA (2015) Enhanced virome sequencing using targeted sequence capture. Genome Res 25(12):1910–1920
40. Noyes NR, Weinroth ME, Parker JK, Dean CJ, Lakin SM, Raymond RA et al (2017) Enrichment allows identification of diverse, rare elements in metagenomic resistome-virulome sequencing. Microbiome 5(1):142
41. Kivioja T, Vaharautio A, Karlsson K, Bonke M, Enge M, Linnarsson S et al (2011) Counting absolute numbers of molecules using unique molecular identifiers. Nat Methods 9(1):72–74
42. Pflug FG, von Haeseler A (2018) TRUmiCount: correctly counting absolute numbers of molecules using unique molecular identifiers. Bioinformatics (Oxford, England)
43. Clarke J, Wu HC, Jayasinghe L, Patel A, Reid S, Bayley H (2009) Continuous base identification for single-molecule nanopore DNA sequencing. Nat Nanotechnol 4(4):265–270
44. Loose M, Malla S, Stout M (2016) Real-time selective sequencing using nanopore technology. Nat Methods 13(9):751–754

45. Naccache SN, Federman S, Veeraraghavan N, Zaharia M, Lee D, Samayoa E et al (2014) A cloud-compatible bioinformatics pipeline for ultrarapid pathogen identification from next-generation sequencing of clinical samples. Genome Res 24(7):1180–1192
46. Wood DE, Salzberg SL (2014) Kraken: ultrafast metagenomic sequence classification using exact alignments. Genome Biol 15(3):R46
47. Truong DT, Franzosa EA, Tickle TL, Scholz M, Weingart G, Pasolli E et al (2015) MetaPhlAn2 for enhanced metagenomic taxonomic profiling. Nat Methods 12:902–903
48. Xavier BB, Das AJ, Cochrane G, De Ganck S, Kumar-Singh S, Aarestrup FM et al (2016) Consolidating and exploring antibiotic resistance gene data resources. J Clin Microbiol 54(4):851–859
49. Zankari E, Hasman H, Cosentino S, Vestergaard M, Rasmussen S, Lund O et al (2012) Identification of acquired antimicrobial resistance genes. J Antimicrob Chemother 67(11):2640–2644
50. McArthur AG, Waglechner N, Nizam F, Yan A, Azad MA, Baylay AJ et al (2013) The comprehensive antibiotic resistance database. Antimicrob Agents Chemother 57(7):3348–3357
51. Martinez JL, Coque TM, Baquero F (2015) What is a resistance gene? Ranking risk in resistomes. Nat Rev Microbiol 13(2):116–123
52. Wallace JC, Port JA, Smith MN, Faustman EM (2017) FARME DB: a functional antibiotic resistance element database. Database 2017
53. Hunt M, Mather AE, Sanchez-Buso L, Page AJ, Parkhill J, Keane JA et al (2017) ARIBA: rapid antimicrobial resistance genotyping directly from sequencing reads. Microbial Genom 3(10):e000131
54. Ruppé E, Lazarevic V, Girard M, Mouton W, Ferry T, Laurent F et al (2017) Clinical metagenomics of bone and joint infections: a proof of concept study. Sci Rep 7:7718
55. Thoendel M, Jeraldo P, Greenwood-Quaintance KE, Yao J, Chia N, Hanssen AD et al (2018) Identification of prosthetic joint infection pathogens using a shotgun metagenomics approach. Clin Infect Dis
56. Wylie KM, Blanco-Guzman M, Wylie TN, Lawrence SJ, Ghobadi A, DiPersio JF et al (2016) High-throughput sequencing of cerebrospinal fluid for diagnosis of chronic Propionibacterium acnes meningitis in an allogeneic stem cell transplant recipient. Transpl Infect Dis 18(2):227–233
57. Wilson MR, O'Donovan BD, Gelfand JM, Sample HA, Chow FC, Betjemann JP et al (2018) Chronic meningitis investigated via metagenomic next-generation sequencing. JAMA Neurol
58. Burton JN, Liachko I, Dunham MJ, Shendure J (2014) Species-level deconvolution of metagenome assemblies with Hi-C-based contact probability maps. G3 (Bethesda, MD) 4(7):1339–1346
59. Grammatico-Guillon L, Baron S, Gettner S, Lecuyer AI, Gaborit C, Rosset P et al (2012) Bone and joint infections in hospitalized patients in France, 2008: clinical and economic outcomes. J Hosp Infect 82(1):40–48
60. Ivy MI, Thoendel MJ, Jeraldo PR, Greenwood-Quaintance KE, Hanssen AD, Abdel MP et al (2018) Direct detection and identification of prosthetic joint infection pathogens in synovial fluid by metagenomic shotgun sequencing. J Clin Microbiol
61. Gyarmati P, Kjellander C, Aust C, Song Y, Ohrmalm L, Giske CG (2016) Metagenomic analysis of bloodstream infections in patients with acute leukemia and therapy-induced neutropenia. Sci Rep 6:23532
62. Parize P, Muth E, Richaud C, Gratigny M, Pilmis B, Lamamy A et al (2017) Untargeted next-generation sequencing-based first-line diagnosis of infection in immunocompromised adults: a multicentre, blinded, prospective study. Clin Microbiol Infect 23(8):574.e1–574.e6
63. Pendleton KM, Erb-Downward JR, Bao Y, Branton WR, Falkowski NR, Newton DW et al (2017) Rapid pathogen identification in bacterial pneumonia using real-time metagenomics. Am J Respir Crit Care Med 196(12):1610–1612
64. Schlaberg R, Queen K, Simmon K, Tardif K, Stockmann C, Flygare S et al (2017) Viral pathogen detection by metagenomics and pan-viral group polymerase chain reaction in children with pneumonia lacking identifiable etiology. J Infect Dis 215(9):1407–1415

65. Langelier C, Zinter MS, Kalantar K, Yanik GA, Christenson S, O'Donovan B et al (2018) Metagenomic sequencing detects respiratory pathogens in hematopoietic cellular transplant patients. Am J Respir Crit Care Med 197(4):524–528
66. Charalampous T, Kay GL, Richardson H, Aydin A, Baldan R, Jeanes C et al (2019) Nanopore metagenomics enables rapid clinical diagnosis of bacterial lower respiratory infection. Nat Biotechnol 37(7):783–792
67. Wilson MR, Naccache SN, Samayoa E, Biagtan M, Bashir H, Yu G et al (2014) Actionable diagnosis of neuroleptospirosis by next-generation sequencing. N Engl J Med 370(25):2408–2417
68. Simner PJ, Miller HB, Breitweiser FP, Pinilla Monsalve G, Pardo CA, Salzberg SL et al (2018) Development and optimization of metagenomic next-generation sequencing methods for cerebral spinal fluid diagnostics. J Clin Microbiol
69. Perlejewski K, Bukowska-Osko I, Nakamura S, Motooka D, Stokowy T, Ploski R et al (2016) Metagenomic analysis of cerebrospinal fluid from patients with multiple sclerosis. Adv Exp Med Biol 935:89–98
70. Greninger AL, Naccache SN, Federman S, Yu G, Mbala P, Bres V et al (2015) Rapid metagenomic identification of viral pathogens in clinical samples by real-time nanopore sequencing analysis. Genome Med 7:99
71. Wilson MR, Shanbhag NM, Reid MJ, Singhal NS, Gelfand JM, Sample HA et al (2015) Diagnosing Balamuthia mandrillaris encephalitis with metagenomic deep sequencing. Ann Neurol 78(5):722–730
72. Ortiz-Alcantara JM, Segura-Candelas JM, Garces-Ayala F, Gonzalez-Duran E, Rodriguez-Castillo A, Alcantara-Perez P et al (2016) Fatal Psychrobacter sp. infection in a pediatric patient with meningitis identified by metagenomic next-generation sequencing in cerebrospinal fluid. Arch Microbiol 198(2):129–135
73. Hasman H, Saputra D, Sicheritz-Ponten T, Lund O, Svendsen CA, Frimodt-Moller N et al (2014) Rapid whole-genome sequencing for detection and characterization of microorganisms directly from clinical samples. J Clin Microbiol 52(1):139–146
74. Schmidt K, Mwaigwisya S, Crossman LC, Doumith M, Munroe D, Pires C et al (2017) Identification of bacterial pathogens and antimicrobial resistance directly from clinical urines by nanopore-based metagenomic sequencing. J Antimicrob Chemother 72(1):104–114
75. Greninger AL, Messacar K, Dunnebacke T, Naccache SN, Federman S, Bouquet J et al (2015) Clinical metagenomic identification of Balamuthia mandrillaris encephalitis and assembly of the draft genome: the continuing case for reference genome sequencing. Genome Med 7:113
76. Charalampous T, Richardson H, Kay GL, Baldan R, Jeanes C, Rae D et al (2018) Rapid diagnosis of lower respiratory infection using nanopore-based clinical metagenomics. bioRxiv:387548

Chapter 9
Advanced Applications of MALDI-TOF MS – Typing and Beyond

Aline Cuénod and Adrian Egli

9.1 General Introduction

Within the last decade, Matrix-Assisted Laser Desorption Ionisation – Time of Flight mass spectrometry (MALDI-TOF MS) has revolutionised species identification in clinical microbiology routine diagnostics and continuous to replace conventional and biochemical methods for bacterial species identification [1]. Bacterial whole cell-mass spectrometry allows identifying microbes to the species level within minutes from cultured single isolates [2]. Several studies have shown that MALDI-TOF MS is a reliable, reproducible and cost-effective method for rapid bacterial and fungal species identification [3–7]. The accuracy of MALDI-TOF MS identification has been investigated by several studies and was determined to be 79.9–93.6% at the species level and 94.5–97.2% at the genus level [8–11]. Since these early validation studies, further database updates and optimisation of the workflows and procedures have additionally increased the accuracy of the method [12]. It has been shown, that for some species e.g. within the family of *Enterobacteriaceae*, identification by MALDI-TOF MS is more accurate than 16S *rRNA* gene sequence analysis which is a highly reputable species identification method [8]. In this chapter, we will discuss the challenges and opportunities of bacterial typing using MALDI-TOF MS, beyond species identification.

A. Cuénod (✉) · A. Egli
Applied Microbiology Research, Department of Biomedicine, University of Basel, Basel, Switzerland

Division of Clinical Bacteriology and Mycology, University Hospital Basel, Basel, Switzerland
e-mail: aline.cuenod@stud.unibas.ch

J. Moran-Gilad, Y. Yagel (eds.), *Application and Integration of Omics-powered Diagnostics in Clinical and Public Health Microbiology*,
https://doi.org/10.1007/978-3-030-62155-1_9

9.1.1 MALDI-TOF MS Functions and Workflow

In diagnostic routine MALDI-TOF MS workflows, a single bacterial colony is placed on the MALDI-TOF MS target plate, air-dried and coated with a specific matrix before measurement. The matrix serves as an energy absorbent organic compound (e.g. alpha- Cyano-4-hydroxycinnamic (CHCA)), which co-crystallises with the bacterial sample. The target plate is then inserted to the mass spectrometer where a vacuum is established. Subsequently, a laser beam (e.g. nitrogen laser) is fired onto the sample-matrix crystals, evaporating and ionising the analytes in the sample, and forming a so-called "ion cloud". The ions are accelerated in electrostatic potential, in a manner dependent on the mass and the charge of the ionised peptides. Ions with a low mass/charge (m/z) ratio are accelerated more easily and reach a higher velocity, whereas ions with a higher m/z ratio fly comparably slower. After acceleration, the ions enter a flight tube without an electrostatic field, where they are further separated according to their m/z ratio and detected at the end of the tube. The m/z ratio of an ion is determined by measuring the time, which is required to travel through the flight tube. According to this time of flight (TOF) information, a spectrum of m/z values is generated. As every sample generates a unique spectrum, these characteristic spectra are called Peptide Mass Fingerprints (PMF). The mass range in which peptides are detected in microbiological routine diagnostics is between 2000 and 20,000 Daltons. Although many aspects of this workflow are standardised, the technical and biological reproducibility of MALDI-TOF MS, and the generation of PMFs are variable both within a laboratory and certainly also in between different laboratories. High reproducibility of PMF quality is the most essential factor allowing typing beyond species identification [13].

9.1.2 MALDI-TOF MS Based Identification Process

For "classical" species identification, the PMF of the unknown sample is compared to a database containing a set of PMF from well-characterised samples [1]. This comparison relies on multiple factors, including the number of characteristic species-specific peaks, which can be detected in both spectra. The two most widely used commercial systems in clinical microbiology routine are the MALDI Biotyper (Bruker Daltonik GmbH, Bremen, Germany) and the Vitek MS (bioMérieux, Marcy l'Etoile, France). A high percentage of the peaks of a spectrum correspond to ribosomal proteins, which are among the most abundant cytosolic proteins [14]. Molecular diagnostic tools such as 16S *rRNA* sequencing also rely on the specific diversity of the bacterial ribosome, making this an ideal target for most species for identification [15]. For (sub)typing applications using MALDI-TOF MS, such as the detection of species within a complex, *in house* developed or modified commercial databases are often used with an increased or optimised set of PMFs of the bacterial species or isolates of interest. As later discussed in more detail, also the

PMF themselves can be compared between bacterial species e.g. in an outbreak scenario.

9.1.3 Resolution of MALDI-TOF MS for Typing

In theory, some differences in the genetic sequence of bacterial strains can be reflected in MALDI-TOF MS. Single point mutations can lead to a change in the amino acid sequence of a protein, which can lead to a shift of mass of the protein. Currently available MALDI-TOF MS systems can detect m/z shifts of about 5–10 daltons. Indeed, the resolution of MALDI-TOF MS has limits and is significantly lower than the resolution of DNA-based sequence analysis, as a SNP in the DNA sequence can only be detected if i.) the mutation leads to a change in the amino acid sequence, or premature truncation of the protein, ii.) the amino acids exchanged have different masses (e.g. glutamine and lysine have almost the same mass) and iii.) the mass of the protein is in the m/z range of the MALDI-TOF MS measurement range [13, 16] and (iv) the protein is expressed in routine growth conditions.

9.2 Current Challenges for Species Identification

Although MALDI-TOF MS is a highly valuable method for species identification in modern microbiological diagnostic laboratories, the technique also has several important technical limitations. Some bacterial species cannot be reliably separated and identified from closely related bacterial species. The low resolution for some species can have various reasons:

(i) *Species classification of highly similar bacterial lineages.* A prominent example of such uncertainty is the inability of MALDI-TOF MS to reliably distinguish between the species *Escherichia coli* and the four species of the genus *Shigella*. Strains of this taxonomic unit are considered as a single species based on DNA relatedness [17, 18]. There are only minor differences in their biochemical profiles and even pathogenicity does not provide an unambiguous classification as strains of the species *E.coli* can cause dysentery-like diarrhoea [19]. This species classification should be understood in a historical context and is maintained because of the severity of shigellosis and to avoid confusion in medical microbiology [18]. Virulence factors which are typically associated with the genus *Shigella* are encoded on plasmids and unfortunately cannot be detected directly with MALDI-TOF MS [20, 21]. Multiple studies have attempted to distinguish strains of the species *E.coli* and strains of the four species of the genus *Shigella* by MALDI-TOF MS using an empirical approach to identify distinctive marker peaks [22, 23]. These suggestions have however not yet been validated and have not been implemented in routine diagnostics [24].

(ii) *Genetic differences are not reflected in the MALDI-TOF MS spectra.* Another example of closely related species which cannot, or can only partially be discriminated by MALDI-TOF MS is the species within the *Enterobacter cloacae*-complex. Here, genome-wide analyses have identified several new species within the last few years [25]. The genetic differences are not reflected in MALDI-TOF MS, to the extent that distinction cannot be implemented with similar algorithms as those for more distantly related species [26]. Other examples where species identification by MALDI-TOF MS is still challenging and not yet implemented in routine microbiology are the viridans group streptococci [27], the *Citrobacter freundii*-complex [11, 28] and the *Bacillus cereus* group [29].

(iii) *Low quality of MALDI-TOF MS spectra.* MALDI-TOF MS resolution increases with spectral quality, as with high-quality spectra the mass range can be enlarged and more peaks can be detected and compared. The quality of MALDI-TOF MS spectra is influenced by several factors such as the sample preparation protocol used, bacterial growth condition (type of liquid or solid media, aerobic or anaerobic incubation) and duration of incubation, quantity of bacterial material applied onto the target plate, the cleanliness of the matrix and the target plate, regular maintenance of device and regular and frequent calibrations [8]. The acquisition of good quality spectra in microbiological routine laboratories is possible, but standardisation of the above-mentioned factors needs to be implemented and an internationally accepted guideline is still missing.

In addition, the quality of MALDI-TOF MS spectra is also strongly dependent on the bacterial species. Bacterial factors which can impede the ionisation of cytosolic proteins include heavy capsule production (eg. hypermucoviscous *K. pneumoniae*), or the thickness of the bacterial cell wall of both *Mycobacterium* species [30] and yeast [31]. Bruker Daltonics have addressed the difficulty to break open the mycobacterial cell wall with the release of a 'MycoEx', a protocol for the sample preparation of *Mycobacterium* isolates and an associated database. Even with this improvement however, several species of the genus *Mycobacterium* remain indistinguishable using MALDI-TOF MS, including the species within the clinically most important *Mycobacterium tuberculosis* complex [32].

9.3 MALDI-TOF Mass Spectrometry-Based Typing

Strains within a bacterial species can have very heterogeneous phenotypes such as virulence, antibiotic resistance or transmissibility. Bacterial typing refers to the differentiation of sub-lineages within a species. The purpose of bacterial typing is to distinguish and recognise subgroups within a species, for example, those associated with a specific phenotype or epidemic strains (from clonal complexes to even single clones). Bacterial typing can help to decipher the population structure of a bacterial species and to follow the development and spread of different subgroups over time [6].

Molecular typing methods such as Multi-Locus-Sequence-Typing (MLST) or Pulse-Field-Gel-Electrophoresis (PFGE) are well established and show a high discriminatory power [33]. Most of the molecular typing methods are expensive, labour-extensive, require specific equipment and expertise. Also, data interoperability and comparability are often reduced: as an example, PFGE typing data cannot be compared easily between different laboratories. Some typing methods were developed particularly for one bacterial species and cannot be adopted to other species, for example, *spa* typing for *S. aureus* [34] or PCR-ribotyping for *C. difficile* [35], as discussed in Chap. 5.

In recent years, some of these problems have been overcome with whole-genome sequencing (WGS), providing the currently highest resolution and highly interoperable, comparable and sharable data formats [36]. However, WGS based typing is still very expensive due to high reagent and equipment costs. Most WGS based workflows currently still require up to 1week for typing in a usual scenario, often too long as time-to-result is a critical element in allowing clinical actions. Due to these obstacles, WGS is only slowly transferring into routine diagnostics.

Therefore, MALDI-TOF MS-based typing may provide a fast and low-cost method to identify isolates for higher resolution typing such as WGS. Indeed, MALDI-TOF MS can be used as a typing method requiring minimal sample preparation, with simple workflows and fast data analysis time. However, closely related species and sub-lineages of the same species provide somewhat ambiguous results when analysed with the conventional approach of pattern matching to reference spectra. Therefore, for bacterial typing with MALDI-TOF MS, different approaches are needed, including high standardisation of spectra acquisition and spectra analysis and interpretation [6, 13].

Multiple studies have shown that MALDI-TOF MS is capable of identifying bacterial strains on a subspecies level. This can especially be useful in the situation of a cluster of infections to identify a potential outbreak [37–39].

Several studies have shown that the sub-lineage identification by MALDI-TOF MS matches phylogenetic classification using MLST. In some instances, it has been shown that phylogenetic units such as Clonal Complexes (CC) [40], Sequence Types (ST) [41] or even single clones [42] can be distinguished using MALDI-TOF MS. The distinction below species level becomes important if the distinguishable units within a species are associated with clinically significant phenotypes such as differences in antibiotic susceptibility, virulence or transmissions in an outbreak context. One meaningful subspecies discrimination is the identification of *E. coli* ST131 which plays an important role in the dissemination of extended-spectrum-β-lactamases (ESBL). It is associated with high transmissibility and antibiotic resistance [43, 44].

In this context, it is easily understandable why bacterial species which often cause infections of public health significance are most often subjected to typing studies. Therefore typing studies considering methicillin-resistant *S. aureus, E.coli, C. difficile and Salmonella* will be discussed in Sect. 9.6.

9.4 Different Typing Approaches

9.4.1 Empirical Identification of Marker Peaks

Multiple studies have come up with empirical study designs, where spectra of the subgroups are compared, and discriminatory peaks are either identified through manual comparison or statistically more elaborated methods [40, 45, 46]. The identification of marker peaks to distinguish the most prevalent STs within methicillin-resistant *Staphylococcus aureus* (MRSA) (ST5, ST59, ST239 ST45) [40] and to distinguish ST131 within the species *E.coli* [43, 47, 48] are successful examples of studies using this approach. Some studies have tried to go further and reach even single clone boundaries. In a study by Egli et al., ESBL-producing *E. coli* could successfully be grouped using MALDI-TOF MS typing in similar clusters to those seen using PFGE [42]. Since then, the reproducibility of these results has been confirmed through a multi-centre study where six diagnostic laboratories characterised the same strains. The technical reproducibility was the most important factor allowing the correct grouping of the bacterial clones [13].

9.4.2 Prediction of Marker Peaks from Bacterial DNA Sequences

Another approach to identify marker peaks is to predict m/z values from predicted genes within bacterial whole genome sequences. As ribosomal subunits are among the most abundant proteins in the bacterial cytosol and have relatively low masses, they can reproducibly be detected in MALDI-TOF MS spectra. Therefore, several studies predicting m/z values from whole-genome sequences have focused on ribosomal subunit proteins [49–53]. Ojima-Kato et al. applied this approach to distinguish serovars within *E. coli* and *Salmonella enterica* subsp. *enterica* [50, 51]. In both studies, m/z values of multiple biomarkers were calculated from WGS data and included in a database called 'Strain Solution', which is available from Shimadzu Corporation (Kyoto, Japan). These biomarkers include ribosomal and other housekeeping proteins. MALDI-TOF MS spectra were analysed by matching the observed m/z profile of the biomarkers to the predicted m/z profiles.

9.4.3 Expansion of the Mass Range to Detect Marker Peaks Otherwise Not Accessible

Another approach is to increase the mass range to capture peptides which have higher or lower m/z values than the routinely detected mass range (2000–20,000 m/z) [54, 55]. One such approach is to combine tryptic digestion and nano-liquid chromatography. The resulting peptide fractions are then identified by MALDI-tandem TOF (MALDI-TOF/TOF) MS; allowing the capture of informative peaks in a lower mass range, making it possible to identify *Salmonella enterica* subspecies [54].

Increasing the mass range to higher m/z values has allowed the successful typing of *C. difficile* isolates, giving MALDI-TOF MS spectral data associated with respective PCR-ribotypes [55].

9.4.4 Self-Learning Classification Algorithms

Recently, several studies have used self-learning algorithms and statistically sophisticated methods such as shrinkage discriminant analysis or supervised neural networks to increase the resolution of conventional MALDI-TOF MS bacterial identification [41, 45, 56].

9.5 Sample Preparation Methods Used for Typing

Integrating bacterial typing in the workflow of routine microbiology laboratories would be a major advantage to rapidly identify infection clusters and transmission within a health care institution.

As resolution increases with spectral quality, many studies use defined sample preparation protocols in order to acquire high-resolution spectra. The most widely applied preparation protocol for high-resolution spectra is the Ethanol-Formic acid protein extraction procedure proposed by Freiwald and Sauer [40, 42, 45, 57, 58]. Beside the sample preparation, the age of the culture is a crucial element, which strongly influences the technical and biological reproducibility [13]. Table 9.1 summarises the methods used in selected MALDI-TOF MS typing studies.

9.6 Examples

9.6.1 Methicillin-Resistant Staphylococcus aureus

Staphylococcus aureus is one of the most frequently isolated bacterial species in clinical microbiological routine laboratories. *S. aureus* infections range from mild superficial skin infections to life-threatening diseases like endocarditis or sepsis [77]. Strains of the species *S. aureus* can harbour multiple different antibiotic

Table 9.1 Summary of selected typing studies and methods applied

Species	Typing unit	Sample preparation method	Identification of distinctive peaks	Ref
S.aureus	CC5, CC8, CC22, CC30, CC45	Ethanol-formic acid protein extraction	Empirical marker detection, hierarchical clustering of profiles	[40]
	CC5, CC8, CC22 and CC398	Direct smear, overlaid with CHAC matrix	Empirical marker detection (statistical tests supported by CLINPROTOOLS (Bruker Daltonics), classification by supervised neural network	[59]
	ST5, ST59, ST239 ST45	Direct smear overlaid with formic acid and CHAC matrix	Machine learning approaches	[41]
	PFGE, *spa* typing	Direct smear, overlaid with CHAC matrix	Clustering of peptide mass fingerprinting	[39]
	CC5, CC22, CC8, CC45, CC30, and CC1, MRSA, MSSA and borderline resistant *S. aureus* (BORSA)	Ethanol-formic acid protein extraction	Empirical marker detection, correlation to mutations in the genomes which cause the peak shift	[58]
	ST239, ST5, ST59, ST45, and 20 MRSA-OST (other clonal lineages)	Ethanol-formic acid protein extraction	Empirical marker detection (statistical tests supported by CLINPROTOOLS (Bruker Daltonics))	[60]
	(PFGE)-types	Direct smear, overlaid with CHAC matrix	Peptide mass fingerprinting (statistical tests supported by Bionumerics software)	[61]
	26 different MALDI-TOF groups comprising 16 MRSA clonal complexes and 89 spa types	Direct smear, matrix: CHAC, ethanol-formic acid protein extraction	Empirical marker detection, visual identification of marker peaks	[62]
	Strain type USA300	Ethanol-formic acid protein extraction	Empirical marker detection (statistical tests supported by CLINPROTOOLS (Bruker Daltonics))	[63]

(continued)

Table 9.1 (continued)

Species	Typing unit	Sample preparation method	Identification of distinctive peaks	Ref
	CC398	Ethanol-formic acid protein extraction	Empirical marker detection (statistical tests supported by CLINPROTOOLS (Bruker Daltonics))	[64]
Staphylococcus	27 strains of the species *S.aureus*, *S. hominis* and *S. epidermidis*	MALDI with CeO_2 (metal oxide laser ionization [MOLI] MS)	Fatty acid components, principal component analysis	[65]
E.coli ESBL	Outbreak strain STEC	Ethanol-formic acid protein extraction	Machine learning algorithm (A.B.O.S. analysis)	[45]
	Outbreak strain STEC	Ethanol-formic acid protein extraction	Empirical marker identification, identification by LC-MS/MS and sequence comparison	[37]
	53 flagellar/H antigen	In house sample preparation protocol, matrices used: CHAC and 2, 5-dihydroxybenzoic acid (DHB), low mass range	Calculation of expected H antigen m/z values from sequence data	[66]
	Serotypes	Solid culture: Direct spotting, liquid culture: *in house* sample preparation protocol, matrices: CHAC, sinapic acid (SA)	Predicting markers from ribosomal proteins encoded by the S10-spc-alpha	[51]
	O157, O26 and O111 Serovars	Direct spotting, matrices: CHAC, sinapic acid (SA)	Predicting markers from WGS data (ribosomal protein S15, L25, acid stress protein H-NS)	[67]
	PFGE-related cluster	Ethanol-formic acid protein extraction	Empirical marker identification, peptide mass fingerprinting, PCA	[42]
	ST131	Ethanol-formic acid protein extraction	Empirical marker identification, (statistical tests supported by CLINPROTOOLS (Bruker Daltonics)	[48]

(continued)

Table 9.1 (continued)

Species	Typing unit	Sample preparation method	Identification of distinctive peaks	Ref
	ST131	Ethanol-formic acid protein extraction, direct spotting	Empirical marker identification, (statistical tests supported by CLINPROTOOLS (Bruker Daltonics)	[43]
Clostridioides difficile	PCR-ribotypes 001, 027 and 126/078	Direct spotting, matrix: CHAH	Empirical marker identification (Superspectra™ function of SARAMIS™ AXIMA)	[46]
	PCR ribotypes 010, 011, 012, 015, 017, 020, 027, 046, 081, SE13a, SE13d, SE20a and SE99/1	Direct spotting, and overlay trans-ferulic acid matrix solution (FeA) and ethanol-formic acid protein extraction with CHAC matrix	High molecular weight typing, peptide mass fingerprinting (statistical tests supported by Bionumerics software)	[55]
Salmonella enterica	Subspecies	Ethanol-formic acid protein extraction with CHAC matrix, followed by tryptic digestion and Nano-LC spotting	Empirical marker identification, in hose excel macro, identification by MALDI-TOF/TOF and NANO-LC	[54]
	Serovars, ribosomal mass profiles	Direct spotting, matrices: CHAC, sinapic acid (SA)	Predicting markers from WGS data, 12 biomarkers, 8 ribosomal proteins 4 non-ribososmal proteins	[50]
Haemophilus influence	Sp., comparison to 16S	Ethanol-formic acid protein extraction with CHAC matrix	Empirical marker identification, peptide mass fingerprinting, PCA	[68]

(continued)

Table 9.1 (continued)

Species	Typing unit	Sample preparation method	Identification of distinctive peaks	Ref
Leptospira	Genomospecies, serovars, (*L. interrogans Hebdomadis L. interrogans Australis L. interrogans Autumnalis L. interrogans Bratislava L. interrogans Canicola L. interrogans Copenhageni L.interrogansHardjo L. interrogans Pomona L. interrogans Pyrogenes L. interrogans Icterohaemorrhagiae L. interrogans Bataviae L. kirschneri Grippotyphosa*)	Ethanol-formic acid protein extraction with CHAC matrix	Empirical marker detection (statistical tests supported by CLINPROTOOLS (Bruker Daltonics)	[69]
Yersinia	*Species*	*In house* tri Fluor acid (TFA) inactivation protocol for highly pathogenic microorganisms	Machine learning algorithms, *in house* Matlab program, identification my MALDI-TOF/TOF	[70]
Yersinia	*Y. pestis*	(TFA) inactivation protocol, direct spotting, DHB matrix	Supervised learning algorithm	[71]
Klebsiella pneumoniae	Hypervirulent strain K1	Ethanol-formic acid protein extraction with CAHC matrix	Empirical marker identification, (statistical tests supported by CLINPROTOOLS (Bruker Daltonics)	[72]
Mycobacterium absesus (sensus lato)	*Mycobacterium absesus (sensu stricto) mycobacteria massiliensis*	Modified mycobacterial preparation protocol suggested by bioMérieux	Empirical marker identification, peptide mass fingerprinting, HCA and PCA, two separate cluster analytical methods, were performed with DataLab software	[73]
Streptococcus pneumoniae	Capsule types 6A, 6B, 6C, 9 N, 9 V or 14.	Formic acid/ Acetonitril premix, overlay with CHCA matrix	Empirical marker identification, peptide mass fingerprinting (statistical tests supported by Bionumerics software)	[74]

(continued)

Table 9.1 (continued)

Species	Typing unit	Sample preparation method	Identification of distinctive peaks	Ref
Bacillus coagulans	26 rep-PCR types	Ethanol-formic acid protein extraction with CAHC matrix	Empirical marker identification, peptide mass fingerprinting, PCA in SARAMIS	[75]
B. pumilus group	Group A and P with species *B. pumilus* and *B. altidunisis*, respectively	Ethanol-formic acid protein extraction with CAHC matrix	SPECLUST analysis, *in house* algorithms	[76]
Rhizobia	*Bradyrhizobium japonicum* strain G49, *Sinorhizobium fredii* strains NGR234 and USDA257	Direct smear, matrix: CHAC, suspension in 25% formic acid, ethanol-formic acid protein extraction with CAHC matrix	Predicting markers from WGS data of 35 ribosomal proteins	[53]

resistance mechanisms and especially methicillin-resistant *S. aureus* (MSRA) strains are a major concern for health care institutions worldwide [78]. Early detection of MRSA spreading clones is essential for the treatment and infection control measures [40]. Several studies have investigated the capability of MALDI-TOF MS to perform this task in order to reduce the need for DNA-based typing methods such as *spa* typing as they are laborious and time-consuming [34].

One way to rapidly detect potential MRSA is to identify the major Clonal Complexes (CC) which are associated with methicillin resistance. This method does not recognise the resistance mechanisms themselves, but rather phylogenetic markers which serve as proxies; although this can be a drawback, as the association of resistances and CC can change over time. Sauget et al. summarised biomarkers which were found to discriminate between the five most frequently detected MRSA CC (CC5, CC8, CC22, CC30, CC45) and the outbreak CC398 [6]. The peak shift from 3876 to 3891 m/z was identified as a characteristic biomarker for CC5, caused by an underlying amino acid exchange in the non-annotated protein SA2420.1 [40, 58, 60, 62]. Two recent studies have approached the distinction at a higher level, by identifying the major ST within MRSA, which are ST5, ST59, ST239 and ST45 [41, 60]. These two studies identified different distinctive signals, an example being ST45 where Zhang et al. identified a discriminatory peak at 4808 m/z and Wang et al. identified a discriminatory low signal at 4813 m/z. This could possibly indicate a peak shift from 4813 m/z to 4808 m/z for ST45. Two further studies have also correlated MALDI-TOF MS profiles to *S.aureus* PFGE types [39, 61].

Whether the absence of a peak can be used as a distinctive marker is questionable as the reproducibility of MALDI-TOF MS marker peaks can be influenced by multiple factors such as spectral quality and growth conditions.

9.6.2 *Escherichia coli*

Strains of the species *Escherichia coli* occur ubiquitously in the environment, being commensals in the human gut, but are also able to cause disease ranging from urinary tract infections to sepsis [19]. As clinical phenotypes of this species are heterogeneous, a distinction below the species level is desirable. The species is classified into several phylogroups, where A and B1 are associated with non-pathogenic colonisation whereas B2 and D are associated with extra-intestinal disease [79]. Within phylogroup B2, the ST131 has drawn special attention as it is associated with ESBL-production and has been the causative agent of multiple nosocomial outbreaks within the last few years [44, 80]. Several typing methods have been adapted to the species *E.coli* such as PFGE, PCR-ribotyping, serotyping and MLST. Many of these have been subjected to studies attempting to correlate them with MALDI-TOF MS findings: Chui et al. distinguished H-antigens, Egli et al. PFGE related clusters, Oijima Serovars and Nakamura et al. as well as Lafolie et al. distinguished ST131 [42, 43, 48, 51, 66]. Christner et al. and Oberle et al. distinguished outbreak-related clones [13, 45].

Multiple distinctive peaks have been identified for ST131 [43, 48] but as Sauget et al. pointed out some of these are not specific to the sequence type level, but rather for all strains of phylogroup B2 [6].

Oberle et al. examined the technical and biological reproducibility of MALDI-TOF MS by analysing 12 closely related ESBL strains representing two nosocomial outbreaks (cluster 1 and cluster 2) from six different centres. Interestingly, PCA analysis showed clustering of the spectra acquired in the same centres, reflecting technical differences between the centres.

Using discriminant analyses, the two outbreak clones could completely be separated. Distinctive peaks were identified empirically using a classifier system from 'AppliedMaths' by Bionumerics [13].

A recently published alternative approach to use MALDI-TOF MS typing for outbreak investigation [45]: 'A.B.O.S.' ('A Better Omics System'; version 1.1.0; Ars Nova AG, Esslingen, Germany) is proposed as an easy to use software for the analysis of omics data. The software uses self-learning algorithms to identify group-specific properties from large datasets by combining various multivariate analysis techniques and predicts the classification of MALDI-TOF MS spectra based on pre-assigned learning groups [45].

As an example, Christner et al. successfully used the A.B.O.S. software for marker peak detection and classification of MALDI-TOF MS spectra from various *E.coli* samples which were isolated during a large Shiga-toxin producing *E.coli* (STEC) outbreak.

Enterobacteriaceae yield relatively high-quality MALDI-TOF MS spectra, even without applying protein extraction protocols. Direct spotting of sample on the MALDI-TOF MS target plate is used in several *E.coli* typing studies, suggesting that a subspecies identification of *E.coli* is possible in microbiological routine settings [45, 51].

9.6.3 *Clostridioides difficile*

Clostridioides difficile are among the most common cause of diarrhoea in hospital settings [81]. Infections with *C. difficile* often occur in the elderly and in patients who have recently been treated with antibiotics. Multiple typing schemes have been established in order to monitor *C. difficile* infections and to rapidly identify outbreaks. Examples are PFGE, multilocus variable-number tandem repeat analysis (MLVA) and PCR-ribotyping [35]. Like many conventional, sequence-based typing methods, these methods are accurate but laborious and time-consuming.

PCR-ribotypes are identified by comparing fragments of the 16S and the 23S *rRNA* genes using capillary gel electrophoresis. One of the first studies to examine the capability of MALDI-TOF MS to distinguish *C. difficile* PCR-ribotypes was conducted by Reil et al. [46], showing that the PCR-ribotypes 001, 027 and 078/126 could be identified.

Rizzardi et al. established a MALDI-TOF MS typing method which could discriminate between 14 different PCR-ribotypes. A higher resolution was accomplished by enlarging the m/z range, by also considering peaks between 30′000 and 50′000 m/z (high molecular weight (HMW) typing method) [55].

In order to detect peaks in this higher m/z range, Ferullic acid (FerA) matrix was used. The higher mass range was empirically selected to improve resolution; in a second step, it was shown that the distinctive peaks detected in this range corresponded mainly to *C. difficile* surface layer proteins and could also give information on the virulence of the strain. The MALDI-TOF MS HMW method has less discriminatory power than the well-established PCR-ribotyping, as 35 HMW profiles were detected from strains including 59 PCR-ribotypes. Due to its easy and cheap implementation, HMW has the potential to complement conventional PCR-ribotyping as it can be applied to a large number of strains with low costs, timewise and financially.

9.6.4 *Salmonella spp.*

In line with MALDI-TOF MS HMW typing, which increases resolution by expanding the mass range to detect higher m/z values, there have been attempts to expand the mass range towards lower m/z values. An example of this is the study by Gekenedis from 2014. In order to identify subspecies-specific peptide biomarkers, the samples were subjected to a standard Ethanol-Formic acid protein extraction [57]. The protein-rich solvent resulting from this procedure is further digested using Trypsin. The peptides were subjected to nano-liquid chromatography and then subsequently identified by MALDI-TOF/TOF MS [54].

This method was found to increase the resolution to subspecies level and assigned biomarkers were identified for the three subspecies *S. enterica* subsp. *arizonae* (n = 17 biomarkers), *S. enterica* subsp. *enterica* (n = 22), and *S. enterica* subsp.

houtenae (n = 29). The number of distinctive biomarkers is relatively high compared to studies which tried to identify biomarkers in the routinely applied mass range. In order to identify the biomarkers amino acid sequence and thereby identity, they were subjected to MALDI-TOF/TOF MS measurements.

Another approach to identifying bacterial strains beyond the species level is to predict distinctive m/z values from sequence data such as whole-genome sequences. Ojima-Kato applied this concept to the subspecies *Salmonella enterica* subsp. *enterica* [50].

Salmonella enterica is an opportunistic pathogen as it can colonise the gut of human and poultry. It regularly causes severe disease outbreaks often associated with the consumption of infected meat, vegetable or fruit [82]. Serotyping is the most widely used typing method. Agglutination assays identify it with specific antibodies of the three surface antigens: the flagellar H, the oligosaccharide O, and the polysaccharide Vi [83]. There are more than 2600 different serotypes described. In 2015 a surveillance study from the US identified the serotypes Typhimurium, Enteritidis, Newport, Heidelberg, and Javiana to be the most common with the first three responsible for almost half of all human infections analysed [84].

Ojima-Kato calculated the m/z values of 12 biomarkers, six of which were ribosomal proteins, from whole-genome sequence data and included them in the software Strain Solution™ ver.2. These theoretically predicted mass profiles were subsequently correlated to serotypes, thereby including the most common outbreak-associated serotypes such as Enteritidis and Typhimurium. From the acquired MALDI TOF MS spectra, the mass profiles of the 12 biomarkers were determined, allowing the correct identification of the serotypes of 109/116 strains. Two of the wrongly classified strains belonged to the serotype 'Typhimurium', two to 'Staintpaul' and three to the serotype 'UN' [50].

This approach can be applied to routine spectral quality but gains resolution with higher spectral quality, as the number of biomarkers identified increases. Theoretically predicting biomarkers allows you to exclusively consider previously predicted marker masses, making this technique independent of growth conditions such as age of the colony and the culture medium used.

9.7 Future Perspectives

Bacterial typing using MALDI-TOF MS is possible, as illustrated with the examples shown. Unfortunately, most of the discussed typing applications are not very straight forward for a routine diagnostic laboratory. Often these methods are used in a research context, are time-consuming and require a significant degree of a machine and/or software knowledge. We anticipate that these applications will require more time in development and translational adaptations for routine use. Easy-to-use and straightforward software modules or databases for analysis would be necessary to allow the expansion of MALDI-TOF MS technology beyond the classical species identification.

In addition, there are discrepancies concerning the reproducibility and feasibility of bacterial typing by MALDI-TOF MS [85, 86]. In part, these discrepancies can possibly be explained by differences in bacterial growth conditions and sample preparation. A desirable development would be the standardisation of growth conditions, sample preparations, maintenance and calibration of the MALDI-TOF MS systems and data analysis methods across routine laboratories [6, 13]. This would allow the comparison and typing of spectra from different centres. An ideal scenario would be to include bacterial typing in the routine workflow. In routine hospital settings, time to species identification can play a crucial role, and sample preparation has to stay as easy as possible and automated data analysis needs to be improved to gain resolution. Another vital step in bacterial typing is a well-curated database [87]. With the introduction of next-generation sequencing (NGS), bacterial taxonomy is changing at a remarkable speed. To keep track with this development, all strains in the database have to be well characterised, most desirably using whole-genome sequencing. MALDI-TOF MS typing can be used to determine that two strains do not belong to the same clonal cluster. On the other hand, high similarity of MALDI-TOF MS spectra does not prove the clonal relationship of the underlying strains, but rather suggests, that the two strains should be typed with higher resolution methods. MALDI-TOF MS can therefore, in the situation of an outbreak, be used as a screening tool to assess a large number of isolates. Based on the spectral similarity or diversity, strains can be identified for subsequent high-resolution sequencing. Using such a screening strategy could help to focus on selected strains, thereby accelerating infection control interventions.

References

1. Angeletti S (2017) Matrix assisted laser desorption time of flight mass spectrometry (MALDI-TOF MS) in clinical microbiology. J Microbiol Methods 138:20–29
2. Dierig A, Frei R, Egli A (2015) The fast route to microbe identification: matrix assisted laser desorption/ionization—time of flight mass spectrometry (maldi-tof Ms). Pediatr Infect Dis J 34(1):97–99
3. Greub G, Moran-Gilad J, Rossen J, Egli A (2017) ESCMID postgraduate education course: applications of MALDI-TOF mass spectrometry in clinical microbiology. Microbes Infect 19(9):433–442
4. Ling H, Yuan Z, Shen J, Wang Z, Xu Y (2014) Accuracy of matrix-assisted laser desorption ionization–time of flight mass spectrometry for identification of clinical pathogenic Fungi: a meta-analysis. J Clin Microbiol 52(7):2573–2582
5. Murray PR (2012) What is new in clinical microbiology—microbial identification by MALDI-TOF mass spectrometry. J Mol Diagn 14(5):419–423
6. Sauget M, Valot B, Bertrand X, Hocquet D (2017) Can MALDI-TOF mass spectrometry reasonably type Bacteria? Trends Microbiol 25(6):447–455
7. Suarez S, Ferroni A, Lotz A, Jolley KA, Guérin P, Leto J et al (2013) Ribosomal proteins as biomarkers for bacterial identification by mass spectrometry in the clinical microbiology laboratory. J Microbiol Methods 94(3):390–396
8. Charretier, Y., Lazarevic, V., Schrenzel, J., Ruppé, E., 2020. Messages from the Fourth International Conference on Clinical Metagenomics. Microbes Infect 22, 635–641. https://doi.org/10.1016/j.micinf.2020.07.007

9. Seng P, Drancourt M, Gouriet F, La Scola B, Fournier P-E, Rolain JM et al (2009) Ongoing revolution in bacteriology: routine identification of bacteria by matrix-assisted laser desorption ionization time-of-flight mass spectrometry. Clin Infect Dis 49(4):543–551
10. Sogawa K, Watanabe M, Sato K, Segawa S, Ishii C, Miyabe A et al (2011) Use of the MALDI BioTyper system with MALDI–TOF mass spectrometry for rapid identification of microorganisms. Anal Bioanal Chem 400(7):1905
11. van Veen SQ, Claas ECJ, Kuijper EJ (2010) High-throughput identification of bacteria and yeast by matrix-assisted laser desorption ionization-time of flight mass spectrometry in conventional medical microbiology laboratories. J Clin Microbiol 48(3):900–907
12. Marín M, Cercenado E, Sánchez-Carrillo C, Ruiz A, Gómez González Á, Rodríguez-Sánchez B, et al. Accurate Differentiation of Streptococcus pneumoniae from other Species within the Streptococcus mitis Group by Peak Analysis Using MALDI-TOF MS. Front Microbiol [Internet]. 2017 [cited 2018 Apr 11];8. Available from: https://www.ncbi.nlm.nih.gov/pmc/articles/PMC5403922/
13. Oberle M, Wohlwend N, Jonas D, Maurer FP, Jost G, Tschudin-Sutter S, et al. The Technical and Biological Reproducibility of Matrix-Assisted Laser Desorption Ionization-Time of Flight Mass Spectrometry (MALDI-TOF MS) Based Typing: Employment of Bioinformatics in a Multicenter Study. PLoS One [Internet]. 2016 [cited 2018 Apr 4];11(10). Available from: https://www.ncbi.nlm.nih.gov/pmc/articles/PMC5087883/
14. Ishihama Y, Schmidt T, Rappsilber J, Mann M, Hartl FU, Kerner MJ et al (2008) Protein abundance profiling of the Escherichia coli cytosol. BMC Genomics 9:102
15. Janda JM, Abbott SL (2007) 16S rRNA gene sequencing for bacterial identification in the diagnostic laboratory: pluses, perils, and pitfalls. J Clin Microbiol 45(9):2761–2764
16. Coluccio ML, Gentile F, Das G, Nicastri A, Perri AM, Candeloro P et al (2015) Detection of single amino acid mutation in human breast cancer by disordered plasmonic self-similar chain. Sci Adv 1(8):e1500487
17. Brenner DJ, Fanning GR, Miklos GV, Steigerwalt AG (1973) Polynucleotide sequence relatedness among Shigella species. Int J Syst Bacteriol 23(1):1–7
18. Pupo GM, Lan R, Reeves PR (2000) Multiple independent origins of Shigella clones of Escherichia coli and convergent evolution of many of their characteristics. PNAS 97(19):10567–10572
19. Croxen MA, Law RJ, Scholz R, Keeney KM, Wlodarska M, Finlay BB (2013) Recent advances in understanding enteric pathogenic Escherichia coli. Clin Microbiol Rev 26(4):822–880
20. He Y, Li H, Lu X, Stratton CW, Tang Y-W (2010) Mass spectrometry Biotyper system identifies enteric bacterial pathogens directly from colonies grown on selective stool culture media. J Clin Microbiol 48(11):3888–3892
21. Martiny D, Busson L, Wybo I, El Haj RA, Dediste A, Vandenberg O (2012) Comparison of the microflex LT and vitek MS systems for routine identification of bacteria by matrix-assisted laser desorption ionization–time of flight mass spectrometry. J Clin Microbiol 50(4):1313–1325
22. Khot PD, Fisher MA (2013) Novel approach for differentiating Shigella species and Escherichia coli by matrix-assisted laser desorption ionization–time of flight mass spectrometry. J Clin Microbiol 51(11):3711–3716
23. Paauw A, Jonker D, Roeselers G, Heng JME, Mars-Groenendijk RH, Trip H et al (2015) Rapid and reliable discrimination between Shigella species and Escherichia coli using MALDI-TOF mass spectrometry. Int J Med Microbiol 305(4):446–452
24. van Belkum A, Welker M, Pincus D, Charrier J-P, Girard V (2017) Matrix-assisted laser desorption ionization time-of-flight mass spectrometry in clinical microbiology: what are the current issues? Ann Lab Med 37(6):475–483
25. Chavda KD, Chen L, Fouts DE, Sutton G, Brinkac L, Jenkins SG et al (2016) Comprehensive genome analysis of carbapenemase-producing enterobacter spp.: new insights into phylogeny, population structure, and resistance mechanisms. mBio 7(6):e02093–e02016

26. Pavlovic M, Konrad R, Iwobi AN, Sing A, Busch U, Huber I (2012) A dual approach employing MALDI-TOF MS and real-time PCR for fast species identification within the Enterobacter cloacae complex. FEMS Microbiol Lett 328(1):46–53
27. La Scola B, Raoult D. Direct Identification of Bacteria in Positive Blood Culture Bottles by Matrix-Assisted Laser Desorption Ionisation Time-of-Flight Mass Spectrometry. PLoS One [Internet]. 2009 [cited 2018 Apr 11];4(11). Available from: https://www.ncbi.nlm.nih.gov/pmc/articles/PMC2777307/
28. Khot PD, Couturier MR, Wilson A, Croft A, Fisher MA (2012) Optimization of matrix-assisted laser desorption ionization–time of flight mass spectrometry analysis for bacterial identification. J Clin Microbiol 50(12):3845–3852
29. Shu L-J, Yang Y-L. Bacillus Classification Based on Matrix-Assisted Laser Desorption Ionization Time-of-Flight Mass Spectrometry—Effects of Culture Conditions. Sci Rep [Internet]. 2017 [cited 2018 Apr 11];7. Available from: https://www.ncbi.nlm.nih.gov/pmc/articles/PMC5686160/
30. Khéchine AE, Couderc C, Flaudrops C, Raoult D, Drancourt M (2011) Matrix-assisted laser desorption/ionization time-of-flight mass spectrometry identification of mycobacteria in routine clinical practice. PLoS One 6(9):e24720
31. Stevenson LG, Drake SK, Shea YR, Zelazny AM, Murray PR (2010) Evaluation of matrix-assisted laser desorption ionization-time of flight mass spectrometry for identification of clinically important yeast species. J Clin Microbiol 48(10):3482–3486
32. Alcaide F, Amlerová J, Bou G, Ceyssens PJ, Coll P, Corcoran D, et al. How to: identify non-tuberculous Mycobacterium species using MALDI-TOF mass spectrometry. Clinical Microbiology and Infection [Internet]. 2017 [cited 2018 Mar 23]; Available from: http://www.sciencedirect.com/science/article/pii/S1198743X17306432
33. Sabat AJ, Budimir A, Nashev D, Sá-Leão R, van Dijl JM, Laurent F et al (2013) Overview of molecular typing methods for outbreak detection and epidemiological surveillance. Eur Secur 18(4):20380
34. Koreen L, Ramaswamy SV, Graviss EA, Naidich S, Musser JM, Kreiswirth BN (2004) Spa typing method for discriminating among Staphylococcus aureus isolates: implications for use of a single marker to detect genetic micro- and macrovariation. J Clin Microbiol 42(2):792–799
35. O'Neill GL, Ogunsola FT, Brazier JS, Duerden BI (1996) Modification of a PCR Ribotyping method for application as a routine typing scheme forClostridium difficile. Anaerobe 2(4):205–209
36. Joensen KG, Scheutz F, Lund O, Hasman H, Kaas RS, Nielsen EM et al (2014) Real-time whole-genome sequencing for routine typing, surveillance, and outbreak detection of Verotoxigenic Escherichia coli. J Clin Microbiol 52(5):1501–1510
37. Christner M, Trusch M, Rohde H, Kwiatkowski M, Schlüter H, Wolters M, et al. Rapid MALDI-TOF Mass Spectrometry Strain Typing during a Large Outbreak of Shiga-Toxigenic Escherichia coli. PLoS One [Internet]. 2014 [cited 2018 Mar 9];9(7). Available from: https://www.ncbi.nlm.nih.gov/pmc/articles/PMC4087019/
38. Sakarikou C, Ciotti M, Dolfa C, Angeletti S, Favalli C. Rapid detection of carbapenemase-producing Klebsiella pneumoniae strains derived from blood cultures by Matrix-Assisted Laser Desorption Ionization-Time of Flight Mass Spectrometry (MALDI-TOF MS). BMC Microbiol [Internet]. 2017 [cited 2018 Mar 9];17. Available from: https://www.ncbi.nlm.nih.gov/pmc/articles/PMC5343375/
39. Steensels D, Deplano A, Denis O, Simon A, Verroken A (2017) MALDI-TOF MS typing of a nosocomial methicillin-resistant Staphylococcus aureus outbreak in a neonatal intensive care unit. Acta Clin Belg 72(4):219–225
40. Wolters M, Rohde H, Maier T, Belmar-Campos C, Franke G, Scherpe S et al (2011) MALDI-TOF MS fingerprinting allows for discrimination of major methicillin-resistant Staphylococcus aureus lineages. Int J Med Microbiol 301(1):64–68
41. Wang H-Y, Lee T-Y, Tseng Y-J, Liu T-P, Huang K-Y, Chang Y-T et al (2018) A new scheme for strain typing of methicillin-resistant Staphylococcus aureus on the basis of matrix-assisted

laser desorption ionization time-of-flight mass spectrometry by using machine learning approach. Becker K, editor. PLOS One 13(3):e0194289
42. Egli A, Tschudin-Sutter S, Oberle M, Goldenberger D, Frei R, Widmer AF. Matrix-Assisted Laser Desorption/Ionization Time of Flight Mass-Spectrometry (MALDI-TOF MS) Based Typing of Extended-Spectrum β-Lactamase Producing E. coli – A Novel Tool for Real-Time Outbreak Investigation. PLoS One [Internet]. 2015 [cited 2018 Mar 9];10(4). Available from: https://www.ncbi.nlm.nih.gov/pmc/articles/PMC4393243/
43. Lafolie J, Sauget M, Cabrolier N, Hocquet D, Bertrand X (2015) Detection of Escherichia coli sequence type 131 by matrix-assisted laser desorption ionization time-of-flight mass spectrometry: implications for infection control policies? J Hosp Infect 90(3):208–212
44. Nicolas Chanoine M-H, Bertrand X, Madec J-Y (2014) Escherichia coli ST131, an intriguing clonal group. Clin Microbiol Rev 27(3):543–574
45. Christner M, Dressler D, Andrian M, Reule C, Petrini O. Identification of Shiga-Toxigenic Escherichia coli outbreak isolates by a novel data analysis tool after matrix-assisted laser desorption/ionization time-of-flight mass spectrometry. PLoS One [Internet]. 2017 [cited 2018 Apr 4];12(9). Available from: https://www.ncbi.nlm.nih.gov/pmc/articles/PMC5587271/
46. Reil M, Erhard M, Kuijper EJ, Kist M, Zaiss H, Witte W et al (2011) Recognition of Clostridium difficile PCR-ribotypes 001, 027 and 126/078 using an extended MALDI-TOF MS system. Eur J Clin Microbiol Infect Dis 30(11):1431–1436
47. Matsumura Y, Yamamoto M, Nagao M, Tanaka M, Machida K, Ito Y et al (2014) Detection of extended-spectrum-β-lactamase-producing Escherichia coli ST131 and ST405 clonal groups by matrix-assisted laser desorption ionization–time of flight mass spectrometry. J Clin Microbiol 52(4):1034–1040
48. Nakamura A, Komatsu M, Kondo A, Ohno Y, Kohno H, Nakamura F et al (2015) Rapid detection of B2-ST131 clonal group of extended-spectrum β-lactamase–producing Escherichia coli by matrix-assisted laser desorption ionization–time-of-flight mass spectrometry: discovery of a peculiar amino acid substitution in B2-ST131 clonal group. Diagn Microbiol Infect Dis 83(3):237–244
49. Hotta Y, Teramoto K, Sato H, Yoshikawa H, Hosoda A, Tamura H (2010) Classification of genus pseudomonas by MALDI-TOF MS based on ribosomal protein coding in S10–spc–alpha operon at strain level. J Proteome Res 9(12):6722–6728
50. Ojima-Kato T, Yamamoto N, Nagai S, Shima K, Akiyama Y, Ota J et al (2017) Application of proteotyping Strain Solution™ ver. 2 software and theoretically calculated mass database in MALDI-TOF MS typing of *Salmonella* serotype. Appl Microbiol Biotechnol 101(23–24):8557–8569
51. Ojima-Kato T, Yamamoto N, Iijima Y, Tamura H (2015) Assessing the performance of novel software Strain Solution on automated discrimination of Escherichia coli serotypes and their mixtures using matrix-assisted laser desorption ionization-time of flight mass spectrometry. J Microbiol Methods 119:233–238
52. Tamura H, Hotta Y, Sato H (2013) Novel accurate bacterial discrimination by MALDI-time-of-flight MS based on ribosomal proteins coding in *S10-spc-alpha* operon at strain level *S10*-GERMS. J Am Soc Mass Spectrom 24(8):1185–1193
53. Ziegler D, Pothier JF, Ardley J, Fossou RK, Pflüger V, de Meyer S et al (2015) Ribosomal protein biomarkers provide root nodule bacterial identification by MALDI-TOF MS. Appl Microbiol Biotechnol 99(13):5547–5562
54. Gekenidis M-T, Studer P, Wüthrich S, Brunisholz R, Drissner D (2014) Beyond the matrix-assisted laser desorption ionization (MALDI) biotyping workflow: in search of microorganism-specific tryptic peptides enabling discrimination of subspecies. Appl Environ Microbiol 80(14):4234–4241
55. Rizzardi K, Åkerlund T (2015) High molecular weight typing with MALDI-TOF MS – a novel method for rapid typing of Clostridium difficile. PLoS One 10(4):e0122457
56. Lasch P, Wahab T, Weil S, Pályi B, Tomaso H, Zange S et al (2015) Identification of highly pathogenic microorganisms by matrix-assisted laser desorption ionization–time of flight mass spectrometry: results of an Interlaboratory ring trial. J Clin Microbiol 53(8):2632–2640

57. Freiwald A, Sauer S (2009) Phylogenetic classification and identification of bacteria by mass spectrometry. Nat Protoc 4(5):732–742
58. Josten M, Reif M, Szekat C, Al-Sabti N, Roemer T, Sparbier K et al (2013) Analysis of the matrix-assisted laser desorption ionization–time of flight mass Spectrum of Staphylococcus aureus identifies mutations that allow differentiation of the Main clonal lineages. J Clin Microbiol 51(6):1809–1817
59. Camoez M, Sierra JM, Dominguez MA, Ferrer-Navarro M, Vila J, Roca I (2016) Automated categorization of methicillin-resistant Staphylococcus aureus clinical isolates into different clonal complexes by MALDI-TOF mass spectrometry. Clin Microbiol Infect 22(2):161.e1–161.e7
60. Zhang T, Ding J, Rao X, Yu J, Chu M, Ren W et al (2015) Analysis of methicillin-resistant Staphylococcus aureus major clonal lineages by Matrix-Assisted Laser Desorption Ionization–Time of Flight Mass Spectrometry (MALDI–TOF MS). J Microbiol Methods 117:122–127
61. Lindgren Å, Karami N, Karlsson R, Åhrén C, Welker M, Moore ERB et al (2018) Development of a rapid MALDI-TOF MS based epidemiological screening method using MRSA as a model organism. Eur J Clin Microbiol Infect Dis 37(1):57–68
62. Østergaard C, Hansen SGK, Møller JK (2015) Rapid first-line discrimination of methicillin resistant Staphylococcus aureus strains using MALDI-TOF MS. Int J Med Microbiol 305(8):838–847
63. Boggs SR, Cazares LH, Drake R (2012) Characterization of a Staphylococcus aureus USA300 protein signature using matrix-assisted laser desorption/ionization time-of-flight mass spectrometry. J Med Microbiol 61(Pt_5):640–644
64. Sauget M, van der Mee-Marquet N, Bertrand X, Hocquet D (2016) Matrix-assisted laser desorption ionization-time of flight Mass spectrometry can detect Staphylococcus aureus clonal complex 398. J Microbiol Methods 127:20–23
65. Saichek NR, Cox CR, Kim S, Harrington PB, Stambach NR, Voorhees KJ (2016) Strain-level Staphylococcus differentiation by CeO2-metal oxide laser ionization mass spectrometry fatty acid profiling. BMC Microbiol 16:72
66. Chui H, Chan M, Hernandez D, Chong P, McCorrister S, Robinson A et al (2015) Rapid, sensitive, and specific Escherichia coli H antigen typing by matrix-assisted laser desorption ionization–time of flight-based peptide mass fingerprinting. J Clin Microbiol 53(8):2480–2485
67. Ojima-Kato T, Yamamoto N, Suzuki M, Fukunaga T, Tamura H (2014) Discrimination of Escherichia coli O157, O26 and O111 from other serovars by MALDI-TOF MS based on the S10-GERMS method. PLoS One 9(11):e113458
68. Månsson V, Gilsdorf JR, Kahlmeter G, Kilian M, Kroll JS, Riesbeck K et al (2018) Capsule typing of Haemophilus influenzae by matrix-assisted laser desorption/ionization time-of-flight mass spectrometry1. Emerg Infect Dis 24(3):443–452
69. Rettinger A, Krupka I, Grünwald K, Dyachenko V, Fingerle V, Konrad R et al (2012) Leptospira spp. strain identification by MALDI TOF MS is an equivalent tool to 16S rRNA gene sequencing and multi locus sequence typing (MLST). BMC Microbiol 12:185
70. Lasch P, Drevinek M, Nattermann H, Grunow R, Stämmler M, Dieckmann R et al (2010) Characterization of yersinia using MALDI-TOF mass spectrometry and chemometrics. Anal Chem 82(20):8464–8475
71. Wittwer M, Heim J, Schär M, Dewarrat G, Schürch N (2011) Tapping the potential of intact cell mass spectrometry with a combined data analytical approach applied to Yersinia spp.: detection, differentiation and identification of Y. pestis. Syst Appl Microbiol 34(1):12–19
72. Huang Y, Li J, Gu D, Fang Y, Chan EW, Chen S, et al. Rapid Detection of K1 Hypervirulent Klebsiella pneumoniae by MALDI-TOF MS. Front Microbiol [Internet]. 2015 [cited 2018 Mar 28];6. Available from: https://www.ncbi.nlm.nih.gov/pmc/articles/PMC4685062/
73. Kehrmann J, Wessel S, Murali R, Hampel A, Bange F-C, Buer J, et al. Principal component analysis of MALDI TOF MS mass spectra separates M. abscessus (sensu stricto) from M. massiliense isolates. BMC Microbiol [Internet]. 2016 [cited 2018 Mar 16];16. Available from: https://www.ncbi.nlm.nih.gov/pmc/articles/PMC4772520/

74. Pinto TCA, Costa NS, Castro LFS, Ribeiro RL, Botelho ACN, Neves FPG, et al. Potential of MALDI-TOF MS as an alternative approach for capsular typing Streptococcus pneumoniae isolates. Sci Rep [Internet]. 2017 [cited 2018 Mar 9];7. Available from: https://www.ncbi.nlm.nih.gov/pmc/articles/PMC5368646/
75. Sato J, Nakayama M, Tomita A, Sonoda T, Hasumi M, Miyamoto T. Evaluation of repetitive-PCR and matrix-assisted laser desorption ionization-time of flight mass spectrometry (MALDI-TOF MS) for rapid strain typing of Bacillus coagulans. PLoS One [Internet]. 2017 [cited 2018 Mar 9];12(10). Available from: https://www.ncbi.nlm.nih.gov/pmc/articles/PMC5636150/
76. Starostin KV, Demidov EA, Bryanskaya AV, Efimov VM, Rozanov AS, Peltek SE (2015) Identification of *Bacillus* strains by MALDI TOF MS using geometric approach. Sci Rep 5:16989
77. Lowy FD (1998) Staphylococcus aureus infections. N Engl J Med 339(8):520–532
78. Cosgrove SE, Sakoulas G, Perencevich EN, Schwaber MJ, Karchmer AW, Carmeli Y (2003) Comparison of mortality associated with methicillin-resistant and methicillin-susceptible Staphylococcus aureus Bacteremia: a meta-analysis. Clin Infect Dis 36(1):53–59
79. Johnson JR, Porter S, Johnston B, Kuskowski MA, Spurbeck RR, Mobley HLT, et al. Host Characteristics and Bacterial Traits Predict Experimental Virulence for Escherichia coli Bloodstream Isolates From Patients With Urosepsis. Open Forum Infect Dis [Internet]. 2015 [cited 2018 Jan 24];2(3). Available from: https://www.ncbi.nlm.nih.gov/pmc/articles/PMC4504731/
80. Downing T (2015) Tackling drug resistant infection outbreaks of global pandemic Escherichia coli ST131 using evolutionary and epidemiological genomics. Microorganisms 3(2):236–267
81. Goudarzi M, Seyedjavadi SS, Goudarzi H, Mehdizadeh Aghdam E, Nazeri S. Clostridium difficile Infection: Epidemiology, Pathogenesis, Risk Factors, and Therapeutic Options. Scientifica (Cairo) [Internet]. 2014 [cited 2018 Mar 28];2014. Available from: https://www.ncbi.nlm.nih.gov/pmc/articles/PMC4058799/
82. Park SH, Ricke SC (2015) Development of multiplex PCR assay for simultaneous detection of Salmonella genus, Salmonella subspecies I, Salm. Enteritidis, Salm. Heidelberg and Salm. Typhimurium. J Appl Microbiol 118(1):152–160
83. Brenner FW, Villar RG, Angulo FJ, Tauxe R, Swaminathan B (2000) Salmonella Nomenclature. J Clin Microbiol 38(7):2465–2467
84. Boore AL, Hoekstra RM, Iwamoto M, Fields PI, Bishop RD, Swerdlow DL (2015) Salmonella enterica infections in the United States and assessment of coefficients of variation: a novel approach to identify epidemiologic Characteristics of individual serotypes, 1996–2011. PLoS One 10(12):e0145416
85. Kang L, Li N, Li P, Zhou Y, Gao S, Gao H et al (2017) MALDI-TOF mass spectrometry provides high accuracy in identification of Salmonella at species level but is limited to type or subtype Salmonella serovars. Eur J Mass Spectrom (Chichester) 23(2):70–82
86. Lasch P, Fleige C, Stämmler M, Layer F, Nübel U, Witte W et al (2014) Insufficient discriminatory power of MALDI-TOF mass spectrometry for typing of Enterococcus faecium and Staphylococcus aureus isolates. J Microbiol Methods 100:58–69
87. Rosselló-Móra R, Amann R (2015) Past and future species definitions for Bacteria and archaea. Syst Appl Microbiol 38(4):209–216

Chapter 10
Advanced Applications of MALDI-TOF: Identification and Antibiotic Susceptibility Testing

Belén Rodríguez-Sánchez and Marina Oviaño

10.1 Introduction

MALDI-TOF (Matrix-Assisted Laser Desorption Ionisation Time of Flight) MS (Mass Spectrometry) has been widely implemented in microbiology laboratories worldwide for the rapid identification of frequent and uncommon bacteria, both aerobic and anaerobic, mycobacteria, yeast and moulds based on their unique pattern of proteins [1–3]. The idea of identifying bacterial isolates by their unique pattern of proteins was already proposed in 1975 [4]. However, the technology was not available until 1985, when Koichi Tanaka, an engineer at Shimadzu Corporation (Kyoto, Japan), developed a soft desorption ionisation method that prevented proteins from being fragmented after laser irradiation [5]. Almost at the same time, Michael Karas and Franz Hillenkamp reported the achievement of soft desorption ionisation by the use of an organic matrix [6]. The procedure described by these authors has been standardised, automatised and adapted to high-throughput by several companies and it is now implemented globally as a rapid and accurate tool for microorganism identification.

Bacterial isolates can be either directly identified from colonies ("whole-cell" method) or submitted to a standard protein extraction procedure with ethanol, formic acid and acetonitrile [7]. The "whole-cell" method and the "on-plate" short protein extraction with formic acid work for most common bacteria but some taxonomic groups such as moulds or mycobacteria require pre-treatment [8, 9]. Either

B. Rodríguez-Sánchez (✉)
Clinical Microbiology and Infectious Diseases Department, Hospital General Universitario Gregorio Marañón and Instituto de Investigación Sanitaria Gregorio Marañón, Madrid, Spain
e-mail: mbelen.rodriguez@iisgm.com

M. Oviaño
Servicio de Microbiología, Complejo Hospitalario Universitario A Coruña, La Coruña, Spain

J. Moran-Gilad, Y. Yagel (eds.), *Application and Integration of Omics-powered Diagnostics in Clinical and Public Health Microbiology*,
https://doi.org/10.1007/978-3-030-62155-1_10

whole cells or the extracted bacterial proteins are mixed with the organic matrix on an *ad-hoc* metallic plate, allowing the co-crystallisation of both elements. Several matrices are used for the identification of microorganisms in routine practice. α-cyano-4-hydroxycinnamic acid (Sigma Aldrich, St. Louis, USA) is the most commonly used, together with 2,5-dihydroxybenzoic acid, sinnapinic acid and ferulic acid (organic matrices are thoroughly detailed by Clark et al., 2013 [2]).

When the mixture is dry, the metallic plate is allocated within the MALDI-TOF instrument and submitted to brief nitrogen laser pulses that trigger the desorption-ionisation reaction: the energy of the laser is transferred to the matrix-analyte mixture that becomes a desorbed and ionised mass of molecules in the gas phase. These analytes are subsequently accelerated and the time they take to cross through a metallic tube under vacuum is measured by a TOF (Time of Flight) mass detector located at the end of the tube [1, 3]. The molecules are separated by their mass-to-charge (*m/z*) ratio and a spectrum is obtained where *m/z* is represented on the x-axis and the intensity of the peaks on the y-axis.

The protein spectra generated are unique for a microbial species and can be used as their fingerprint. Thus, protein spectra from well-characterised microorganisms are used as Main Spectra Profiles (MSP) for the identification of other isolates from the same species. This method has been shown to be highly reliable for the microbial species most commonly encountered in the clinical practice [10–14]. Commercial databases are available from different companies -bioMérieux (Marcy-l'Étoile, France), Bruker Daltonik GmbH, Bremen, Germany). Moreover, in-house databases have been reported for taxonomic groups under-represented or lacking in the commercial databases [15, 16].

A search in PubMed (https://www.ncbi.nlm.nih.gov/pubmed) with the keywords "MALDI-TOF" and "identification" from 2008 to the time of writing yielded over 3500 articles describing the rapid and accurate identification of a wide range of microorganisms using MALDI-TOF MS. This technology has demonstrated to be cost-effective, user-friendly and very flexible. The implementation of in-house libraries or the application of MALDI-TOF for direct identification of microorganisms from clinical samples, typing or antibiotic susceptibility testing demonstrate this fact and the great potential of this technology [16–18].

10.2 Identification of Microorganisms Directly from Clinical Samples (Fig. 10.1)

The high rate of satisfactory species-level identification of microorganism from single colonies provided by MALDI-TOF encouraged researchers to implement this technology directly from clinical samples in order to reduce the turnaround time (TAT) to identify the causative agent of infection and provide prompt and directed therapy.

Since bloodstream infections are associated with severe disease and high mortality and morbidity, the implementation of MALDI-TOF for the identification of

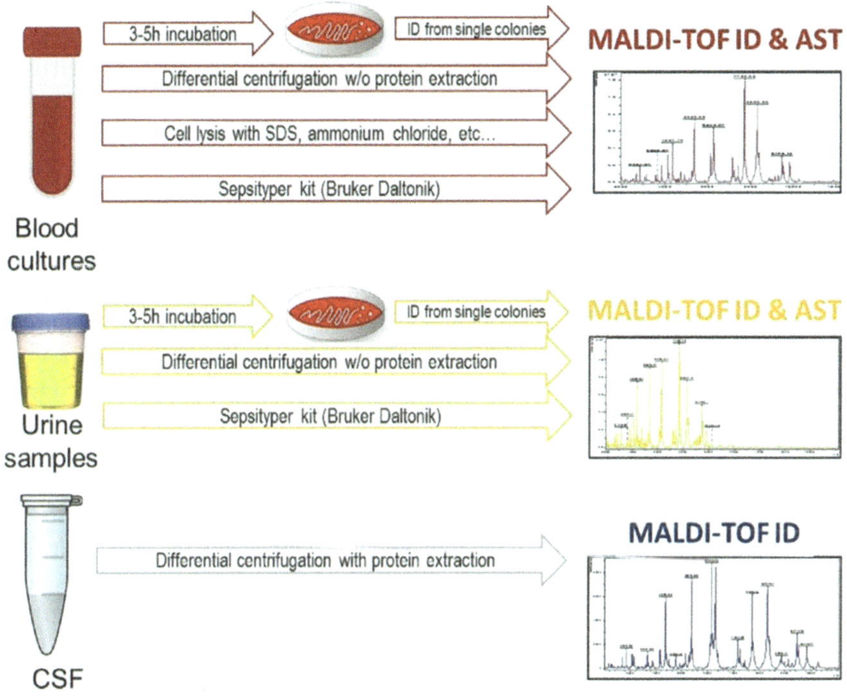

Fig. 10.1 Sample processing methods described for the identification of microorganisms present in blood cultures, urine samples and CSF

microorganisms directly from blood cultures became a priority. The first studies on this topic demonstrated that the microorganisms present in positive blood cultures could be concentrated by differential centrifugation and, once the blood cells and other components present in the broth were eliminated using different proceedings, the resulting pellet could be identified by MALDI-TOF (Fig. 10.1) [19, 20]. Between 78.7% and 87.0% of the analysed isolates were successfully identified at the species level. The method was reported to work equally well for aerobic and anaerobic microorganisms but its performance was poorer for Gram-positive isolates [19]. The explanation for these results could be the close relatedness of the streptococci and coagulase-negative staphylococci species analysed and the resistance of the cell wall from Gram-positive bacteria to be lysed.

Further studies have corroborated these initial results: either the identification from the bacterial pellet [21, 22] or after 3–5 hours incubation on agar plates [23, 24] provided high rate of satisfactory species-level identification especially for Gram-negative rods (97.0–100%) but authors reportedly showed lower percentages of Gram-positive cocci and yeasts successfully identified at the species level [22, 23, 25]. The use of the Sepsityper kit (Bruker Daltonik) has allowed a more robust identification of Gram-positive bacteria with approximately 81.0% accurate species assignment [26, 27]. A meta-analysis published in 2015 by Morgenthaler and

Kostrzewa nicely reviewed the available studies on the use of the Sepsityper kit on blood cultures and emphasised the need for an improved method that allows reliable yeast identification since the highest rate of species-level identification reached only 62.5% of the analysed isolates [28]. The use of 20% SDS for cell lysis was reported by Bidart et al. to perform better than the Sepsityper kit on blood culture bottles spiked with *Candida*, *Cryptococcus* and *Saccharomyces* species [29]. In their study, 88.8% of the isolates were accurately identified using their in-house method versus 81.7% using the Sepsityper kit. These results have been recently confirmed by Jeddi et al. who compared the Sepsityper kit and the SDS method in a head-to-head study that included 71 clinical samples and established 1.7 as the cut-off result score for accurate identification [30]. The authors achieved 95.6% correct identifications using the SDS method in comparison with 66.6% using the Sepsityper kit.

In conclusion, the available sample processing methods allow a rapid and reliable identification of most bacteria and yeasts present in blood cultures using MALDI-TOF. The incomplete identification of the different species present in polymicrobial samples remains one of the few flaws of this method nowadays [22]. The tremendous impact of MALDI-TOF implementation has already been reported: it has been shown to reduce the overall time to optimal treatment, the hospital length of stay, and when coupled with on-site antimicrobial stewardship intervention, both reduces the exposure to unnecessary antibiotics when the detected pathogen is considered a contaminant and facilitates the onset of the optimal therapy in less than 24 hours after blood culture positivity [31–33]. The importance of these improvements is even higher in the case of pediatric blood cultures, where it has been shown that combining MALDI-TOF and AST (Antibiotic Susceptibility Testing) of the identified microorganisms (both by conventional methods or using MALDI-TOF as explained later in this chapter) reduces the time to optimal therapy as well as the use of unnecessary antibiotics, improving hospital costs without compromising patients' outcomes [34, 35].

MALDI-TOF has also been implemented for the identification of microorganisms present in urine samples (Fig. 10.1) [36, 37]. Although the impact of MALDI-TOF for this application is clearly less cost-efficient than for blood cultures since urine samples are not incubated and therefore the number of microorganisms present may be below the detection limit, it can be used to discriminate positive from negative samples, optimising the initiation of adequate therapy and patient outcomes [37]. Moreover, rapid identification of the pathogen using MALDI-TOF enables obtaining AST results 18–24 earlier than standard methods when using the disc diffusion test and even as fast as 90 minutes when applying MALDI-TOF for this purpose [38, 39]. Despite the advantages of MALDI-TOF, polymicrobial samples are not always optimally identified and one or more pathogens may be missed. Besides, a sorting method to discriminate positive from negative urine samples (Gram staining or flow cytometry, for instance) is needed, complicating the implementation of this method in the routine of the microbiology laboratory.

Finally, recent studies have reported the use of MALDI-TOF for the identification of pathogens directly from normally sterile sites such as cerebrospinal fluid

(CSF) (Fig. 10.1). Applying this method enabled the rapid identification of the causative microorganisms in a highly important clinical sample [40]. The authors reported 81.0% correct identification of Gram-negative bacteria analysed. Although the overall rate of identification was 38.6% (17/44) and limited to the Gram negatives (yeasts and Gram-positive cocci were either misidentified or not identified) the fact that MALDI-TOF can identify microorganisms present in only 1 ml of CSF holds a great potential for the rapid and reliable identification of microorganisms from CSF and other sterile sites if sample processing methods can be optimised.

In summary, the application of MALDI-TOF for the direct identification of microorganisms from blood cultures is currently a widely implemented method in most routine laboratories around the world. Although the achieving and processing of the bacterial pellet may vary among centres, the basics of this methodology is well established and the method is known to produce successful identification of microorganisms present in the blood culture sample, reaching almost 100% accuracy for Gram-negative bacteria but still lower rates for Gram positives and yeasts (Fig. 10.1). The acquisition of a high rate of satisfactory identifications will clearly impact the management and outcome of bacteremic patients. Moreover, the methodology developed for blood cultures has now been extended to other samples such as urine and CSF. Although the impact of rapid identification of microorganisms from urine samples is clearly lower than from blood cultures, the implementation of MALDI-TOF directly on samples of normally sterile sites could potentially have a great impact on the management of critically ill patients.

10.3 Proteomic Approaches to Detect Antibiotic Susceptibility by MALDI-TOF MS

A major factor enabling the application of MS to the identification of bacteria and other microorganisms was the advent of nonfragmenting or "soft ionisation" techniques, including MALDI-TOF MS which facilitates the analysis of large macromolecules, including nucleic acids and proteins [41]. This application has long been used for identification in clinical microbiology laboratories; however, other applications can be developed with MALDI-TOF MS, for example, antibiotic resistance detection. Several techniques for achieving this purpose have already been reported, such as the detection of antibiotic molecules and their hydrolysis products due to enzymatic activity, the analysis of bacterial cell components, the measurement of bacterial growth in the presence of an antimicrobial agent and the detection of mutations with mini-sequencing. The reason for applying MALDI-TOF MS for detection of resistance is not only using the same user-friendly platform that we use for identification for another important purpose, but also for rapidly achieving antibiotic resistance detection, without the need for expert personnel.

10.3.1 Detection of Antibiotic Susceptibility by Measuring Enzymatic Activity in MALDI-TOF MS

In contrast to the mass spectrometric approaches that try to find a characteristic "resistance peak pattern", detection of enzymatic activity by MALDI-TOF MS is a functional assay that monitors how the presence of hydrolytic enzymes impact the integrity of the antibiotic applied. In this manner, this method resembles traditional biochemical resistance tests [18]. The capability of MALDI-TOF mass spectrometry to precisely detect mass changes in small molecules is one of the most promising applications. Analysis of antibiotics and their degradation products usually takes place in a mass range between 100 and 1000 Da, much lower than for identification purposes.

10.3.1.1 Detection of β-Lactamase Activity

Assays based on direct monitoring of β-lactamase activity on the β-lactam antibiotics are the starting point of all resistance detection assays by MALDI-TOF MS [42]. β-lactam antibiotics are inactivated by hydrolysis of the amide bond in the β-lactam ring, mediated by a water molecule. This molecule of water is added to the new structure formed, giving place to a new structure with a higher molecular weight, +18 Da. This mass change is what differentiates susceptible from resistant isolates and allows the detection of this reaction by MALDI-TOF MS.

Direct detection of β-lactamase activity is similarly performed in all published assays. A fresh bacterial culture is resuspended in an antibiotic buffer and incubated at 37 °C under agitation. After incubation, the sample is centrifuged, and the supernatant is analyzed with a proper matrix, usually α-cyano-4-hydroxy-cinnamic acid, HCCA. Once dried, the MALDI-TOF MS target is ready for analysis.

The first two studies reporting the detection of β-lactamase activity were published in 2011. Hrabak et al. used a series of 124 samples, including *Enterobacteriaceae* and *Pseudomonas aeruginosa* for detecting carbapenem resistance using meropenem as an indicator [43]. The carbapenemases represented were bla_{IMP}, bla_{VIM}, bla_{NDM} and bla_{KPC}. Bacteria were incubated for 3 h in a buffer containing a solution of meropenem. In this case, the matrix used for detection of hydrolysis products was the acid 2,5-dihidroxybenzoic (DHB). However, the use of DHB results in heterogeneous preparations, complicating automated acquisition of spectra [44]. This assay has been improved in two occasions, the first one with the addition of 0.01% dodecylsulfate sodic (SDS), that permits diminishing bacterial concentrations [45], and incorporating NH_4HCO_3 to the buffer solution, that allows detection of hydrolysis products by $bla_{\mathrm{OXA\text{-}48}}$ enzymes without decreasing sensitivities for other enzymes [46]. Burckhardt et al. used an incubation step with meropenem to detect resistance to carbepenems. The isolates carried $bla_{\mathrm{NDM\text{-}1}}$, $bla_{\mathrm{VIM\text{-}1}}$, $bla_{\mathrm{VIM\text{-}2}}$, $bla_{\mathrm{KPC\text{-}2}}$ and bla_{IMP}-type enzymes. Incubation time was between 1 and 2.5 hours [47].

In 2012 Sparbier et al. [48] determined the structure and the mass peaks corresponding to ampicillin, piperacillin, cefotaxime, ceftazidime, ertapenem, imipenem, meropenem and their hydrolysis products. The incubation time was set at 3 hours for all antibiotics. In 2017, Oviaño et al. improved the detection of β-lactamase activity using ceftriaxone as the antibiotic marker for ESBL or AmpC resistance [49]. Hydrolysis detection of ceftriaxone yielded 70% more positive results than cefotaxime, 80% more than ceftazidime and 20% more than cefpodoxime, with 100% specificity. The use of cefepime yielded 100% sensitivity, but only 27% specificity. β-lactamase resistance was detected only after 30 min of incubation with 100% sensitivity and specificity, considerably improving the time to results compared with previous studies [49] –see Table 10.1.

Table 10.1 Antibiotic resistance mechanisms detected by MALDI-TOF MS

MALDI-TOF MS principle	Application	Commercially available	Advantage	Disadvantage
Detection of enzymatic activity	Detection of beta-lactamases (ESBL, AmpC...)	No	Easy and fast technique (aprox. 30 min)	No identification of the b-lactamase type
	Detection of carbapenemases	Yes (MBT STAR-Carba Kit IVD)	Detects resistance directly from the clinical sample Does not require further interpretation of spectra	Does not provide information regarding the MIC
	Detection of the AAC(6′)-Ib-cr enzyme	No	Easy and fast technique (aprox. 30 min)	Does not provide information regarding the MIC
			More simple than the reference method (molecular technique)	Developed skills for interpretation of spectra
Measure of growth in the presence of an antibiotic	Detection of resistance by measuring protein synthesis in the presence of isotope-labelled amino acids	No	Provides information of susceptibility and resistance to different antibiotics	Requirement of isotope-labelled medium Labour-intensive methodology Time-consuming methodology (aprox. 3 hours to deliver results)
	Detection of resistance by semi-quantification of bacterial growth at the breakpint concentration	No	Provides information of susceptibility and resistance to different antibiotics Easy-handling using the microdoplet assay	Time-consuming methodology (aprox. 3 hours to deliver results)

Regarding carbapenemase detection, Lasserre et al. used a 20 min incubation step with imipenem to detect carbapenemase producers using an MS ratio (mass peaks of metabolite/ imipenem + metabolite) cut-off that statistically determined the classification of strains as carbapenemase producers (MS ratio of ≥0.82) and yielded 100% sensitivity and specificity [50]. Monteferrante et al. designed a protocol for carbapenemase detection, starting from a defined amount of bacterial cells (3.0 McFarland) followed by hydrolysis of imipenem or ertapenem [51]. The comparison between the antibiotics revealed 100% sensitivity and specificity for imipenem and a higher hydrolysis rate. However, imipenem is less stable and its peaks show low intensity. Standardisation of the number of cells makes this method less technician dependent, but much more laborious and time-consuming.

10.3.1.2 Detection of Resistant Strains Directly from Clinical Samples

The possibility to use MALDI-TOF MS for detecting antimicrobial resistance directly from clinical samples is one of the most important applications of this method, limited only by its sensitivity. Different studies report that the minimal amount of cellular material required for MALDI-TOF MS analysis is around 10^5 CFU/ml [52, 53]. Despite this limitation, direct-from-sample resistance detection has been performed from positive blood cultures and urine samples. Jung et al. [54] and Oviaño et al. [55] evaluated the possibility of detection of extended-spectrum β-lactamases in blood cultures in 90 min, using cefotaxime or cefotaxime and ceftazidime plus clavulanic acid, respectively. Both studies lead to very similar results, with sensitivity and specificity close to 100%. Continuing with the aim of detecting carbapenemase activity, the authors developed a universal method for detecting carbapenemase producers in Gram-negative bacilli including *Enterobacteriaceae*, *Pseudomonas* spp. and *Acinetobacter* spp. within 30 min, using imipenem [56]. The overall sensitivity and specificity was 98% and 100%, respectively. Hydrolysis results were interpreted by the STAR-BL (Selective Testing of β-Activity) module of MALDI-TOF Biotyper® Compass software (Bruker Daltonik GmbH), which automatically provides a result of susceptibility or resistance, calculated as the logRQ or ratio of hydrolysis of the antibiotic, turning interpretation of results into an easy task ready to be performed by non-expert users of mass spectrometry. This methodology unifies carbapenemase detection for all types of Gram-negative bacilli, without the need for different protocols for different bacteria.

Regarding carbapenemase detection directly from urine samples, Oviaño et al. developed a method combining flow cytometry, using a cut-off value of $\geq 1.5 \times 10^5$ bacteria/mL, and bacterial protein extraction with the Sepsityper Kit (Bruker Daltonik GmbH) [39]. Imipenem was used as the antibiotic marker of the presence of carbapenemase enzymes, and the results were delivered by the MALDI-TOF Biotyper® Compass software (Bruker Daltonik GmbH). The assay allowed reliable identification of 91% (503/553) of the samples and showed 100% sensitivity (30/30) and specificity (454/454) for detecting carbapenemase activity. The main advantage of this methodology is that the turnaround time is 24–48 hours earlier than conventional methods. However, it also has disadvantages as the high volume (10 ml) of urine required, the need for bacterial counts higher than 1.5×10^{5} bacteria/mL, and the exclusive identification of monomicrobial infections.

Although MALDI-TOF MS is highly automated for identification of microorganisms, it is not so for the detection of resistance mechanisms so far. As a result of the growing demand for the implementation of these techniques in clinical laboratories and the shortage of trained personnel, in 2017 the STAR-CARBA diagnostic kit (Bruker Daltonik GmbH) was launched (Table 10.1). It can be used for the detection of carbapenemases from *Enterobacteriaceae*, *Acinetobacter* spp., and *Pseudomonas aeruginosa*. This is the first kit based on MALDI-TOF MS read-out and the first mass spectrometry resistance test [57]. The kit uses imipenem as the antibiotic marker for carbapenem resistance and the only pre-requisites are an optimal calibration and the presence of positive and negative control in every assay.

10.3.1.3 Detection of the AAC(6′)-Ib-Cr Enzyme

Similar to the detection of β-lactamase activity by MALDI-TOF MS, detection of other resistance mechanisms that produce a mass change in the antibiotics can be performed. The detection of the AAC(6′)-Ib-cr enzyme highlights this ability of MALDI-TOF MS [58–60]. This functional assay is based on the acetylation reaction of the fluoroquinolones ciprofloxacin and norfloxaxin, increasing the mass of the previously exposed antibiotics by 43 Da. Oviaño et al. developed a method to detect the presence of the AAC(6′)-Ib-cr enzyme in clinical isolates by visual inspection of the mass peaks of ciprofloxacin and norfloxacin [58]. Clear differentiation between AAC(6′)-Ib-cr-producing isolates and non-producing isolates was seen after an incubation time of 30 min (Table 10.1). Presence of other determinants of quinolone resistance had no impact on the acetylation reaction and therefore on the results obtained by MALDI-TOF MS. This assay was further improved by automating the processing of spectra analysis and the release of results, using the MALDI Biotyper Peak Shift Prototype (Bruker Daltonik GmbH) [59]. Norfloxacin was found to be the best marker for detecting the AAC(6′)-Ib-cr enzyme as it suffered enhanced acetylation. Pardo et al. reached similar conclusions using norfloxacin and 4 hours of incubation. The analysis was performed in this case using the VITEK MS RUO (Shimadzu Corporation, Kyoto, Japan) [60].

10.3.2 Detection of Antibiotic Susceptibility by MALDI-TOF MS through Other Techniques

Detection of antibiotic susceptibility besides a particular drug inactivation mechanism, cannot be done by the previously described methodologies. Further approaches have been developed in order to avoid this limitation. An example is a method developed by Sparbier et al. [61] that uses the measuring of protein synthesis in cells supplemented with isotope labelled amino acids. The incorporation of the heavy amino acids increases the molecular weight of the newly synthesised proteins, causing peak shifts in the mass spectral profiles which can be detected automatically by a software algorithm. Resistance could be detected in about 2 hours by this method (Table 10.1). Another recent test much related to conventional AST is the MALDI Biotyper Antibiotic Susceptibility Test Rapid Assay (MBT-ASTRA) [62]. This test quantifies the relative growth of a microorganism after incubation with an antibiotic at the breakpoint concentration. While the antibiotic suppresses the growth of a susceptible organism, a resistant one will grow and therefore give proper signals, enough to perform identification with MALDI-TOF MS. Different antibiotics have been tested so far [63] and on different microorganisms [64–67]. A micro-droplet assay has been developed with an in-target incubation that eliminates the centrifugations steps so that the methodology is simplified and able to be introduced into routine laboratories [68]. However, these assays require a certain growth of bacteria, so at least a few hours are still needed to deliver results, in contrast to enzymatic assays which can be performed in minutes.

To conclude, we believe that in the future, MALDI-TOF MS will play a significant role in the detection of antibiotic resistance in clinical microbiology laboratories, as it currently has for identification of bacteria. However, there is still a need for standardised procedures, cost-effectiveness studies and commercially available kits and software applications in order for this method to be widely used for routine clinical purposes.

10.4 Automation in Clinical Practice

The integration of MALDI-TOF in the routine of the clinical microbiology laboratory is a fact nowadays. Different laboratories have established this instrument either as a central service where qualified staff analyse isolates and samples from different origin or as a service freely applied by different users to their specific isolates of interest. In both cases, a high number of isolates are manually spotted and processed daily for MALDI-TOF to identify. In order to optimise these proceedings, several automated approaches have been developed.

Fifteen years ago, the first attempts to automatise sample preparation were reported [69]. An automated sample spotting technique was developed for the analysis of synthetic polymers using a commercially available robot. The automated sample preparation system allowed the authors to integrate MALDI-TOF analysis in their research routine. The same approach is currently used commercially by companies such as Copan (Brescia, Italy), that developed the Colibrí™ system, a colony picker integrated in their WASPlab™ system for agar plate management, incubation and detection of positive samples by imaging acquisition (http://www.copanusa.com/products/automation/colibri-universal-colony-picker/). The Colibrí™ system allows the users to select the colonies grown on agar plates and the robotised system transfers them to the MALDI target plates and covers them with matrix, rendering the target plate ready to be analysed by MALDI-TOF.

One of the most recent efforts towards the automation of sample processing for MALDI-TOF analysis is the prototype developed by Broyer et al. that allows the rapid preparation of blood cultures for analysis by MALDI-TOF [70]. The system is based on a filter wand where the bacteria present in the blood culture broth get fixed under vacuum conditions. They are subsequently transferred to the MALDI target plate by gently tapping the wand on the target surface. Matrix pipetting can also be robotised, rendering the target plate ready to be analysed by MALDI-TOF.

The development and commercialisation of these automated systems will help reduce the time to identification of a large number of microorganisms. This path will surely lead microbiology laboratories to provide rapid and accurate information about the identity of pathogens and their susceptibility pattern to the most commonly used antibiotics, allowing optimised management of patients.

References

1. Croxatto A, Prod'hom G, Greub G (2012) Applications of MALDI-TOF mass spectrometry in clinical diagnostic microbiology. FEMS Microbiol Rev 36:380–407
2. Clark AE, Kaleta EJ, Arora A, Wolk DM (2013) Matrix-assisted laser desorption ionization-time of flight mass spectrometry: a fundamental shift in the routine practice of clinical microbiology. Clin Microbiol Rev 26:547–603
3. Patel R (2015) MALDI-TOF MS for the diagnosis of infectious disease. Clin Chem 61:100–111
4. Anhalt J, Fenselau C (1975) Identification of bacteria using mass spectrometry. Anal Chem 47:219–225
5. Tanaka K, Waki H, Ido Y, Akita S, Yoshida Y, Yoshida T (1988) Protein and polymer analyses up to m/z 100 000 by laser ionization time-of-flight mass spectrometry. Rapid Commun Mass Spectrom 2:151–153
6. Karas M, Hillenkamp F (1988) Laser desorption ionization of proteins with molecular masses exceeding 10000 Daltons. Anal Chem 60:2299–2301
7. Schmitt BH, Cunningham SA, Dailey AL, Gustafson DR, Patel R (2013) Identification of anaerobic bacteria by Bruker Biotyper matrix-assisted laser desorption ionization-time of flight mass spectrometry with on-plate formic acid preparation. J Clin Microbiol 51:782–786
8. Alcaide F, Amlerová J, Bou G, Ceyssens PJ, Coll P, Corcoran D et al (2017) How to: identify non-tuberculous Mycobacterium species using MALDI-TOF mass spectrometry. Clin Microbiol Infect. pii: S1198-743X(17)30643-2
9. Lau AF, Drake SK, Calhoun LB, Henderson CM, Zelazny AM (2013) Development of a clinically comprehensive database and a simple procedure for identification of molds from solid media by matrix-assisted laser desorption ionization-time of flight mass spectrometry. J Clin Microbiol 51:828–834
10. Seng P, Drancourt M, Gouriet F, La Scola B, Fournier PE, Rolain JM, Raoult D (2009) Ongoing revolution in bacteriology: routine identification of bacteria by matrix-assisted laser desorption ionization time-of-flight mass spectrometry. Clin Infect Dis 49:543–551
11. Veloo AC, Erhard M, Welker M, Welling GW, Degener JE (2011) Identification of gram-positive anaerobic cocci by MALDI-TOF mass spectrometry. Syst Appl Microbiol 34(1):58–62
12. Buckwalter SP, Olson SL, Connelly BJ, Lucas BC, Rodning AA, Walchak RC, Deml SM, Wohlfiel SL, Wengenack NL (2016) Evaluation of matrix-assisted laser desorption ionization-

time of flight mass spectrometry for identification of *Mycobacterium* species, *Nocardia* species, and other aerobic actinomycetes. J Clin Microbiol 54:376–384
13. Quiles-Melero I, García-Rodríguez J, Gómez-López A, Mingorance J (2012) Evaluation of matrix-assisted laser desorption/ionisation time-of-flight (MALDI-TOF) mass spectrometry for identification of *Candida parapsilosis*, *C. orthopsilosis* and *C. metapsilosis*. Eur J Clin Microbiol Infect Dis 31:67–71
14. Posteraro B, De Carolis E, Vella A, Sanguinetti M (2013) MALDI-TOF mass spectrometry in the clinical mycology laboratory: identification of fungi and beyond. Expert Rev Proteomics 10(2):151–164
15. L'Ollivier C, Cassagne C, Normand AC, Bouchara JP, Contet-Audonneau N, Hendrickx M, Fourquet P, Coulibaly O, Piarroux R, Ranque S (2013) A MALDI-TOF MS procedure for clinical dermatophyte species identification in the routine laboratory. Med Mycol 51:713–720
16. Zvezdanova ME, Escribano P, Ruiz A, Martínez-Jiménez MC, Peláez T, Collazos A, Guinea J, Bouza E, Rodríguez-Sánchez B (2018) Increased species-assignment of filamentous fungi using MALDI-TOF MS coupled with a simplified sample processing and an in-house library. Med Mycol. https://doi.org/10.1093/mmy/myx154
17. Oberle M, Wohlwend N, Jonas D, Maurer FP, Jost G, Tschudin-Sutter S, Vranckx K, Egli A (2016) The technical and biological reproducibility of matrix-assisted laser desorption ionization-time of flight mass spectrometry (MALDI-TOF MS) based typing: employment of bioinformatics in a multicenter study. PLoS One 11:e0164260
18. Kostrzewa M, Sparbier K, Maier T, Schubert S (2013) MALDI-TOF MS: an upcoming tool for rapid detection of antibiotic resistance in microorganisms. Proteomics Clin Appl 7(11–12):767–778
19. Prod'hom G, Bizzini A, Durussel C, Bille J, Greub G (2010) Matrix-assisted laser desorption ionization time of flight mass spectrometry for direct bacterial identification from positive blood culture pellets. J Clin Microbiol 48:1481–1483
20. Christner M, Rohde H, Wolters M, Sobottka I, Wegscheider K, Aepfelbacher M (2010) Rapid identification of bacteria from positive blood culture bottles by use of matrix-assisted laser desorption-ionization time of flight mass spectrometry fingerprinting. J Clin Microbiol 48(5):1584–1591
21. Hoyos-Mallecot Y, Miranda-Casas C, Cabrera-Alvargonzalez JJ, Gómez-Camarasa C, Pérez-Ramirez MD, Navarro-Marí JM (2013) Bacterial identification from blood cultures by a rapid matrix-assisted laser desorption-ionisation time-of-flight mass spectrometry technique. Enferm Infecc Microbiol Clin 31:152–155
22. Rodríguez-Sánchez B, Sánchez-Carrillo C, Ruiz A, Marín M, Cercenado E, Rodríguez-Créixems M, Bouza E (2014) Direct identification of pathogens from positive blood cultures using matrix-assisted laser desorption-ionization time-of-flight mass spectrometry. Clin Microbiol Infect 20:O421–O427
23. Verroken A, Defourny L, Lechgar L, Magnette A, Delmée M, Glupczynski Y (2015) Reducing time to identification of positive blood cultures with MALDI-TOF MS analysis after a 5-h subculture. Eur J Clin Microbiol Infect Dis 34:405–413
24. Idelevich EA, Schüle I, Grünastel B, Wüllenweber J, Peters G, Becker K (2014) Rapid identification of microorganisms from positive blood cultures by MALDI-TOF mass spectrometry subsequent to very short-term incubation on solid medium. Clin Microbiol Infect 20:1001–1006
25. Kohlmann R, Hoffmann A, Geis G, Gatermann S (2015) MALDI-TOF mass spectrometry following short incubation on a solid medium is a valuable tool for rapid pathogen identification from positive blood cultures. Int J Med Microbiol 305:469–479
26. Schieffer KM, Tan KE, Stamper PD, Somogyi A, Andrea SB, Wakefield T, Romagnoli M, Chapin KC, Wolk DM, Carroll KC (2014) Multicenter evaluation of the Sepsityper extraction kit and MALDI-TOF MS for direct identification of positive blood culture isolates using the BD BACTEC FX and VersaTREK diagnostic blood culture systems. J Appl Microbiol 116:934–941

27. Martinez RM, Bauerle ER, Fang FC, Butler-Wu SM (2014) Evaluation of three rapid diagnostic methods for direct identification of microorganisms in positive blood cultures. J Clin Microbiol 52:2521–2529
28. Morgenthaler NG, Kostrzewa M (2015) Rapid identification of pathogens in positive blood culture of patients with sepsis: review and meta-analysis of the performance of the sepsityper kit. Int J Microbiol 2015:827416
29. Bidart M, Bonnet I, Hennebique A, Kherraf ZE, Pelloux H, Berger F, Cornet M, Bailly S, Maubon D (2015) An in-house assay is superior to Sepsityper for direct matrix-assisted laser desorption ionization-time of flight (MALDI-TOF) mass spectrometry identification of yeast species in blood cultures. J Clin Microbiol 53:1761–1764
30. Jeddi F, Yapo-Kouadio GC, Normand AC, Cassagne C, Marty P, Piarroux R (2017) Performance assessment of two lysis methods for direct identification of yeasts from clinical blood cultures using MALDI-TOF mass spectrometry. Med Mycol 55:185–192
31. French K, Evans J, Tanner H, Gossain S, Hussain A (2016) The clinical impact of rapid. Direct MALDI-ToF Identification of Bacteria from Positive Blood Cultures PLoS One 11:e0169332
32. Beganovic M, Costello M, Wieczorkiewicz SM (2017) Effect of matrix-assisted laser desorption ionization-time of flight mass spectrometry (MALDI-TOF MS) alone versus MALDI-TOF MS combined with real-time antimicrobial stewardship interventions on time to optimal antimicrobial therapy in patients with positive blood cultures. J Clin Microbiol 55:1437–1445
33. Osthoff M, Gürtler N, Bassetti S, Balestra G, Marsch S, Pargger H, Weisser M, Egli A (2017) Impact of MALDI-TOF-MS-based identification directly from positive blood cultures on patient management: a controlled clinical trial. Clin Microbiol Infect 23:78–85
34. Malcolmson C, Ng K, Hughes S, Kissoon N, Schina J, Tilley PA, Roberts A (2017) Impact of matrix-assisted laser desorption and ionization time-of-flight and antimicrobial stewardship intervention on treatment of bloodstream infections in hospitalized children. J Pediatric Infect Dis Soc 6:178–186
35. Reuter CH, Palac HL, Kociolek LK, Zheng XT, Chao YY, Patel R, Patel SJ (2018) Ideal and actual impact of rapid diagnostic testing and antibiotic stewardship on antibiotic prescribing and clinical outcomes in children with positive blood cultures. Pediatr Infect Dis J 38:131–137. https://doi.org/10.1097/INF.0000000000002102
36. Íñigo M, Coello A, Fernández-Rivas G, Rivaya B, Hidalgo J, Quesada MD, Ausina V (2016) Direct identification of urinary tract pathogens from urine samples, combining urine screening methods and matrix-assisted laser desorption ionization-time of flight mass spectrometry. J Clin Microbiol 54:988–993
37. Burillo A, Rodríguez-Sánchez B, Ramiro A, Cercenado E, Rodríguez-Créixems M, Bouza E (2014) Gram-stain plus MALDI-TOF MS (matrix-assisted laser desorption ionization-time of flight mass spectrometry) for a rapid diagnosis of urinary tract infection. PLoS One 9:e86915
38. Zboromyrska Y, Rubio E, Alejo I, Vergara A, Mons A, Campo I, Bosch J, Marco F, Vila J (2016) Development of a new protocol for rapid bacterial identification and susceptibility testing directly from urine samples. Clin Microbiol Infect 22:561.e1–561.e6
39. Oviaño M, Ramírez CL, Barbeyto LP, Bou G (2017) Rapid direct detection of carbapenemase-producing Enterobacteriaceae in clinical urine samples by MALDI-TOF MS analysis. J Antimicrob Chemother 72:1350–1354
40. Bishop B, Geffen Y, Plaut A, Kassis O, Bitterman R, Paul M, Neuberger A (2018) The use of matrix-assisted laser desorption/ionization time-of-flight mass spectrometry for rapid bacterial identification in patients with smear-positive bacterial meningitis. Clin Microbiol Infect 24:171–174
41. Sauer S, Kliem M (2010) Mass spectrometry tools for the classification and identification of bacteria. Nat Rev Microbiol 8:74–82
42. Hrabák J, Chudácková E, Walková R (2013) Matrix-assisted laser desorption ionization-time of flight (MALDI-TOF) mass spectrometry for detection of antibiotic resistance mechanisms: from research to routine diagnosis. Clin Microbiol Rev 26:103–114

43. Hrabák J, Walková R, Studentová V, Chudácková E, Bergerová T (2011) Carbapenemase activity detection by matrix-assisted laser desorption ionization-time of flight mass spectrometry. J Clin Microbiol 49:3222–3227
44. Horneffer V, Strupat K, Hillenkamp F (2006) Localization of noncovalent complexes in MALDI-preparations by CLSM. J Am Soc Mass Spectrom 17:1599–1604
45. Hrabák J, Studentová V, Walková R, Zemlicková H, Jakubu V, Chudácková E et al (2012) Detection of NDM-1, VIM-1, KPC, OXA-48, and OXA-162 carbapenemases by matrix-assisted laser desorption ionization-time of flight mass spectrometry. J Clin Microbiol 50:2441–2443
46. Papagiannitsis CC, Študentová V, Izdebski R, Oikonomou O, Pfeifer Y, Petinaki E et al (2015) Matrix-assisted laser desorption ionization-time of flight mass spectrometry meropenem hydrolysis assay with NH4HCO3, a reliable tool for direct detection of carbapenemase activity. J Clin Microbiol 53:1731–1735
47. Burckhardt I, Zimmermann S (2011) Using matrix-assisted laser desorption ionization-time of flight mass spectrometry to detect carbapenem resistance within 1 to 2.5 hours. J Clin Microbiol 49:3321–3324
48. Sparbier K, Schubert S, Weller U, Boogen C, Kostrzewa M (2012) Matrix-assisted laser desorption ionization-time of flight mass spectrometry-based functional assay for rapid detection of resistance against β-lactam antibiotics. J Clin Microbiol 50:927–937
49. Oviaño M, Gómara M, Barba MJ, Revillo MJ, Barbeyto LP, Bou G (2017) Towards the early detection of β-lactamase-producing Enterobacteriaceae by MALDI-TOF MS analysis. J Antimicrob Chemother 72:2259–2262
50. Lasserre C, De Saint ML, Cuzon G, Bogaerts P, Lamar E, Glupczynski Y et al (2015) Efficient detection of carbapenemase activity in *Enterobacteriaceae* by matrix-assisted laser desorption ionization-time of flight mass spectrometry in less than 30 minutes. J Clin Microbiol 53:2163–2171
51. Monteferrante CG, Sultan S, Ten Kate MT, Dekker LJ, Sparbier K, Peer M et al (2016) Evaluation of different pretreatment protocols to detect accurately clinical carbapenemase-producing *Enterobacteriaceae* by MALDI-TOF. J Antimicrob Chemother 71:2856–2867
52. Ferreira L, Sanchez-Juanes F, Gonzalez-Avila M, Cembrero-Fucinos D, Herrero-Hernandez A, Gonzalez-Buitrago JM et al (2010) Direct identification of urinary tract pathogens from urine samples by matrix-assisted laser desorption ionization-time of flight mass spectrometry. J Clin Microbiol 48:2110–2115
53. Yan Y, He Y, Maier T, Quinn C, Shi G, Li H et al (2011) Improved identification of yeast species directly from positive blood culture media by combining Sepsityper specimen processing and Microflex analysis with the matrix-assisted laser desorption ionization Biotyper system. J Clin Microbiol 49:2528e32
54. Jung JS, Popp C, Sparbier K, Lange C, Kostrzewa M, Schubert S (2014) Evaluation of matrix-assisted laser desorption ionization-time of flight mass spectrometry for rapid detection of β-lactam resistance in Enterobacteriaceae derived from blood cultures. J Clin Microbiol 52:924–930
55. Oviaño M, Fernández B, Fernández A, Barba MJ, Mouriño C, Bou G (2014) Rapid detection of Enterobacteriaceae producing extended spectrum beta-lactamases directly from positive blood cultures by matrix-assisted laser desorption ionization-time of flight mass spectrometry. Clin Microbiol Infect 20:1146–1157
56. Oviaño M, Sparbier K, Barba MJ, Kostrzewa M, Bou G (2016) Universal protocol for the rapid automated detection of carbapenem-resistant Gram-negative bacilli directly from blood cultures by matrix-assisted laser desorption/ionisation time-of-flight mass spectrometry (MALDI-TOF/MS). Int J Antimicrob Agents 48:655–660
57. Kostrzewa M (2018) Application of the MALDI Biotyper to clinical microbiology: progress and potential. Expert Rev Proteomics. https://doi.org/10.1080/14789450.2018.1438193

58. Oviaño M, Rodríguez-Martínez JM, Pascual Á, Bou G (2017) Rapid detection of the plasmid-mediated quinolone resistance determinant AAC(6′)-Ib-cr in *Enterobacteriaceae* by MALDI-TOF MS analysis. J Antimicrob Chemother 72:1074–1080
59. Oviaño M, Gómara M, Barba MJ, Sparbier K, Pascual Á, Bou G (2017) Quantitative and automated MALDI-TOF MS-based detection of the plasmid-mediated quinolone resistance determinant AAC(6′)-Ib-cr in *Enterobacteriaceae*. J Antimicrob Chemother 72:2952–2954
60. Pardo CA, Tan RN, Hennequin C, Beyrouthy R, Bonnet R, Robin F (2016) Rapid detection of AAC(6′)-Ib-cr production using a MALDI-TOF MS strategy. Eur J Clin Microbiol Infect Dis 35:2047–2051
61. Sparbier K, Lange C, Jung J, Wieser A, Schubert S, Kostrzewa M (2013) MALDI biotyper-based rapid resistance detection by stable-isotope labeling. J Clin Microbiol 51:3741–3748
62. Lange C, Schubert S, Jung J, Kostrzewa M, Sparbier K (2014) Quantitative matrix-assisted laser desorption ionization-time of flight mass spectrometry for rapid resistance detection. J Clin Microbiol 52:4155–4162
63. Jung JS, Hamacher C, Gross B, Sparbier K, Lange C, Kostrzewa M et al (2016) Evaluation of a semiquantitative matrix-assisted laser desorption ionization-time of flight mass spectrometry method for rapid antimicrobial susceptibility testing of positive blood cultures. J Clin Microbiol 54:2820–2824
64. Maxson T, Taylor-Howell CL, Minogue TD (2017) Semi-quantitative MALDI-TOF for antimicrobial susceptibility testing in Staphylococcus aureus. PLoS One 12:e0183899
65. Ceyssens PJ, Soetaert K, Timke M, Van den Bossche A, Sparbier K, De Cremer K et al (2017) Matrix-assisted laser desorption ionization-time of flight mass spectrometry for combined species identification and drug sensitivity testing in mycobacteria. J Clin Microbiol 55:624–634
66. De Carolis E, Vella A, Florio AR, Posteraro P, Perlin DS, Sanguinetti M et al (2012) Use of matrix-assisted laser desorption ionization-time of flight mass spectrometry for caspofungin susceptibility testing of *Candida* and *Aspergillus species*. J Clin Microbiol 50:2479–2483
67. Vella A, De Carolis E, Vaccaro L, Posteraro P, Perlin DS, Kostrzewa M et al (2013) Rapid antifungal susceptibility testing by matrix-assisted laser desorption ionization-time of flight mass spectrometry analysis. J Clin Microbiol 51:2964–2969
68. Idelevich EA, Sparbier K, Kostrzewa M, Becker K (2017) Rapid detection of antibiotic resistance by MALDI-TOF mass spectrometry using a novel direct-on-target microdroplet growth assay. Clin Microbiol Infect. pii: S1198-743X(17)30578-5
69. Meier MA, Hoogenboom R, Fijten MW, Schneider M, Schubert US (2003) Automated MALDI-TOF-MS sample preparation in combinatorial polymer research. J Comb Chem 5:369–374
70. Broyer P, Perrot N, Rostaing H, Blaze J, Pinston F, Gervasi G, Charles MH, Dachaud F, Dachaud J, Moulin F, Cordier S, Dauwalder O, Meugnier H, Vandenesch F (2018) An automated sample preparation instrument to accelerate positive blood cultures microbial identification by MALDI-TOF mass spectrometry (Vitek® MS). Front Microbiol 9:911

Chapter 11
Fourier Transform Infrared Spectroscopy (FT-IR) for Food and Water Microbiology

Ângela Novais and Luísa Peixe

11.1 Introduction

An extraordinary development of sensitive, rapid and increasingly precise physical techniques with applications in microbiology occurred during the 1980s and the 1990s. It included mass spectrometry (MS), molecular spectroscopy (including fluorescence, Fourier-transform infrared (FTIR), and Raman spectroscopy], flow cytometry and high-resolution separation techniques [1]. In parallel, there was a significant advance in genotypic methods for bacterial characterisation based on partial (particular genotypic markers) or whole genomes, that became more affordable and sensitive [2]. In fact, the high resolution provided by whole genome sequencing (WGS) demonstrated the insufficient discriminatory potential of methods used in the past 30 years and is the current gold-standard diagnostic tool [3]. Combining the increasing knowledge available at the genomic level with that provided by reliable and high-throughput biophysical methods is an excellent opportunity to increase our knowledge on bacterial evolution and pathogenicity, as well as to develop cost-effective methods for bacterial characterisation [4].

Currently, a diagnostic method in microbiology needs to be reliable and accurate, but also rapid, low-cost, and user-friendly, in order to meet clinical or food microbiology demands and contribute to efficient decisions in several areas such as bacterial identification, outbreak control, microbial source tracking, amongst others. In addition, the stability of results over time, portability and appropriate software for both data storage and automated interpretation are a plus to guarantee standardised data and international coverage [2, 3]. While traditionally developed and used for

Â. Novais · L. Peixe (✉)
UCIBIO, Laboratory of Microbiology, Biological Sciences Department, Faculty of Pharmacy, University of Porto, Porto, Portugal
e-mail: lpeixe@ff.up.pt

J. Moran-Gilad, Y. Yagel (eds.), *Application and Integration of Omics-Powered Diagnostics in Clinical and Public Health Microbiology*,
https://doi.org/10.1007/978-3-030-62155-1_11

chemical analysis, matrix-assisted laser desorption-ionisation time-of-flight mass spectrometry (MALDI-TOF MS) became the first-line tool in clinical microbiology laboratories around the world because it simplified and speeded-up microbial identification [5, 6].

On their turn, vibrational spectroscopy methods such as infrared (IR) and Raman spectroscopy (RS) might have a place in the microbiology armamentarium since they provide a very attractive performance due to their high-throughput, speed, low cost and simplicity [7]. They are complementary techniques providing a biochemical fingerprint of the bacterial cell. Even though Raman spectroscopy and its derivative surface-enhanced Raman scattering (SERS) have been considered useful for outbreak detection or characterisation of bacterial strains, it presents comparatively lower sensitivity, reproducibility and resolution than Fourier transform infrared (FT-IR) spectroscopy. The latter has been profusely used during the 1990s for variable purposes in the microbiology field, but especially for bacterial discrimination purposes at different taxonomic levels [8, 9]. However, the inaccuracy in bacterial classification systems at that time (either by blurred taxonomic positioning or by the insufficiency of the methodology used for bacterial typing), the lack of consistent databases thereof and standardised protocols, as well as the development of genotypic-based methods motivated their growing abandonment. Nevertheless, improvements in bacterial taxonomy and cell biomolecules knowledge, mostly derived by whole genome sequencing, support FT-IR-based bacterial characterisations and might bring back vibrational spectroscopy to the spotlight.

11.2 Fourier-Transform Infrared (FT-IR) Spectroscopy in Microbiology: An Overview

The analysis of biological materials by infrared (IR) spectroscopy was firstly suggested by W. W. Coblentz at the beginning of the twentieth century [10]. Later in the 1950s and 1960s, IR spectroscopy has been profusely applied to microorganisms, for differentiation and identification purposes [11, 12], but at that time the process was complicated, time-consuming and lacked reproducibility. It was only after the development of modern interferometers and Fourier-Transform techniques providing an improved time of analysis, reproducibility and sensitivity, together with new hardware and data analysis algorithms that it was possible to revive the technique for microbiological applications. The definitive establishment of FT-IR for analysis of microorganisms occurred after the developments of Naumann and collaborators at the Robert Koch Institute in the late 1980s, in collaboration with Bruker Optics [13]. These experiments were fundamental to settle the experimental conditions for sample preparation and data analysis [9, 13, 14].

Since then, FT-IR has been applied in diverse microbiological contexts and for different purposes. One of the most explored applications is for discrimination, classification and identification of bacteria at different taxonomic levels (genera, species) or even at serotype/serogroup and strain level. Many different reviews have explored the topic in the past 10 years, and all of them agree that this methodology is a quick and low-cost alternative for bacterial typing [8, 15, 16]. However, only very recently, the fundamentals for FT-IR discrimination have been clarified and explored for different bacterial species [17–20]. In some of these studies, the discriminatory ability at the strain level was related to variations on surface bacterial structures, especially on the saccharidic somatic (O) and capsular (K) antigens, and thus with correspondence with traditional serotyping. The antigen formula (O:H:K) comprising the somatic, flagellar (H) and capsular antigens is still often used as a strain signature useful for identification of outbreaks or certain pathogenic strains (e.g. *Escherichia coli* O157:H7) [21]. It is known that several of these surface antigens are composed of variable types and combinations of saccharide units that determine the final composition and structure of oligo and polysaccharide chains, that can vary between and within species. A comprehensive correlation between FT-IR-based assignments, sugar-based bacterial coating structures and genotypic features reflecting strain evolution has been recently established for different bacterial species, which settles the principles for bacterial typing by FT-IR [4].

FT-IR also has many applications in the food industry and food safety microbiology. One of the most explored is the detection of food spoilage by *Enterobacterales*, lactic acid bacteria or others present in different food matrices [22–25]. The detection and differentiation of contaminants along food production lines (microbiological source tracking) or in cosmetics' products is also a useful application, in the latter case developed with a library-independent approach for detection of fungal contamination, that can also be used for bacteria [26, 27]. There are also descriptions of the detection and source-tracking of outbreak strains from different bacterial species involved in foodborne disease either as pure colonies or directly from food matrices, with a highly attractive time to response (usually around 24 h) [28–30]. FT-IR-based approaches have also been tested for differentiation of the physiological state (viable, dead or injured cells) of food-related bacteria such as *E. coli*, *Salmonella enterica* or *Listeria monocytogenes* or in mixed bacterial populations when exposed to different stress elements (e.g. those used in food-processing chain for bacterial elimination) [31–36] or for evaluation of the success of spore inactivation in spore-forming bacteria [37, 38]. Along with the detection of stress-injured micro-organisms in food-related environments, the methodology has also been used to assess mechanisms of bacterial inactivation/response for exposure to stresses (heat, acids, tolerants) or antimicrobial compounds, monitoring membrane properties in changing environments, dynamic changes in bacterial populations and the study of spore ecology [39–41]. Studies using FT-IR for monitoring of biomass composition or even for quantification of bacteria, by measuring some of its components (cell lipids, polysaccharides) or their by-products (e.g. polypolyhydroxykalkanoate, dipicolinic acid, lactic acid) including in food matrices also showed promising results [38, 42–53]. In most cases, variability was observed using the whole spectra or some spectral regions, whereas in specific cases, particular biomarker peaks were devised as discriminatory [38, 54].

Other types of application include the evaluation of metabolic signatures associated with host adaptation in *L. monocytogenes* or characterisation of surface structures (teichoic wall acids) that are relevant for *Staphylococcus aureus* pathogenesis [55, 56], which opens new avenues on the study of host-pathogen interactions. Finally, the potential of FT-IR to detect metabolic changes associated with the expression of antibiotic resistance genes has also been tested [57, 58], which is an ambitious goal. Nevertheless, these studies lack robustness and correlation with reference methods or well-characterized collections of isolates thus hindering an accurate association of spectral changes with specific antibiotic resistance mechanisms.

11.3 Principles of FT-IR Spectroscopy Applied to Bacterial Cells

Fourier Transform Infrared (FT-IR) spectroscopy is a biophysical technique traditionally used in chemistry to determine the molecular composition of diverse sample types. It is a rapid, non-destructive, simple, reagentless, low-cost and high-throughput analytical tool, associated with very low running costs. The principle of the technique is that the absorption of the infrared (IR) radiation by a given sample type (chemical, microbiological, etc.) causes a change in the vibrational modes of the chemical bonds present on the sample. When applied to bacterial cells, the interaction of the IR radiation with the different cellular components (nucleic acids, proteins, lipids, and carbohydrates) creates a complex spectrum reflecting the relative abundance of the different functional groups at different wavenumber ranges [9, 15]. Peaks represent a superposition of contributions from all its biomolecules, and as such, each micro-organism has a highly specific infrared spectrum signature that can be used for bacterial identification when compared with reference databases (Fig. 11.1). Established correlations between band frequencies (peak positions cm^{-1}, peak intensities and peak width) and known biochemical structures (functional groups) can be used to tentatively assign particular IR absorption bands to a specific molecular bond. Some of these are available in several publications [8, 59], but it is well recognised that the absence of specific information on discriminatory molecular biomarkers is a limitation of the method [15].

Usually, the mid-infrared region of the electromagnetic spectrum comprising wavenumbers 4000–400 cm^{-1} is used to acquire bacterial FT-IR spectra. Helm et al. [14] settled the parameters of FT-IR for identification and classification of bacteria, by dividing this region in five spectral windows corresponding to the absorption expressed in wavenumbers (cm^{-1}) of: (i) lipids (window 1, 3000–2800 cm^{-1}, dominated by vibrations of functional groups usually present in fatty acids); (ii) proteins and peptides (window 2, 1800-1500 cm^{-1}, dominated by vibrations of amide I and amide II bands); (iii) a mixed region (window 3, 1500-1200 cm^{-1}, with information of proteins, fatty acids and phosphate-carrying compounds); (iv) carbohydrates (window 4, 1200-900 cm^{-1}, a fingerprint-like region with absorption bands of the carbohydrates); and (v) a fingerprint region (window 5, 900-700 cm^{-1}, showing some remarkably specific spectral patterns, not yet assigned to particular cellular components or functional groups) (Fig. 11.1). Windows 3 and 4 have been consistently pointed out as the most discriminatory for routine bacterial identification purposes [14, 19, 60, 61].

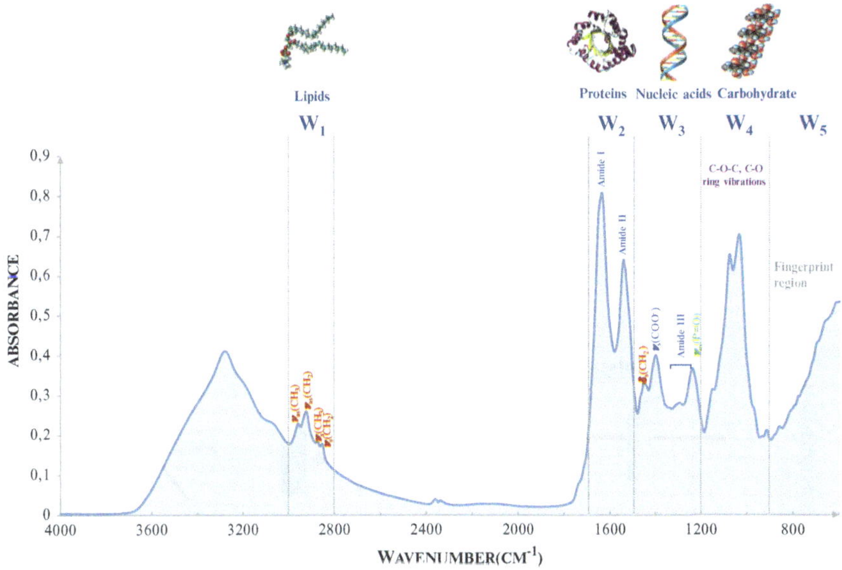

Fig. 11.1 Typical spectrum from a *Salmonella enterica* isolate showing spectral windows as defined by Helm et al. [14] and main band assignments. The spectrum was acquired using a Perkin Elmer Spectrum BXFT-IR System spectrophotometer in the ATR mode with a PIKE Technologies Gladi ATR accessory from 4000–600 cm-1 and a resolution of 4 cm^{-1} and 32 scan co-additions. *ν* stretching vibrations, δ bending vibrations, s symmetric vibrations, as asymmetric vibrations

11.4 Experimental Details and Data Analysis

Baker et al. described in detail the experimental conditions for using FT-IR to analyse biological samples [62]. A FT-IR experiment comprises *sample preparation*, *spectra acquisition* and *data analysis* as essential steps. Variations have been described by different authors, which are summarised in Fig. 11.2.

Sample preparation is non-destructive and minimal since only one bacterial colony or a bacterial suspension is used, a choice that can vary according to the FT-IR acquisition mode (see below). Bacteria cultivation conditions (culture medium, temperature or time of incubation) are established according to the micro-organism studied, that in most cases described herein are adapted to aerobic rapid-growth bacteria and standardised for spectral database development. It has been widely accepted that FT-IR bacterial spectra can vary according to the culturing conditions due to variable metabolic activity [63]. It has been assumed that variations in culturing conditions (primarily the culture media and the incubation time) can significantly affect reproducibility between spectra, at least when the whole spectra is considered [64].

To minimise the influence of experimental or biological variation, spectral data are usually processed (e.g. derivatised) and compartmentalised to focus on specific regions of the spectra. With this approach, preliminary data from our group support

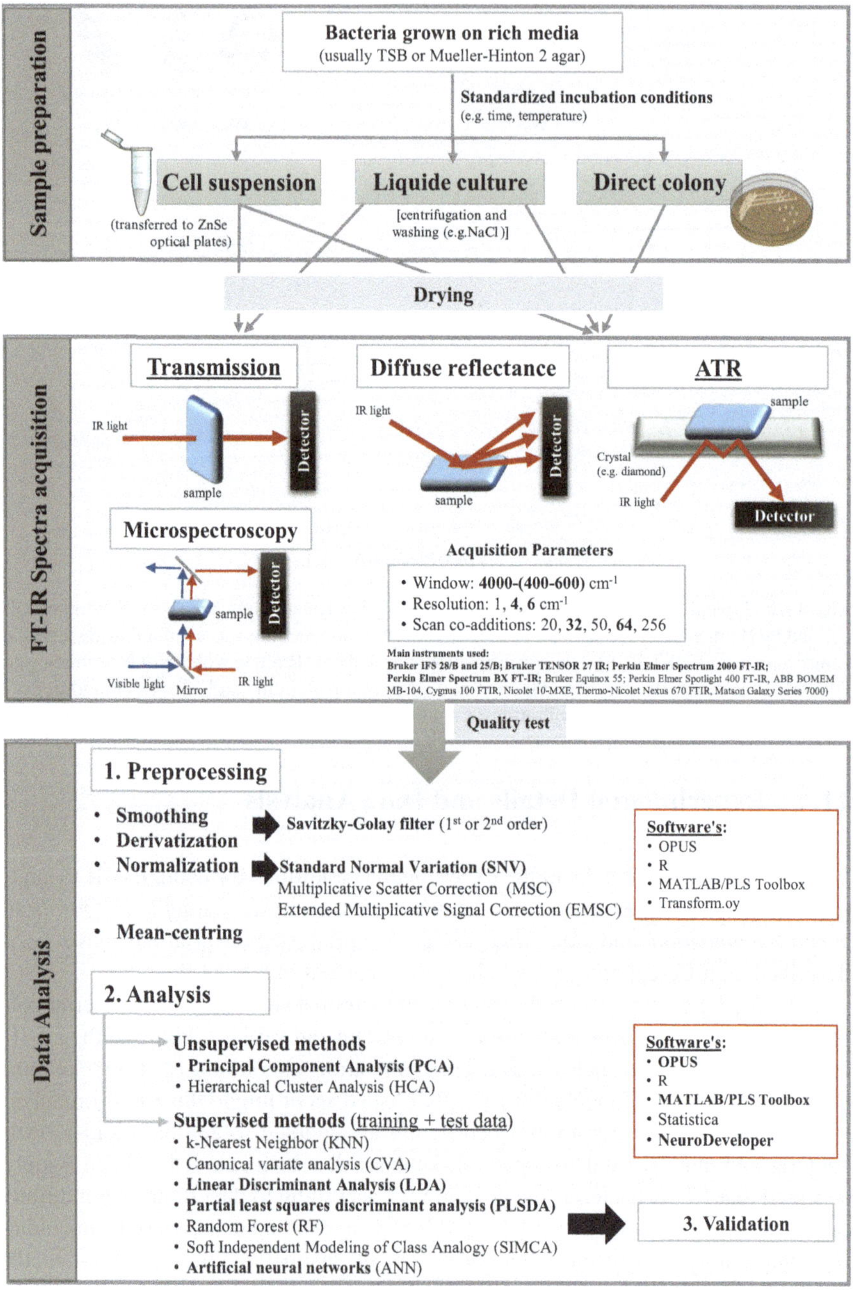

Fig. 11.2 Variations on steps required for a typical FT-IR experiment

previous observations [65, 66] that small variations in the incubation time (+/− 4 h) and growth in different non-selective culture media are not detrimental to reproducibility when evaluated against spectral databases obtained in similar conditions. Some spectral regions may be less affected by differences in culture conditions such as one of the most commonly used for bacterial typing purposes such as the 1200–900 cm^{-1} region, dominated by polysaccharides.

The most common *spectra acquisition* modes for bacterial characterisation are transmittance, attenuated total reflectance (ATR) and diffuse reflectance [62]. One of the earliest and most commonly used methods is the transmittance mode, where the sample is placed on the path of an IR beam and scanned. This was the technique used in the prototype developed by Bruker and researchers from Robert Koch Institute [13], and subsequently by many other researchers in a wide diversity of bacterial species [55, 65]. In transmission mode, the bacterial cells are suspended in water or saline and then transferred to water-insoluble optical plates (frequently from ZnSe). The samples must be uniformly dried, usually at 50–60 °C during 1-2 h or under vacuum before measurement. The increased signal-to-noise ratio, increased time of sample preparation and the spectra variability due to differences in sample amount and thickness are commonly recognised disadvantages. Another frequently used mode is the attenuated total reflectance (ATR), where one colony or a small fraction of a bacterial suspension is directly placed on an optically dense crystal, the IR beam creates an evanescent wave that is absorbed by the sample and the reflected radiation is passed to the detector. It is an equally inexpensive and low-cost method associated with greater simplicity (little or no sample preparation) and higher reproducibility. Till very recently, this method allowed only one single sample per measurement which is an important caveat for routine applications. However, the development and commercialisation of a high-throughput ATR accessory for multiple sample analysis now allows testing several samples simultaneously [67]. Finally, the diffuse reflectance mode is used to analyse low amounts of solid, powder or freeze-dried samples, however, due to higher costs, this technique is barely applied to bacterial cells [68]. Further details regarding instruments, technical specifications and data analysis tools are precisely explained in these comprehensive technical reviews [15, 62].

Data analysis of the complexity of chemical information generated is performed using chemometrics tools, a series of mathematical and statistical methods to interpret and model chemical phenomena and get identification or classification results. For classification purposes, these tools need to be sufficiently efficient to detect class-specific features and exclude intraclass variation, and their choice often depends on a trial and error approach [14, 62, 69]. Different software analysis tools are available, differing on the availability and cost (open source or commercial), ease of operation and need of trained personnel. Some of them are designed to operate in specific IR systems (e.g. OPUS from Bruker or OMNIC for ThermoScientific) and hence represent closed systems, with low manoeuvrability [62].

After guaranteeing the quality of the spectra (absorption signal, signal to noise ratio and level of water vapour), data analysis first starts with preprocessing of bacterial spectra to correct issues related to spectra acquisition. Spectra pre-processing is used to scale-up differences in spectra acquired in similar conditions. It is essential in order to: (i) minimise variation associated with sampling conditions (e.g. sample amount, relative humidity), (ii) amplify chemically-based spectral differences to facilitate interpretation and analysis, and (iii) assure the robustness and accuracy of the multivariate pattern recognition methods used subsequently. The most common pre-processing methods used in raw spectral data are smoothing (reduces background noise), derivatisation (improves the resolution of complex and overlapping bands) and normalisation (compensates differences associated with sample size or thickness). One of the most commonly used algorithms is the one developed by Savitzky and Golay that simultaneously derivates (first or second-order) and smooth's the spectra [70]. Usually, multivariate data analysis is performed on a subset of the spectral data, corresponding to the spectral window with the highest discriminatory potential. This pre-selection is useful to reduce data amount and simplify classification methods.

In the second step, the pre-processed spectra are subsequently interpreted by multivariate methods (unsupervised or supervised) to evaluate large numbers of spectral features at the same time. These methods can be used for classification and/or identification of bacterial isolates [16, 60, 69]. The choice of the analysis workflow is guided not only by the sensitivity and specificity of the method but also by its ease of use, software availability and experience of the operator. Usually, each research group uses its own workflow, whenever a reliable one is available.

Unsupervised methods are used when spectra differentiation is performed without *a priori* class assignment, *i.e.*, species or strain information. Isolates relationships' are established according to the similarity between spectra using pattern recognition techniques for classification of groups of related isolates. Several methods can be used, but the most commonly applied are: (i) the hierarchical cluster analysis (HCA) method, that generates dendrograms representing distances between the spectra; (ii) principal component analysis (PCA) where variation is captured in principal components and expressed in score plots; (iii) canonical variant analysis (CNV), a variant of PCA, where similarities between spectra are represented in graphics. They are extremely useful to assess data complexity, similarity and heterogeneity of data with unknown composition. They can also be used to check reproducibility and groups' assignment of known datasets or to identify and exclude outliers before a supervised analysis.

On the other hand, *supervised methods* require *a priori* knowledge of the sample and its classes, and this information is used as learning information to build models (training data) and generate a classification function that will subsequently be used to predict unknown datasets. Several methods are used, such as the partial least squares discriminant analysis (PLS-DA), soft independent modelling by class analogy (SIMCA) or artificial neural networks (ANN). ANN is by far the most popular machine-(self)learning technique, which is highly adaptable but also requires large training sets and specialised personnel [17, 71, 72]. To the best of our knowledge, commercial versions are available for ANN (NeuroDeveloper, Synthon-Analytics)

[73] and SIMCA [74]. Importantly, the robustness of each supervised analysis is only assured when the generated model is validated with an unknown sample.

Identification of unknown isolates requires a comparison of spectra with those from a reference database containing representative spectra of isolates covering the intrinsic variability of a given taxa (spectral libraries). Methods based on machine learning techniques such as partial least squares discriminant analysis (PLS-DA) or artificial neural networks (ANN) are known to be powerful to extract specific spectral signatures [16, 17, 19], but they have only been tested and validated on specific and limited collections of isolates and require skills of specialised personnel. There are commercial databases to assist FT-IR-based species identification of a wide range of bacterial species and yeasts (Bruker Optik), whereas there are no consistent reference datasets for FT-IR-based infraspecies typing. Some research groups have their own reference libraries and have been using them especially for bacterial identification at the genus and species level [4, 8, 16, 65]. Particular applications for common source outbreak investigations in human or veterinary medicine or identification of contaminants in food products have also been reported [28, 30, 75, 76].

Routine applications based on FT-IR require standardised conditions to assure reproducibility and data compatibility within and/or between instruments, and the establishment of reference databases created judiciously according to the purpose. Many of these issues have been evaluated during the '90s [13], the golden era of FT-IR-based microbiological characterisations, but to date, there is no universally accessible spectra database [16, 62]. In 2016, Bruker launched a bench FT-IR-based equipment (IR Biotyper®) for bacterial typing for clinical microbiology routines that could facilitate data analysis, automation of the whole process and interlaboratory comparisons (https://www.bruker.com/applications/microbiology/strain-typing with ir-biotyper/overview html).

11.5 FT-IR Based Typing at the Species and Infra-Species Level of Bacteria in Food and Water Microbiology

There is accumulating evidence that FT-IR spectra of bacterial cells possess discriminatory features that can be useful for determination of species, serogroups, serotypes and clones in a wide range of Gram-positive and Gram-negative bacteria if appropriately extracted and compared by multivariate data analysis methods. These studies demonstrated that FT-IR has considerable potential as a rapid high-throughput screening method that can be useful for bacterial identification, outbreak detection, microbial source tracking, and identification of pathogenic strains in contexts that are relevant for the food industry and environmental safety. While some of them can be considered "proof-of-principle" studies that need further support by comprehensive and representative collections of isolates with adequate biological diversity, others include extensive and well-characterised collections of isolates characterised by reference genotypic methods, strengthening both methodology and purpose. It is essential to highlight that FT-IR-based identification at the species level demands the construction of calibration models including strains representing the natural biodiversity of the species and generally requires the use of wider spectral windows covering information from different types of biomolecules (Table 11.1). At the infra-species level, discrimination is possible when associated with particular cell biomolecules usually from the cell surface. Thus, to further support and substantiate the discriminatory patterns obtained, we need to increase the molecular knowledge on bacterial phylogeny and bacterial surface structures delineated by the most reliable genotypic methods (ideally by whole genome sequencing). A comprehensive analysis of existing knowledge on FT-IR for bacteria differentiation at the strain level is included in the recently published review by Novais et al. [4]. Herein we will review the current accumulated experience with FT-IR typing of bacterial species of relevance in the context of food and water microbiology.

11.5.1 *Bacillus sp.*

Bacillus sp. are rod-shaped and spore-forming bacteria that are important foodborne pathogens and frequent spoilage agents in food products, but they are also abundant in several environments such as water, soil and air. More than 70 species have been described, many of them in recent years, driving a significant re-structuration of the genera and the recognition of difficulties in differentiating closely related species. Two studies published in 1998 assessed the differentiation between *Bacillus* species type strains using FT-IR in the transmission mode. The study by Lin et al. demonstrated a reliable differentiation of *Bacillus cereus* from other species by a characteristic peak in the amide I region (1738–1740 cm^{-1}) in different culture media [54]. Beattie et al. showed a reliable identification of closely related *Bacillus cereus*, *Bacillus mycoides* and *Bacillus thuringiensis* using a wider spectral window (2000–500 cm^{-1}) [77]. Lucking et al. identified a broad set of isolates derived from food spoilage in diverse dairy products and industrial processing environments belonging to different species (e.g. *B. subtilis*, *Bacillus sporothermodurans*, *Bacillus amyloliquefaciens*) [78]. Approaches using FT-IR in the ATR mode, were also successful in distinguishing *B. cereus* from *Bacillus subtilis* [68] and also several closely related *Bacilllus pumilus* group species [79] using different spectral windows (4000–900 cm^{-1} *vs* 1200–900 cm^{-1}), although only the type strains were tested.

Regarding differentiation at the strain level within *Bacillus* species, it has been pointed out that FT-IR ATR might provide a higher resolution for strain typing since it reflects cell surface biochemistry and thus, potential strain-specific markers [68]. Some studies have attempted to evaluate FT-IR-based strain typing in species from the *B. cereus* group in comparison with molecular or phenetic methods, but problems associated with the low number of strains used, the identification of closely related species and/or lack of knowledge regarding variation at the cell surface, hinder the accuracy of groupings obtained [80, 81]. However, the nature and

Table 11.1 Fourier-Transform Infrared spectroscopy of bacterial species of relevance in food and water microbiology

Species	Strains included (N°)[a]	FT-IR mode	Spectral region used for analysis (cm^{-1})	Discriminatory peaks (cm^{-1})	Multivariate data analysis methods used[b]	Reference
Gram positive						
Bacillus cereus group	T (6)	Transmission	1800–1500	1738–1740	–	53
	T (6) + N (38)		2000–500	–	CVA	76
Bacillus pumilus group	T (5)	ATR	1200–900	–	PCA	78
	N (10)		4000–900	–	PCA, DFA	67
Listeria sensu stricto	T (5)	Transmission	2000–750	947, 985	CVA	83
	T (7) + N (24)		3000–500	–		65
	N (5)	Transmission	1400–720	–	CDA + PLSDA	84
	T (245)+N (520)	Transmission	1800–700 + 3100–2800	–	ANN	85
					(neurodeveloper)	71
Lactobacillus	T+N (56)	Transmission	1400–720	–	PLS, SIMCA, KNN	98
	T (12) + N (42)		3000–2800 + 1800-	–	HCA	
			1600 + 1200–700			99
Staphylococcus	T (94) + N (85)	transmission	2888–2868	–	SVD + PCA + SOM + KNN	91
			1800–750	–	PCA	92
Enterococcus	N (19)	Diffuse reflectance	4000–600	–	ANN	102
	T+N (18)	Transmission/ATR	4000–500	–	HCA	101
	N (33)	Microspectroscopy	3000–2820 + 1800–750	–	HCA + PLS	93
	T+N (22)	Transmission	3000–2800 + 1800–700	–	ANN	100
Gram negative						
Escherichia	T+N (4)	ATR	3068–2941 + 1780-	–	PCA + SIMCA	105
			1695 + 1523–673			

(continued)

Table 11.1 (continued)

Species	Strains included (N°)[a]	FT-IR mode	Spectral region used for analysis (cm^{-1})	Discriminatory peaks (cm^{-1})	Multivariate data analysis methods used[b]	Reference
Campylobacter	T+N (26)	Transmission	3000–2800 + 1800–700	–	ANN	120
	T (11)		1500–500	Several	PCA + DFA	121
Yersinia	T+N (847)	Transmission	1800–500	–	ANN	70
	N (198)		3000–2800 + 1500–800	–		119
Vibrio	T (3)	ATR	3068–2941 + 1780-		PCA + SIMCA	105
			1695 + 1523–673			
Mycobacterium	T+N (28)	Microspectroscopy	1700–900	–	HCA	125

[a]*T* type strains, *N* Non-type strains
[b]CVA = canonical variant analysis; PCA = principal component analysis; DFA = discriminant function analysis; CDA = canonical discriminate analysis; PLSDA = partial least square discriminant analysis; ANN = artificial neural networks; SIMCA = soft independent modelling of class analogy; KNN = multivariate k-nearest neighbor; HCA = hirarchical clustering analysis; SVD = singular value decomposition; SOM = self-organizing maps

composition of surface structures such as capsules or S-layers is only barely known for *B. cereus* and shows some promising differential features [82].

11.5.2 *Listeria sp.*

Listeria sp. are non-spore producing rod-shaped bacteria, that might be found in several environments and geographical regions. Only *Listeria monocytogenes* and *Listeria ivanova* are well-recognised pathogens, transmitted through contaminated food. These species belong to the recently recognised *Listeria* sensu strictu (together with *Listeria seeligeri*, *Listeria welshimeri* and *Listeria innocua*) whereas other 11 species belong to the *Listeria* sensu lato group and are thought to represent different genera [83]. These changes in the taxonomy can have a significant impact in the food industry and the strategies for detection of the human pathogen *L. monocytogenes*. Holt et al. and Lefier et al. have primarily shown reliable identification of *Listeria* sensu strictu species with very similar approaches [66, 84]. These studies, published in the late 1990s, used FT-IR in the transmission mode and compared the whole spectra (3000–500 cm^{-1} or 2000–750 cm^{-1}) of type strains (one isolate/species) by CVA, and maximised the influence of particular peaks (mainly around 947 cm^{-1} and 985 cm^{-1}). These results were corroborated later in a study including a higher number of strains per species [85]. Nevertheless, two further studies have established a more robust and semi-automated workflow for identification of *Listeria* sensu strictu species using FT-IR in a high-throughput transmission mode and commercial software based on the powerful artificial neural networks (ANN) machine learning method, that provided results in 25 h [72, 86]. In these studies, a high number of strains (n = 245 or n = 520) was tested to construct a calibrated model that was subsequently validated with test strains, identified by phenotypic and molecular methods, with correct identifications reaching 96–99% for all species and 96.6%–99.2% for *L. monocytogenes*. The adoption of a rapid analytical tool such as FT-IR by food industry could be useful to prescreen potentially persistent *L. monocytogenes* contaminants in food products. In fact, it seems to be useful in the identification of persistent *L. monocytogenes* strains in cheese from different producers in several countries [87].

The ability of FT-IR to differentiate at the infra-species level was only tested for *L. monocytogenes* in 1997 [66]. Other studies followed and showed reliable discrimination between serogroups (correct identifications up to 98.8%) and main serotypes (up to 96.6%), including those frequently implicated in human listeriosis (1/2a and 4b) [66, 72, 87–89]. All of them are generally concordant with the most discriminatory region, dominated by polysaccharides (1200–900 cm^{-1}), which presumably reflect differences in the composition of wall teichoic acids (WTAs) that seem to be specific for each serotype [90]. Comparisons of FT-IR-based strain assignments with those obtained by standard molecular methods revealed generally congruent results but diverse spectral regions and genotypic methods were used as a reference, hindering the ability to establish a reliable correspondence [34, 89, 91].

11.5.3 *Staphylococcus sp.*

Many species of *Staphylococcus* sp. are natural inhabitants of skin and mucosa of mammals, as well as the causative agents of opportunistic infections, and can roughly be subdivided into coagulase-negative staphylococci (CNS) and coagulase-positive, the latter including the pathogenic *Staphylococcus aureus*. This species is a major causative agent of foodborne disease due to enterotoxin production, and its reliable and quick identification is primordial to detect contaminated food or to trace potential outbreaks.

For this reason, several studies directed FT-IR-based approaches to differentiate *S. aureus* from CNS species, including the earlier studies by Helm et al. [8, 14, 92, 93]. These studies included reference/Type strains and a high number of isolates and species, in most cases identified by phenotypic and reliable molecular methods, and created models that were subsequently validated with isolates from food products. They used several spectral windows or a widened region (1500–780 cm^{-1}), though a narrow spectral region also seemed to be reliable [92]. However, the potential to differentiate other *Staphylococcus* species remains to be clarified [14, 94]. FT-IR has also demonstrated utility in the reliable detection and source-tracking of *S. aureus* food-related outbreaks in a shorter time than reference methods [29, 76]. Moreover, FT-IR based subtyping focusing on *S. aureus* demonstrated reliable results considering mainly the polysaccharide region (1200–900 cm^{-1}), which correlated with the *cap* specific locus and the glycostructural composition of different *S. aureus* capsular types, using bacteria isolated from different sources. Furthermore, discrimination was also specifically associated with variation in WTAs or other glycopolymers of the cell wall such as peptidoglycan and lipoteichoic acid [17, 95].

11.5.4 *Lactobacillus sp.*

Diverse *Lactobacillus* species are ubiquitous in the environment and are also widely used as starter cultures during fermentation processes or as probiotics in the food industry, while some species can be frequently encountered as spoilage in different food products, making reliable species identification of great interest. The genus comprises a large number of species (around 170), that are divided according to their carbohydrate fermentation pathways into obligate homofermentative, facultative heterofermentative or obligate heterofermentative lactobacilli, though it is currently under taxonomic restructuring [96].

Different studies assessed the potential of FT-IR (all of them in the transmission mode) for differentiation of lactobacilli isolated from different food products (beer, cheese, meat and kefir). The species analysed included in most cases supposed homofermentative species, but it is currently known that species identification is problematic in this genus, which might compromise the reliability of some of the results presented. Two of these studies reported a good resolution (up to 94% correct identifications at the species level) using a combination of three spectral regions

(1200–700 cm^{-1}), but they included only a small number of strains per species and/or a few species [97, 98]. Oust et al. revealed a high discriminatory potential among four closely related homofermentative species using a similar region (1400–720 cm^{-1}) and supervised methods (PLS or KNN), where the experimental approach allowed the re-identification of a few misidentified strains [99]. Studies from Bosch et al. and Wenning et al. included more representative samples of both hetero- and homofermentative strains, including Type strains, and demonstrated a consistent species identification using a combination of several spectral regions in a step-wise process or ANN [100, 101] but with lower resolution for closely related species possibly associated with their misidentification by conventional methods.

11.5.5 *Enterococcus sp.*

Enterococci also belong to the group of lactic acid bacteria and are important in the food context since they are used as probiotics, as fermentative agents or are involved in food spoilage, as well as they are also common agents of infections in humans and animals. Their ability to produce bacteriocins is useful to control foodborne pathogens in food products, and considering the variable pathogenicity associated with the different species or strains, there is a great need for reliable and cost-effective tools to support their identification in the food industry.

Some potential for FT-IR to discriminate *Enterococcus* species has been suggested by studies using variable FT-IR analysis workflows, though all of them included a low number of species (mainly *E. faecium* and *E. faecalis*), isolates (from clinical or food origin) or reference strains [94, 101–103]. These preliminary results need to be substantiated with more diverse and representative collections of isolates, but they represent important proof-of-principle studies evaluating diverse acquisition modes (one of the very few using diffuse reflectance), analysis methods (including ANN) and reproducibility among laboratories. The potential for discrimination at the infra-species level remains to be clarified since the few studies published (focusing only in *E. faecium*) either used non-representative collections or compared with methods that are nowadays recognized as inadequate for strain typing [101, 104]. Moreover, there is a lack of knowledge on the cell surface polysaccharide variation of *E. faecium*, though preliminary studies suggest there is a potential for discrimination between main lineages of clinical interest (Freitas AR et al. unpublished data).

11.5.6 *Other Gram-Positive Bacteria*

Few information exists on other relevant bacteria in the contexts of food and water microbiology as those belonging to the genus *Clostridium* (some of them responsible for severe foodborne diseases such as *C. perfringens* or *C. botulinum*) or other lactic acid bacteria such as *Lactococcus* sp. The studies available evaluated only a few strains and therefore lacked appropriate validation [14, 101, 105].

11.5.7 *Escherichia sp.*

Species from the genus *Escherichia* are common inhabitants of the gastrointestinal tract of humans and animals and belong to the order *Enterobacterales*. *E. coli* includes opportunistic strains that cause extraintestinal infections or pathogens that cause intestinal infections which can be transmitted through contaminated food. Also, the detection of *E. coli* has long been considered an indicator of poor water quality attributed to faecal contamination.

There are several reports on the differentiation of *E. coli* from other *Enterobacterales* or other non-fermentative Gram-negative bacteria using the whole spectra or only lipid cellular components [94, 106]. Nevertheless, they include only a small number of strains (mainly Type strains) requiring further optimisation of protocols and validation in more extensive collections of isolates. In the first studies by Helm et al. [14], *E. coli* strains were grouped in a small number of serogroups, and the discrimination according to their O-antigenic structure was afterwards corroborated [107], both studies considering the 1200–900 cm^{-1} polysaccharides region. The ability of FT-IR to selectively detect the *E. coli* O157:H7 serotype, a particularly relevant foodborne pathogen, has been tested using different approaches including directly from several food matrices (ground beef, apple juice), or its differentiation from other closely related serotypes or non-pathogenic strains [30, 108–110], which is of high relevance for food quality control. Another useful application is the possibility to detect *E. coli* in drinking water (even in mixed culture) [111] and quantify *E. coli* at a limit of detection of 100 CFU/ml, as was done for example in baby spinach leaves [112]. Discrimination of clinically-relevant clones frequently involved in urinary tract infections with high reliability (up to 91–100%) was also demonstrated, though the correlation with variation at surface antigens still remains to be precisely clarified [61, 113].

11.5.8 *Salmonella sp.*

There are only two species described in the genus *Salmonella*, *Salmonella enterica* and *Salmonella bongori*, *S. enterica* further divided into six subspecies and over 2600 serotypes defined based on the somatic (O) and flagellar (H) antigens. *S. enterica* is distributed in the environment, in humans and animals, and particular serotypes are the leading cause of foodborne illness worldwide (salmonellosis). Some serotypes can also cause invasive disease and even life-threatening infections in certain geographic areas and patient populations. Thus, quick and reliable differentiation of *S. enterica* serotypes, epidemiological and food-source tracking investigation is crucial. Conventional microbiological methods for detection and identification of *Salmonella* sp. are usually labour-demanding and time-consuming (results in 5–7 days), whereas molecular methods have disadvantages including inadequate sensitivity and high-cost and the fact that these methods are not able to differentiate live from dead cells.

The discriminatory ability of *S. enterica* at the serotype level was initially assessed in a low number of strains or serotypes (mainly B, C, D or E) in studies that validated the methodology using various experimental conditions and analysis workflows [74, 114, 115]. These results were corroborated using more extensive collections of isolates from different serogroups, where specific peaks were considered to be discriminatory [116]. More recently, a comprehensive analysis of a vast and well representative collection of isolates (n = 325 from 15 serogroups and 57 serotypes) previously typed by up-to-date methods settled the correlation between FT-IR based assignments and variation on O-units structures and available molecular data [19]. Some previous studies showed the ability to identify and differentiate particular serotypes (Typhimurium or Enteritidis, including live and dead cells) from complex food matrices (chicken breast and minced meat) even in a few hours, which is an additional advantage of this method [32, 117, 118].

11.5.9 *Yersinia sp.*

The genus *Yersinia* comprises 17 species, three of which are well-known human pathogens: *Y. pestis*, *Y. pseudotuberculosis*, and *Y. enterocolitica*. *Y. enterocolitica* causes acute enteric disease, originating from contamination of food from animal origin (especially pork), triggering the analysis of food samples and veterinary controls in livestock for epidemiological purposes. However, domestic animals are thought to be possible reservoirs, and *Y. enterocolitica* is also widespread in soil and water environments. *Y. enterocolitica* sensu lato has been recently subdivided into *Y. enterocolitica* sensu strictu (including several biotypes and serotypes, some of which are pathogenic) and other species, which may be misidentified by conventional cultural and/or biochemical methods.

A very robust study assessed FT-IR potential for differentiating *Yersinia* at the species and subspecies levels [71]. The authors included Type strains from different species, representative collections of isolates including a large number of well-characterised *Yersinia* isolates (~200) from different sources for calibration and validations models, as well as other Gram-negative bacteria (>600). Using the spectral region 1800–500 cm^{-1} and a complex ANN model, they demonstrated very high levels of correct predictions at the genus level (91.5%) and subspecies level (98.3% of correct identification of bioserotypes) while lower at the species level (77.9%). This model was subsequently optimised for identification of *Y. ruckeri* from diseased salmonid fish, allowing to increase the proportion of correct identification of *Y. ruckeri* specifically (97.4%, including sorbitol fermenting/non-fermenting strains) as well as other species (87.1%) [119]. The same approach was subsequently used to reliably identify *Y. enterocolitica* and their corresponding bioserotypes isolated from the faeces of companion animals (dogs and cats) [120]. In all of them, pathogenic strains from diverse bioserotypes were also discriminated by FT-IR (>90% correct predictions), though there was no information regarding the discriminatory spectral region.

11.5.10 *Campylobacter sp.*

Species belonging to the *Campylobacter* genus are causative agents of foodborne illness worldwide, most commonly originating from poultry as a source of human diarrheal disease. *Campylobacter jejuni* and *Campylobacter coli* are the most common species, though there are other species with pathogenic potential. Quick and reliable identification is important to understand the epidemiology of the disease as well as for prevention and control. However, available methods do not appropriately distinguish some of the most relevant species.

FT-IR based discrimination of *C. jejuni* and *C. coli* between each other and from other much less represented species (e.g. *C. fetus*, *C. lari*, *C. concisus*), and also the subspecies of the less common *C. fetus* (*C. fetus* subsp. *fetus* and *C. fetus* subsp. *venerealis*) was tested, with rates of correct predictions reaching 99%. Some of these studies were based on ANN models, that need to be enriched with a higher number of strains and species to increase robustness [121, 122]. Strains within *C. jejuni* and *C. coli* were also reliably distinguished using the 1200–900 cm^{-1} region in accordance with the highly discriminatory enterobacterial repetitive intergenic consensus (ERIC)-PCR typing method, though the robustness and generalizability of these results are unclear [123].

11.5.11 *Legionella spp.*

More than 58 species and 70 serotypes from the genus *Legionella* are known, some of them (and particularly some *L. pneumophila* serotypes) cause disease, and are commonly distributed in soil, natural or industrial/domestic aquatic environments. The potential for discrimination of species remains to be elucidated but it was already demonstrated in the late 1980's that it was possible to differentiate *L. pneumophila* isolates from different serogroups in the polysaccharide region (1200–900 cm^{-1}) [124], a study that was partially complemented by Helm et al. in 1991, that attributed differentiation indices to the production of poly-β-hydroxy fatty acids [14]. However, these constitute only preliminary observations that were not further explored.

11.5.12 *Vibrio sp.*

Vibrio sp. are usual inhabitants of marine coastal waters, estuaries, lakes and streams. Some species cause disease and are transmitted through contaminated water (*V. cholera*) or by contaminated seefood (*V. parahaemolyticus* and *V. vulnificus*). Mainly *V. parahaemolyticus* strains are pathogenic and can cause acute gastroenteritis due to the production of toxins, and their detection is an important food-safety issue. Representative Type strains of these species were distinguished from *Enterobacterales* by FT-IR using fatty acid methyl ester profiles, though these preliminary results need to be substantiated with larger collections [106]. Very recently, it has been shown that FT-IR can help distinguish pathogenic from non-pathogenic *V. parahaemolyticus* strains presumptively on the basis of the production of particular toxins in comparison with reference strains [125]. However, it remains to be clarified if the basis for that discrimination is related to toxin production or strains' relatedness.

11.5.13 *Mycobacterium sp.*

Mycobacterium organisms are ubiquitous and globally found in drinking water systems. The genus includes over 180 species and, the most clinically important of them is *M. tuberculosis* which causes tuberculosis. It is followed by nontuberculosis mycobacteria (NTM) and *M. leprae* causing most commonly pulmonary disease or leprae, respectively. Differentiation of NTM species by FT-IR microspectroscopy has been demonstrated using a set of reference and test strains from 10 different species, showing the potential to broaden the spectrum of identification to a broader set of strains with higher biological diversity [126]. Additionally, a high congruence was observed between FT-IR-based assignments with those obtained by the reference genotyping methods (spoligotyping and variable number of tandem repeats units, VNTR) for certain *M. bovis* isolates in a step-wise discriminatory workflow, showing promise for intraspecies differentiation that needs to be substantiated with molecular-based knowledge [127].

11.6 Future Perspectives

The potential of FT-IR spectroscopy to accurately discriminate biologically significant bacterial groups at different taxonomic levels at a quick, low cost and high-throughput rate is evident. The current availability of bacterial genomes from a wide range of bacteria and bioinformatic tools for data analysis is an opportunity to establish significant and comprehensive correlations between genotypic features and cell biomolecules that will support FT-IR-based assignments. One can envision a wide range of applications of such a versatile methodology in diverse areas of food and water microbiology, including food safety, food industry or quality control. Settled on a vast list of proof-of-concept studies, the implementation of FT-IR to facilitate routine microbiological procedures depends on a proper validation of bacterial reference databases and reproducible experimental workflows between instruments and laboratories, and, finally, on the adaptation of the method for a non-specialist user.

Acknowledgments This work was supported by the Applied Molecular Biosciences Unit - UCIBIO which is financed by national funds from FCT (UIDB/04378/2020). This work was also funded by national funds through FCT – Foundation for Science and Technology, I.P., in the frame of the Transitional Norm - DL57/2016/CP1346/CT0032. The authors would like to thank the European Society of Clinical Microbiology and Infectious Diseases (ESCMID) Food- and Water-borne Infections Study Group (EFWISG) for scientific support of activities related with the topic, and Ana R. Freitas and Carla Rodrigues for critical reading.

References

1. Naumann D (2006) Infrared spectroscopy in microbiology. In: Encyclopedia of analytical chemistry [Internet]. Chichester, Wiley. Available from: http://doi.wiley.com/10.1002/9780470027318.a0117
2. Sabat AJ, Budimir A, Nashev D, Sá-Leão R, van Dijl JM, Laurent F et al (2013) Overview of molecular typing methods for outbreak detection and epidemiological surveillance. Euro Surveill 18(4):20380
3. Nadon C, Van Walle I, Gerner-Smidt P, Campos J, Chinen I, Concepcion-Acevedo J et al (2017) PulseNet international: vision for the implementation of whole genome sequencing (WGS) for global food-borne disease surveillance. Eur Secur 22(23):30544
4. Novais Â, Freitas AR, Rodrigues C, Peixe L (2018) Fourier transform infrared spectroscopy: unlocking fundamentals and prospects for bacterial strain typing. Eur J Clin Microbiol Infect Dis 38(3):427–448
5. Kostrzewa M, Sparbier K, Maier T, Schubert S (2013) MALDI-TOF MS: an upcoming tool for rapid detection of antibiotic resistance in microorganisms. Proteomics Clin Appl 7(11–12):767–778
6. Singhal N, Kumar M, Kanaujia PK, Virdi JS (2015) MALDI-TOF mass spectrometry: an emerging technology for microbial identification and diagnosis. Front Microbiol 6:791
7. Ellis DI, Goodacre R (2006) Metabolic fingerprinting in disease diagnosis: biomedical applications of infrared and Raman spectroscopy. Analyst 131(8):875
8. Maquelin K, Kirschner C, Choo-Smith L-P, van den Braak N, Endtz HP, Naumann D et al (2002) Identification of medically relevant microorganisms by vibrational spectroscopy. J Microbiol Methods 51(3):255–271
9. Naumann D, Helm D, Labischinski H (1991) Microbiological characterizations by FT-IR spectroscopy. Nature 351(6321):81–82
10. Coblentz WW (1873) Investigations of infrared spectra. Nabu Press, Place of publication not identified, p 2010
11. Goulden JD, Sharpe ME (1958) The infra-red absorption spectra of lactobacilli. J Gen Microbiol 19(1):76–86
12. Randall HM, Smith DW, Colm AC, Nungester WJ (1951) Correlation of biologic properties of strains of *Mycobacterium* with infra-red spectrums. I. Reproducibility of extracts of *M. tuberculosis* as determined by infra-red spectroscopy. Am Rev Tuberc 63(4):372–380
13. Naumann D (2008 [cited 2018 Dec 21]) FT-IR spectroscopy of microorganisms at the Robert Koch Institute: experiences gained during a successful project. In: Mahadevan-Jansen A,

Petrich W, Alfano RR, Katz A (eds) , San Jose, p 68530G. Available from: http://proceedings.spiedigitallibrary.org/proceeding.aspx?doi=10.1117/12.761698

14. Helm D, Labischinski H, Schallehn G, Naumann D (1991) Classification and identification of bacteria by Fourier-transform infrared spectroscopy. J Gen Microbiol 137(1):69–79
15. Lasch P, Naumann D (2015) Infrared spectroscopy in microbiology. In: Encyclopedia of analytical chemistry [Internet]. Wiley, Chichester, pp 1–32. Available from: http://doi.wiley.com/10.1002/9780470027318.a0117.pub2
16. Wenning M, Scherer S (2013) Identification of microorganisms by FTIR spectroscopy: perspectives and limitations of the method. Appl Microbiol Biotechnol 97(16):7111–7120
17. Grunert T, Wenning M, Barbagelata MS, Fricker M, Sordelli DO, Buzzola FR et al (2013) Rapid and reliable identification of *Staphylococcus aureus* capsular serotypes by means of artificial neural network-assisted fourier transform infrared spectroscopy. J Clin Microbiol 51(7):2261–2266
18. Sahu RK, Mordechai S, Pesakhov S, Dagan R, Porat N (2006) Use of FTIR spectroscopy to distinguish between capsular types and capsular quantities in *Streptococcus pneumoniae*. Biopolymers 83(4):434–442
19. Campos J, Sousa C, Mourão J, Lopes J, Antunes P, Peixe L (2018) Fourier transform infrared spectroscopy with attenuated total for non-typhoidal *Salmonella* clinically relevant serogroups and serotypes discrimination: a comprehensive analysis. Int J Food Microbiol 285(July):34–41
20. Rodrigues C, Sousa C, Lopes JA, Novais Â, Peixe L (2020) A front line on Klebsiella pneumoniae capsular polysaccharide knowledge: fourier transform infrared spectroscopy as an accurate and fast typing tool. mSystems 5(2):e00386-19
21. Henriksen SD (1978) Serotyping of bacteria. In: Methods in microbiology, pp 1–13
22. Rahman U, Sahar A, Pasha I, Rahman S, Ishaq A (2018) Assessing the capability of Fourier transform infrared spectroscopy in tandem with chemometric analysis for predicting poultry meat spoilage. PeerJ 6:e5376
23. Saraiva C, Vasconcelos H, de Almeida JMMM (2017) A chemometrics approach applied to Fourier transform infrared spectroscopy (FTIR) for monitoring the spoilage of fresh salmon (Salmo salar) stored under modified atmospheres. Int J Food Microbiol 241:331–339
24. Panagou EZ, Mohareb FR, Argyri AA, Bessant CM, Nychas G-JE (2011) A comparison of artificial neural networks and partial least squares modelling for the rapid detection of the microbial spoilage of beef fillets based on Fourier transform infrared spectral fingerprints. Food Microbiol 28(4):782–790
25. Nicolaou N, Goodacre R (2008) Rapid and quantitative detection of the microbial spoilage in milk using Fourier transform infrared spectroscopy and chemometrics. Analyst 133(10):1424–1431
26. Shapaval V, Møretrø T, Wold Åsli A, Suso HP, Schmitt J, Lillehaug D et al (2017) A novel library-independent approach based on high-throughput cultivation in bioscreen and fingerprinting by FTIR spectroscopy for microbial source tracking in food industry. Lett Appl Microbiol 64(5):335–342
27. Dieckmann R, Hammerl JA, Hahmann H, Wicke A, Kleta S, Dabrowski PW et al (2016) Rapid characterisation of *Klebsiella oxytoca* isolates from contaminated liquid hand soap using mass spectrometry, FTIR and Raman spectroscopy. Faraday Discuss 187:353–375
28. Eisenberg T, Mauder N, Contzen M, Rau J, Ewers C, Schlez K et al (2015) Outbreak with clonally related isolates of *Corynebacterium ulcerans* in a group of water rats. BMC Microbiol 15:42
29. Fetsch A, Contzen M, Hartelt K, Kleiser A, Maassen S, Rau J et al (2014) *Staphylococcus aureus* food-poisoning outbreak associated with the consumption of ice-cream. Int J Food Microbiol 187:1–6
30. Davis R, Irudayaraj J, Reuhs BL, Mauer LJ (2010) Detection of *E. coli* O157:H7 from ground beef using Fourier transform infrared (FT-IR) spectroscopy and chemometrics. J Food Sci 75(6):M340–M346

31. Lu X, Liu Q, Wu D, Al-Qadiri HM, Al-Alami NI, Kang D-H et al (2011) Using of infrared spectroscopy to study the survival and injury of *Escherichia coli* O157:H7, *Campylobacter jejuni* and *Pseudomonas aeruginosa* under cold stress in low nutrient media. Food Microbiol 28(3):537–546
32. Davis R, Burgula Y, Deering A, Irudayaraj J, Reuhs BL, Mauer LJ (2010) Detection and differentiation of live and heat-treated *Salmonella enterica* serovars inoculated onto chicken breast using Fourier transform infrared (FT-IR) spectroscopy. J Appl Microbiol 109(6):2019–2031
33. Davis R, Deering A, Burgula Y, Mauer LJ, Reuhs BL (2012) Differentiation of live, dead and treated cells of *Escherichia coli* O157:H7 using FT-IR spectroscopy. J Appl Microbiol 112(4):743–751
34. Nyarko EB, Puzey KA, Donnelly CW (2014) Rapid differentiation of *Listeria monocytogenes* epidemic clones III and IV and their intact compared with heat-killed populations using Fourier transform infrared spectroscopy and chemometrics: *Listeria monocytogenes* epidemic clones. J Food Sci 79(6):M1189–M1196
35. Al-Qadiri HM, Lin M, Al-Holy MA, Cavinato AG, Rasco BA (2008) Detection of sublethal thermal injury in *Salmonella enterica* Serotype Typhimurium and *Listeria monocytogenes* using Fourier Transform Infrared (FT-IR) Spectroscopy (4000–600 cm $^{-1}$). J Food Sci 73(2):M54–M61
36. Toziou P-M, Barmpalexis P, Boukouvala P, Verghese S, Nikolakakis I (2018) Quantification of live *Lactobacillus acidophilus* in mixed populations of live and killed by application of attenuated reflection Fourier transform infrared spectroscopy combined with chemometrics. J Pharm Biomed Anal 154:16–22
37. Subramanian A, Ahn J, Balasubramaniam VM, Rodriguez-Saona L (2006) Determination of spore inactivation during thermal and pressure-assisted thermal processing using FT-IR spectroscopy. J Agric Food Chem 54(26):10300–10306
38. Goodacre R, Shann B, Gilbert RJ, Timmins EM, McGovern AC, Alsberg BK et al (2000) Detection of the dipicolinic acid biomarker in *Bacillus* spores using Curie-point pyrolysis mass spectrometry and Fourier transform infrared spectroscopy. Anal Chem 72(1):119–127
39. Alvarez-Ordóñez A, Mouwen DJM, López M, Prieto M (2011) Fourier transform infrared spectroscopy as a tool to characterize molecular composition and stress response in foodborne pathogenic bacteria. J Microbiol Methods 84(3):369–378
40. Hlaing MM, Wood BR, McNaughton D, Rood JI, Fox EM, Augustin MA (2018) Vibrational spectroscopy combined with transcriptomic analysis for investigation of bacterial responses towards acid stress. Appl Microbiol Biotechnol 102(1):333–343
41. Gurbanov R, Simsek Ozek N, Gozen AG, Severcan F (2015) Quick discrimination of heavy metal resistant bacterial populations using infrared spectroscopy coupled with chemometrics. Anal Chem 87(19):9653–9661
42. Brandes Ammann A, Brandl H (2011) Detection and differentiation of bacterial spores in a mineral matrix by Fourier transform infrared spectroscopy (FTIR) and chemometrical data treatment. BMC Biophys 4:14
43. Gupta MJ, Irudayaraj J, Debroy C (2004) Spectroscopic quantification of bacteria using artificial neural networks. J Food Prot 67(11):2550–2554
44. Pistorius AMA, DeGrip WJ, Egorova-Zachernyuk TA (2009) Monitoring of biomass composition from microbiological sources by means of FT-IR spectroscopy. Biotechnol Bioeng 103(1):123–129
45. Santos MI, Gerbino E, Tymczyszyn E, Gomez-Zavaglia A (2015) Applications of Infrared and Raman spectroscopies to probiotic investigation. Foods 4(3):283–305
46. Millan-Oropeza A, Rebois R, David M, Moussa F, Dazzi A, Bleton J et al (2017) Attenuated total reflection Fourier transform infrared (ATR FT-IR) for rapid determination of microbial cell lipid content: correlation with gas chromatography-mass spectrometry (GC-MS). Appl Spectrosc 71(10):2344–2352
47. Isak I, Patel M, Riddell M, West M, Bowers T, Wijeyekoon S et al (2016) Quantification of polyhydroxyalkanoates in mixed and pure cultures biomass by Fourier transform infrared spectroscopy: comparison of different approaches. Lett Appl Microbiol 63(2):139–146

48. Correa E, Sletta H, Ellis DI, Hoel S, Ertesvåg H, Ellingsen TE et al (2012) Rapid reagentless quantification of alginate biosynthesis in *Pseudomonas fluorescens* bacteria mutants using FT-IR spectroscopy coupled to multivariate partial least squares regression. Anal Bioanal Chem 403(9):2591–2599
49. Nicolaou N, Xu Y, Goodacre R (2011) Fourier transform infrared and Raman spectroscopies for the rapid detection, enumeration, and growth interaction of the bacteria *Staphylococcus aureus* and *Lactococcus lactis* ssp. cremoris in milk. Anal Chem 83(14):5681–5687
50. Schäwe R, Fetzer I, Tönniges A, Härtig C, Geyer W, Harms H et al (2011) Evaluation of FT-IR spectroscopy as a tool to quantify bacteria in binary mixed cultures. J Microbiol Methods 86(2):182–187
51. Arcos-Hernandez MV, Gurieff N, Pratt S, Magnusson P, Werker A, Vargas A et al (2010) Rapid quantification of intracellular PHA using infrared spectroscopy: an application in mixed cultures. J Biotechnol 150(3):372–379
52. Marcotte L, Kegelaer G, Sandt C, Barbeau J, Lafleur M (2007) An alternative infrared spectroscopy assay for the quantification of polysaccharides in bacterial samples. Anal Biochem 361(1):7–14
53. Oberreuter H, Mertens F, Seiler H, Scherer S (2000) Quantification of micro-organisms in binary mixed populations by Fourier transform infrared (FT-IR) spectroscopy. Lett Appl Microbiol 30(1):85–89
54. Lin SF, Schraft H, Griffiths MW (1998) Identification of *Bacillus cereus* by Fourier transform infrared spectroscopy (FTIR). J Food Prot 61(7):921–923
55. Grunert T, Monahan A, Lassnig C, Vogl C, Müller M, Ehling Schulz M (2014) Deciphering host genotype-specific impacts on the metabolic fingerprint of *Listeria monocytogenes* by FTIR spectroscopy. Boneca IG, editor. *PLoS One* 9(12):e115959
56. Grunert T, Jovanovic D, Sirisarn W, Johler S, Weidenmaier C, Ehling-Schulz M et al (2018) Analysis of *Staphylococcus aureus* wall teichoic acid glycoepitopes by Fourier transform infrared spectroscopy provides novel insights into the staphylococcal glycocode. Sci Rep 8(1).1889
57. Adamus-Bialek W, Lukasz L, Kubiak-Szeligowska AB, Wawszczak M, Kamińska E, Chrapek M (2017) A new look at the drug-resistance investigation of uropathogenic *E. coli* strains. *Molc Biol Rep* 44(1):191–202
58. Sharaha U, Rodriguez-Diaz E, Riesenberg K, Bigio IJ, Huleihel M, Salman A (2017) Using infrared spectroscopy and multivariate analysis to detect antibiotics' resistant *Escherichia coli* Bacteria. Anal Chem 89(17):8782–8790
59. Coates J (2006) Interpretation of infrared spectra, a practical approach. In: encyclopedia of analytical chemistry [internet]. Chichester, Wiley. Available from: http://doi.wiley.com/10.1002/9780470027318.a5606
60. Davis R, Mauer LJ (2010) Fourier tansform infrared (FT-IR) spectroscopy: a rapid tool for detection and analysis of foodborne pathogenic bacteria. In: Méndez-Vilas A (ed) Current research, technology and education topics in applied microbiology and microbial biotechnology [Internet], 2nd edn. Formatex Research Center, pp 1582–1594. Available from: https://www.researchgate.net/profile/Reeta_Davis/publication/257781807_Fourier_Transform_Infrared_FT-IR_Spectroscopy_A_Rapid_Tool_for_Detection_and_Analysis_of_Foodborne_Pathogenic_Bacteria/links/5614e8f108ae4ce3cc649412.pdf
61. Sousa C, Novais Â, Magalhães A, Lopes J, Peixe L (2013) Diverse high-risk B2 and D *Escherichia coli* clones depicted by Fourier Transform Infrared Spectroscopy. Sci Rep 3:3278
62. Baker MJ, Trevisan J, Bassan P, Bhargava R, Butler HJ, Dorling KM et al (2014) Using Fourier transform IR spectroscopy to analyze biological materials. Nat Protoc 9(8):1771–1791
63. Choo-Smith LP, Maquelin K, van Vreeswijk T, Bruining HA, Puppels GJ, Ngo Thi NA et al (2001) Investigating microbial (micro)colony heterogeneity by vibrational spectroscopy. Appl Environ Microbiol 67(4):1461–1469

64. Helm D, Labischinski H, Naumann D (1991) Elaboration of a procedure for identification of bacteria using Fourier-Transform IR spectral libraries: a stepwise correlation approach. J Microbiol Methods 14(2):127–142
65. Wenning M, Breitenwieser F, Konrad R, Huber I, Busch U, Scherer S (2014) Identification and differentiation of food-related bacteria: a comparison of FTIR spectroscopy and MALDI-TOF mass spectrometry. J Microbiol Methods 103:44–52
66. Lefier D, Hirst D, Holt C, Williams AG (1997) Effect of sampling procedure and strain variation in *Listeria monocytogenes* on the discrimination of species in the genus Listeria by Fourier transform infrared spectroscopy and canonical variates analysis. FEMS Microbiol Lett 147(1):45–50
67. Sykora L, Müller A (2017) ATR-FTIR microplate reader and micromachined ATR silicon crystals. In: 11th Workshop FT-IR spectroscopy in microbiological and medical diagnostic [Internet], Berlin. Available from: http://www.ftir-workshop.org/workshop/posters2017/Sykora.pdf
68. Winder CL, Goodacre R (2004) Comparison of diffuse-reflectance absorbance and attenuated total reflectance FT-IR for the discrimination of bacteria. Analyst 129(11):1118–1122
69. Lasch P (2012) Spectral pre-processing for biomedical vibrational spectroscopy and microspectroscopic imaging (2012) Chemometrics and intelligent laboratory systems. Chemom Intell Lab Syst 117:100–114
70. Savitzky A, Golay MJE (1964) Smoothing and differentiation of data by simplified least squares procedures. Anal Chem 36(8):1627–1639
71. Kuhm AE, Suter D, Felleisen R, Rau J (2009) Identification of *Yersinia enterocolitica* at the species and subspecies levels by fourier transform infrared spectroscopy. Appl Environ Microbiol 75(18):5809–5813
72. Romanolo KF, Gorski L, Wang S, Lauzon CR (2015) Rapid identification and classification of *Listeria* spp. and serotype assignment of *Listeria monocytogenes* using Fourier Transform-Infrared Spectroscopy and Artificial Neural Network Analysis. *PLoS One* 10(11):e0143425
73. Udelhoven T, Naumann D, Schmitt J (2000) Development of a hierarchical classification system with artificial neural networks and FT-IR spectra for the identification of bacteria. Appl Spectrosc 54(10):1471–1479
74. Baldauf NA, Rodriguez-Romo LA, Yousef AE, Rodriguez-Saona LE (2006) Differentiation of selected *Salmonella enterica* serovars by Fourier transform mid-infrared spectroscopy. Appl Spectrosc 60(6):592–598
75. Silva L, Rodrigues C, Lira A, Leão M, Mota M, Lopes P et al (2020) Fourier transform infrared (FT-IR) spectroscopy typing: a real-time analysis of an outbreak by carbapenem-resistant *Klebsiella pneumoniae*. Eur J Clin Microbiol Infect Dis 39(12):2471–2475.
76. Johler S, Tichaczek-Dischinger PS, Rau J, Sihto H-M, Lehner A, Adam M et al (2013) Outbreak of staphylococcal food poisoning due to SEA-producing *Staphylococcus aureus*. Foodborne Pathog Dis 10(9):777–781
77. Beattie SH, Holt C, Hirst D, Williams AG (1998) Discrimination among *Bacillus cereus*, *B. mycoides* and *B. thuringiensis* and some other species of the genus *Bacillus* by Fourier transform infrared spectroscopy. *FEMS Microbiol Lett* 164(1):201–206
78. Lücking G, Stoeckel M, Atamer Z, Hinrichs J, Ehling-Schulz M (2013) Characterization of aerobic spore-forming bacteria associated with industrial dairy processing environments and product spoilage. Int J Food Microbiol 166(2):270–279
79. Branquinho R, Sousa C, Osório H, Meirinhos-Soares L, Lopes J, Carriço JA et al (2014) *Bacillus invictae* sp. nov., isolated from a health product. Int J Syst Evol Microbiol 64(Pt 11):3867–3876
80. Mietke H, Beer W, Schleif J, Schabert G, Reissbrodt R (2010) Differentiation between probiotic and wild-type *Bacillus cereus* isolates by antibiotic susceptibility test and Fourier transform infrared spectroscopy (FT-IR). Int J Food Microbiol 140(1):57–60
81. Ehling-Schulz M, Svensson B, Guinebretiere M-H, Lindbäck T, Andersson M, Schulz A et al (2005) Emetic toxin formation of *Bacillus cereus* is restricted to a single evolutionary lineage of closely related strains. Microbiology (Reading, England) 151(Pt 1):183–197

82. Leoff C, Saile E, Sue D, Wilkins P, Quinn CP, Carlson RW et al (2008) Cell wall carbohydrate compositions of strains from the *Bacillus cereus* group of species correlate with phylogenetic relatedness. J Bacteriol 190(1):112–121
83. Orsi RH, Wiedmann M (2016) Characteristics and distribution of *Listeria* spp., including *Listeria* species newly described since 2009. Appl Microbiol Biotechnol 100(12):5273–5287
84. Holt C, Hirst D, Sutherland A, MacDonald F (1995) Discrimination of species in the genus *Listeria* by Fourier transform infrared spectroscopy and canonical variate analysis. Appl Environ Microbiol 61(1):377–378
85. Janbu AO, Møretrø T, Bertrand D, Kohler A (2008) FT-IR microspectroscopy: a promising method for the rapid identification of *Listeria* species. FEMS Microbiol Lett 278(2):164–170
86. Rebuffo CA, Schmitt J, Wenning M, von Stetten F, Scherer S (2006) Reliable and rapid identification of *Listeria monocytogenes* and *Listeria* species by artificial neural network-based Fourier transform infrared spectroscopy. Appl Environ Microbiol 72(2):994–1000
87. Stessl B, Fricker M, Fox E, Karpiskova R, Demnerova K, Jordan K et al (2014) Collaborative survey on the colonization of different types of cheese-processing facilities with *Listeria monocytogenes*. Foodborne Pathog Dis 11(1):8–14
88. Rebuffo-Scheer CA, Schmitt J, Scherer S (2007) Differentiation of *Listeria monocytogenes* serovars by using artificial neural network analysis of Fourier-transformed infrared spectra. Appl Environ Microbiol 73(3):1036–1040
89. Davis R, Mauer LJ (2011) Subtyping of Listeria monocytogenes at the haplotype level by Fourier transform infrared (FT-IR) spectroscopy and multivariate statistical analysis. Int J Food Microbiol 150(2–3):140–149
90. Nelson KE, Fouts DE, Mongodin EF, Ravel J, DeBoy RT, Kolonay JF et al (2004) Whole genome comparisons of serotype 4b and 1/2a strains of the food-borne pathogen *Listeria monocytogenes* reveal new insights into the core genome components of this species. Nucleic Acids Res 32(8):2386–2395
91. Oust A, Møretrø T, Naterstad K, Sockalingum GD, Adt I, Manfait M et al (2006) Fourier transform infrared and raman spectroscopy for characterization of *Listeria monocytogenes* strains. Appl Environ Microbiol 72(1):228–232
92. Amiali NM, Mulvey MR, Sedman J, Simor AE, Ismail AA (2007) Epidemiological typing of methicillin-resistant *Staphylococcus aureus* strains by Fourier transform infrared spectroscopy. J Microbiol Methods 69(1):146–153
93. Lamprell H, Mazerolles G, Kodjo A, Chamba JF, Noël Y, Beuvier E (2006) Discrimination of *Staphylococcus aureus* strains from different species of *Staphylococcus* using Fourier transform infrared (FTIR) spectroscopy. Int J Food Microbiol 108(1):125–129
94. Sandt C, Madoulet C, Kohler A, Allouch P, De Champs C, Manfait M et al (2006) FT-IR microspectroscopy for early identification of some clinically relevant pathogens. J Appl Microbiol 101(4):785–797
95. Johler S, Stephan R, Althaus D, Ehling-Schulz M, Grunert T (2016) High-resolution subtyping of *Staphylococcus aureus* strains by means of Fourier-transform infrared spectroscopy. Syst Appl Microbiol 39(3):189–194
96. De Angelis M, Gobbetti M (2016) Lactobacillus SPP.: General Characteristics ☆. In: Reference module in food science [Internet]. Elsevier. [cited 2018 Dec 21]. Available from: https://linkinghub.elsevier.com/retrieve/pii/B9780081005965008519
97. Curk MC, Peledan F, Hubert JC (1994) Fourier transform infrared (FTIR) spectroscopy for identifying *Lactobacillus* species. FEMS Microbiol Lett 123(3):241–248
98. Amiel C, Mariey L, Curk-Daubié M-C, Pichon P, Travert J (2000) Potentiality of Fourier Transform Infrared Spectroscopy (FTIR) for discrimination and identification of dairy lactic acid bacteria. Lait 80(4):445–459
99. Oust A, Møretrø T, Kirschner C, Narvhus JA, Kohler A (2004) FT-IR spectroscopy for identification of closely related lactobacilli. J Microbiol Methods 59(2):149–162

100. Bosch A, Golowczyc MA, Abraham AG, Garrote GL, De Antoni GL, Yantorno O (2006) Rapid discrimination of lactobacilli isolated from kefir grains by FT-IR spectroscopy. Int J Food Microbiol 111(3):280–287
101. Wenning M, Büchl NR, Scherer S (2010) Species and strain identification of lactic acid bacteria using FTIR spectroscopy and artificial neural networks. J Biophotonics 3(8–9):493–505
102. Kirschner C, Maquelin K, Pina P, Ngo Thi NA, Choo-Smith L-P, Sockalingum GD et al (2001) Classification and identification of enterococci: a comparative phenotypic, genotypic, and vibrational spectroscopic study. J Clin Microbiol 39(5):1763–1770
103. Goodacre R, Timmins EM, Rooney PJ, Rowland JJ, Kell DB. Rapid identification of *Streptococcus* and *Enterococcus* species using diffuse reflectance-absorbance Fourier transform infrared spectroscopy and artificial neural networks. FEMS Microbiol Lett 1996;140(2–3):233–9
104. Preisner O, Lopes JA, Guiomar R, Machado J, Menezes JC (2007) Fourier transform infrared (FT-IR) spectroscopy in bacteriology: towards a reference method for bacteria discrimination. Anal Bioanal Chem 387(5):1739–1748
105. Samelis J, Bleicher A, Delbès-Paus C, Kakouri A, Neuhaus K, Montel M-C (2011) FTIR-based polyphasic identification of lactic acid bacteria isolated from traditional Greek Graviera cheese. Food Microbiol 28(1):76–83
106. Whittaker P, Mossoba M, Al-Khaldi S, Fry F, Dunkel V, Tall B et al (2003) Identification of foodborne bacteria by infrared spectroscopy using cellular fatty acid methyl esters. J Microbiol Methods 55(3):709–716
107. Beutin L, Wang Q, Naumann D, Han W, Krause G, Leomil L et al (2007) Relationship between O-antigen subtypes, bacterial surface structures and O-antigen gene clusters in *Escherichia coli* O123 strains carrying genes for Shiga toxins and intimin. J Med Microbiol 56(Pt 2):177–184
108. Al-Qadiri HM, Lin M, Cavinato AG, Rasco BA (2006) Fourier transform infrared spectroscopy, detection and identification of *Escherichia coli* O157:H7 and Alicyclobacillus strains in apple juice. Int J Food Microbiol 111(1):73–80
109. Davis R, Paoli G, Mauer LJJ (2012) Evaluation of Fourier transform infrared (FT-IR) spectroscopy and chemometrics as a rapid approach for sub-typing *Escherichia coli* O157:H7 isolates. Food Microbiol 31(2):181–190
110. Al-Holy MA, Lin M, Cavinato AG, Rasco BA (2006) The use of Fourier transform infrared spectroscopy to differentiate *Escherichia coli* O157:H7 from other bacteria inoculated into apple juice. Food Microbiol 23(2):162–168
111. Al-Qadiri HM, Al-Holy MA, Lin M, Alami NI, Cavinato AG, Rasco BA (2006) Rapid detection and identification of *Pseudomonas aeruginosa* and *Escherichia coli* as pure and mixed cultures in bottled drinking water using fourier transform infrared spectroscopy and multivariate analysis. J Agric Food Chem 54(16):5749–5754
112. Wang J, Kim KH, Kim S, Kim YS, Li QX, Jun S (2010) Simple quantitative analysis of *Escherichia coli* K-12 internalized in baby spinach using Fourier Transform Infrared spectroscopy. Int J Food Microbiol 144(1):147–151
113. Dawson SE, Gibreel T, Nicolaou N, AlRabiah H, Xu Y, Goodacre R et al (2014) Implementation of Fourier transform infrared spectroscopy for the rapid typing of uropathogenic *Escherichia coli*. Eur J Clin Microbiol Infect Dis 33(6):983–988
114. Baldauf NA, Rodriguez-Romo LA, Männig A, Yousef AE, Rodriguez-Saona LE (2007) Effect of selective growth media on the differentiation of *Salmonella enterica* serovars by Fourier-Transform Mid-Infrared Spectroscopy. J Microbiol Methods 68(1):106–114
115. Männig A, Baldauf NA, Rodriguez-Romo LA, Yousef AE, Rodríguez-Saona LE (2008) Differentiation of *Salmonella enterica* serovars and strains in cultures and food using infrared spectroscopic and microspectroscopic techniques combined with soft independent modeling of class analogy pattern recognition analysis. J Food Prot 71(11):2249–2256
116. Preisner OE, Menezes JC, Guiomar R, Machado J, Lopes JA (2012) Discrimination of *Salmonella enterica* serotypes by Fourier transform infrared spectroscopy. Food Res Int 45(2):1058–1064

117. Koluman A, Celik G, Unlu T (2012) *Salmonella* identification from foods in eight hours: a prototype study with *Salmonella* Typhimurium. Iran J Microbiol 4(1):15–24
118. Sundaram J, Park B, Hinton A, Yoon SC, Windham WR, Lawrence KC (2012) Classification and structural analysis of live and dead *Salmonella* cells using Fourier transform infrared spectroscopy and principal component analysis. J Agric Food Chem 60(4):991–1004
119. Wortberg F, Nardy E, Contzen M, Rau J (2012) Identification of *Yersinia ruckeri* from diseased salmonid fish by Fourier transform infrared spectroscopy: FT-IR for the identification of Yersinia ruckeri. J Fish Dis 35(1):1–10
120. Stamm I, Hailer M, Depner B, Kopp PA, Rau J (2013) *Yersinia enterocolitica* in diagnostic fecal samples from European dogs and cats: identification by fourier transform infrared spectroscopy and matrix-assisted laser desorption ionization-time of flight mass spectrometry. J Clin Microbiol 51(3):887–893
121. Mouwen DJM, Capita R, Alonso-Calleja C, Prieto-Gómez J, Prieto M (2006) Artificial neural network based identification of *Campylobacter* species by Fourier transform infrared spectroscopy. J Microbiol Methods 67(1):131–140
122. Muhamadali H, Weaver D, Subaihi A, AlMasoud N, Trivedi DK, Ellis DI et al (2016) Chicken, beams, and *Campylobacter*: rapid differentiation of foodborne bacteria via vibrational spectroscopy and MALDI-mass spectrometry. Analyst 141(1):111–122
123. Mouwen DJM, Weijtens MJBM, Capita R, Alonso-Calleja C, Prieto M (2005) Discrimination of enterobacterial repetitive intergenic consensus PCR types of *Campylobacter coli* and *Campylobacter jejuni* by Fourier transform infrared spectroscopy. Appl Environ Microbiol 71(8):4318–4324
124. Horbach I, Naumann D, Fehrenbach FJ (1988) Simultaneous infections with different serogroups of *Legionella pneumophila* investigated by routine methods and Fourier transform infrared spectroscopy. J Clin Microbiol 26(6):1106–1110
125. Li Z, Chen S, Xu C, Ju L, Li F (2018) Rapid subtyping of pathogenic and nonpathogenic *Vibrio parahaemolyticus* by fourier transform infrared spectroscopy with chemometric analysis. J Microbiol Methods 155:70–77
126. Rebuffo-Scheer CA, Kirschner C, Staemmler M, Naumann D (2007) Rapid species and strain differentiation of non-tubercoulous mycobacteria by Fourier-Transform Infrared microspectroscopy. J Microbiol Methods 68(2):282–290
127. Winder CL, Gordon SV, Dale J, Hewinson RG, Goodacre R (2006) Metabolic fingerprints of *Mycobacterium bovis* cluster with molecular type: implications for genotype-phenotype links. Microbiology (Reading, England) 152(Pt 9):2757–2765

Chapter 12
Omics for Forensic and Post-Mortem Microbiology

Amparo Fernández-Rodríguez, Fernando González-Candelas, and Natasha Arora

12.1 Introduction

Forensic microbiology, also known as microbial forensics, is a relatively new scientific field resulting from the interaction of several disciplines (Fig. 12.1), which converge in a set of specific problems that are approached with different methodologies and conceptual backgrounds. The term "microbial forensics" was introduced about 15 years ago to denote the investigation of criminal acts in which spores of the bacterium *Bacillus anthracis* were used to kill several people who received letters impregnated with the deadly anthrax agent [1]. Later, the two synonymous terms have been extended to include the analysis of microorganisms in any forensic application such as establishing the cause of death (COD), the study of pathogen transmission between donor-recipient pairs, the source(s) of outbreaks, the identification of body fluids or the identification of individuals through the microbial composition of their remains, to name just a few of the current applications. In some of these applications, the ultimate goal of forensic microbiology is to determine who or what is involved in a criminal event and to reconstruct the criminal activities.

Although necessarily rooted in human (clinical) microbiology, forensic microbiology is closely related to other disciplines (Fig. 12.1). The field has expanded as new technologies, especially those derived from high-throughput sequencing (HTS) methods, have allowed the culture-independent investigation of the microbial composition of fluids, surfaces, soil and almost any type of sample. The tremendous level of resolution that these new technologies provide has led to an unprecedented opportunity to characterise the composition of complex samples. However, this

A. Fernández-Rodríguez (✉) · F. González-Candelas · N. Arora
National Institute of Toxicology and Forensic Sciences,
José Echegaray 4, Las Rozas de Madrid 28232, España
e-mail: amparo.fernandezrodriguez@justicia.es

J. Moran-Gilad, Y. Yagel (eds.), *Application and Integration of Omics-powered Diagnostics in Clinical and Public Health Microbiology*,
https://doi.org/10.1007/978-3-030-62155-1_12

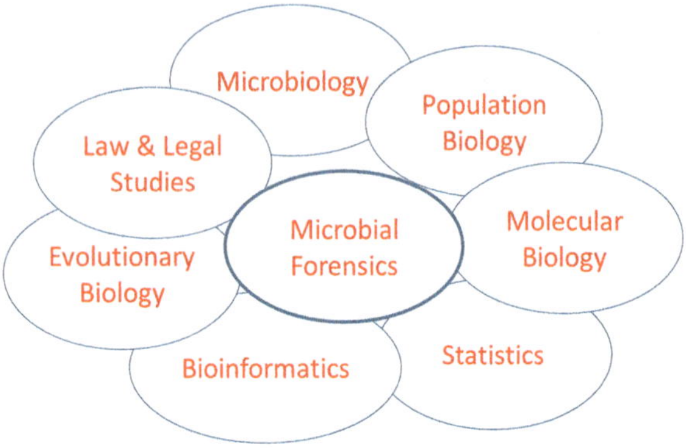

Fig. 12.1 The interdisciplinary nature of microbial forensics is reflected in the different scientific and social disciplines involved in this field

same resolution is the basis for uncertainties and further development and validation is needed to harness these new methods. These challenges appear in many applications of "omics" methodologies but they deserve special consideration when these results are applied in a forensic context.

Forensic microbiology is a diverse field because the problems it deals with encompass a wide range of organisms, methodologies, techniques and questions. It has wide applications and can be aimed both at improving the prevention of infectious diseases as well as improving the fight against crime. In the following sections, we describe some of the major topics in which forensic microbiology utilises "omics" data.

12.2 Omics, Databases and Detection of Biological Agents

HTS studies aimed at the detection of pathogens involved in criminal acts, including bioterrorism events, are one of the prime investigation lines in microbial forensics. As outlined in the introduction, one of the goals of forensic microbiology is the identification of the person responsible of a criminal act [2]. In this process, the first step is the individualisation of the microorganism used to cause harm, that is, the weapon. This individualisation is done through the comparison of strains, which requires reference collections representing the maximum possible genetic diversity, spanning geographical and seasonal variability; accompanying information (metadata) on the strains should also be included [2, 3]. Among the different techniques in use to identify these biothreats, whole-genome sequencing (WGS) is the ultimate strategy to compare strains [4]. This is a very robust and powerful method, regularly applied in clinical microbiology [5], allowing downstream analyses to both discriminate between closely related strains and to reconstruct accurate phylogenies. As WGS is increasingly used, more and more complete genome sequences are being deposited into databases. It is important to not only include data from strains from repository laboratories but also the complete information about the isolates

[3]. Building these databases is an ongoing and cumbersome process, and standardisation of computational pipelines is a must [4].

Bioinformatics analyses have to consider all the complexity in the population structure of the investigated agent, and as this requires considerable research efforts, it should be a matter of investment [6]. Moreover, the information recovered from natural cases involving potential biological agents should be coupled with experimental assays. In light of this need, the application of WGS to biothreats requires international efforts. Apart from the drawbacks derived from not publishing specific WGS due to private or commercial interests, continuous support to the maintenance of these databases is required. Another aspect to consider is that a database, in order to be acceptable in the forensic arena, has to comply with internationally accepted standard criteria. Additional requirements of databases aimed at the forensic identification of biological agents are limited access to specifically authorised staff and compliance with the specific regulations for international data exchange.

12.3 Analysis of Pathogen Transmission and Outbreaks

Another core application in Forensic microbiology is the analysis of transmission events of infectious microorganisms. Most of the cases involve the transmission of human immunodeficiency vi-rus (HIV) or hepatitis C virus (HCV), usually between regular or occasional sexual partners but not restricted to these. These problems are investigated through comparing the nucleotide sequences derived from viruses sampled from the persons involved along with those from population controls not related to the transmission event under investigation. Each of these cases is a complex problem and, although the analysis of sequences is a well-established field in evolutionary biology, the application of the same methods to forensic cases is not as straightforward as it might appear at first sight [7]. There are technical difficulties in the experimental procedures before a sequence is obtained and these occasionally lead to interpretation issues. In addition, viruses are constantly evolving and two persons receiving an infectious fluid from a common source will develop somewhat different viral populations. Sequence identity between source and recipient is hardly ever observed and the differences observed will vary depending on, among other factors, the technique used for sequencing, the time elapsed since the transmission event occurred and the samples were obtained, or the region in the viral genome targeted for analysis [8].

The fast and constant evolution of RNA viruses rapidly creates heterogeneous populations after infection of a new host. These populations, sometimes denoted as "quasi-species" [9], usually become more complex through time, but additional processes such as selection in response to the host's immune system or antiviral treatment, compartmentalisation and genetic drift may transiently reduce population complexity. Nevertheless, because of the particular changes experienced by the virus infecting each host, we should not expect that the sequences derived from each of the corresponding populations be identical. In consequence, the forensic analysis of transmissions must incorporate the analysis and comparison of intrinsically heterogeneous populations. High-throughput sequencing technologies represent the best alternative currently available for the characterisation of viral populations but its use in forensics has some limitations worth considering.

HTS technologies were introduced in 2005 when Roche released the first system based on pyrosequencing, the 454 Genome Sequencer. Further developments by Illumina and Applied BioSystems led to second-generation high throughput sequencing in which a very large number of very short sequencing reads were generated and analysed in parallel, generating huge volumes of sequence information. More recently, different companies such as Oxford Nanopore and Pacific BioSciences have introduced third-generation sequencers such as MinION and PacBio, which are capable of retrieving large (up to hundreds of kilobases) sequences from single molecules. All these technologies offer several advantages over traditional, Sanger-based sequencing such as: (i) there is no need to clone DNA fragments, (ii) they can parallelise thousands, even millions, of reactions, and (iii) there is no need for electrophoresis separation to detect the output of the sequencing reactions [10]. But these clear advantages come with a price. For instance, second-generation technologies generate reads that are too short to unambiguously represent the structure of even relatively simple bacterial genomes and researchers have to use sophisticated, and as of yet non-standardized bioinformatics tools to analyse their huge outputs. A more relevant concern for forensic microbiology is the relatively high error-rates intrinsic to these technologies because they are in the same range as the natural mutation rates of RNA viruses, thus complicating the distinction between true genetic variants and technical errors. Nevertheless, many forensics applications are currently using HTS as the technology of choice for data acquisition [10].

In the case of the analysis of transmission events, the preferred method consists of comparing the nucleotide sequence of a portion of the viral or bacterial genome obtained from samples of the potential source(s) and recipient(s) with those of a set of reference samples. Next, molecular phylogenetic analysis is performed to establish the existence of a most recent common ancestor between the sequences derived from the source and the recipient than that between the source/recipient and the reference samples [11]. Even if complete genome sequences are used for these analyses, Sanger-sequencing offers excellent results in terms of costs, accuracy, reproducibility and speed for some analyses, mainly at a preliminary stage of a forensic study. However, unless limiting dilution or sequencing of cloned PCR-products are incorporated, with the concomitant increases in costs and time, Sanger-sequencing will not provide information on the underlying variability of the viral popula-tions obtained from different samples (as also discussed in Chap. 6). An immediate solution to this problem is to apply NGS to those samples. This can be done in two different ways. Firstly, the intrapatient variability of the viral population is estimated through targeted PCR amplification of a selected genetic region and subsequent HTS of the products. This approach was applied by Campo et al. [12] to analyse the hypervariable region 1 (HVR1) of the hepatitis C virus from several hundreds of samples of diverse HCV genotypes. The results of pyrosequencing PCR-products were analysed phylogenetically to establish which sets of samples shared common ancestors and these were then matched with previous information on HCV outbreaks. The HTS approach provided excellent resolution in establishing which samples belonged to each outbreak. In addition, in the 8 outbreaks, the viral population from the source showed higher levels of genetic variability (on average, 6.2 times more) than any of the corresponding incident cases, thus allowing the inference of the directionality of the transmission. A similar approach with different genome regions of HCV was used in other studies employing either 454-pyrosequencing [13, 14] or Illumina MiSeq [15] with NS5B amplicons. Using a fraction of the genome may not provide enough resolution to analyse transmission events. Consequently, Gonçalves et al. reported an improved resolution of HCV transmission events by analysing both HVR1 and a fragment of NS5A, or by obtaining complete genome sequences [13]. However, this approach is not straightforward with first or second-generation sequencing methodologies, because read-lengths are much shorter than the length of the viral genomes (around 10 kb in the case of HIV and HCV). The output consists of many thousands of small fragments which cannot be assembled to confirm individual genome sequences. Representative sequences can be obtained by the application of consensus methods, which bring us back to Sanger sequencing. Presently, actual applications of this methodology in forensic analyses of transmissions have not been reported yet, but this approach has been successfully applied in the on-the-field molecular epidemiology analysis of recent Ebola [16] and Zika [17] viral outbreaks.

Phylogenetic analysis of sequences can be applied not only to establish transmission chains or clusters, occasionally including their directionality, but also to infer approximate times of transmission events [8]. Although genome sequences accumulate changes at a roughly constant rate – which constitutes the well-known molecular clock hypothesis [18], there is a very large variance to this rate, which prevents its direct application to infer, for instance, times since infection or transmission episodes. Apart from the intrinsic variability resulting from the stochastic nature of mutational processes, additional factors, especially those related to natural selection and genetic drift, contribute to local or regional deviations of the clock rate even among relatively close sequences. However, these deviations can be accommodated and analysed by using several more advanced molecular clock models [19], most of which are implemented in software packages such as BEAST [20, 21].

The calibration of molecular clocks can benefit from using asynchronous sequences, that is, sequences derived from samples taken at different times, not simultaneously, with differences relevant to the time-scale and evolutionary rate of the problem under consideration [22, 23]. When this data is complemented with additional calibration points, such as known dates of events related to internal nodes in a phylogeny, estimates of other time points of interest in a phylogenetic reconstruction can be very accurate, even under varying evolutionary rates (relaxed clock models). For example, González-Candelas et al. used this approach to establish likely dates of infection of more than 250 patients in the context of a large outbreak of HCV from a single donor, using only sequence information, sampling dates and known dates of infection of 24 patients [8]. The estimates and their confidence intervals were highly congruent with those derived independently from hospital records and other documentation.

12.4 Omics to Determine Infection as the Cause of Death (COD)

Determining the COD is an important aim in forensic medicine. Forensic pathologists have to deal with many unascertained deaths, some of them of criminal origin, and some others which are sudden or unexpected. Their protocols include ancillary analyses that help them understand autopsy findings. One of these types of analyses is post-mortem microbiology (PMM), which is used to detect pathogens responsible for an infectious COD [24]. In the scenario of sudden unexpected death (SD), forensic microbiology can be used as a synonym of PMM. From now on, these two terms will be equally used in this epigraph. Both bacteria and viruses have been described as aetiological agents of SD. In fact, in infancy and childhood, an infection can be either a COD or a predisposing factor to death and, therefore, microbiology should be included in autopsy analytical protocols [25]. In young adults, sudden cardiac death can be due to viral myocarditis [26], and in adults of any age, fulminant infections can lead to a fatal outcome [24]. Many of these situations have to be investigated by forensic pathologists, as there are no clinical symptoms and some of them occur at home and are unwitnessed. In addition, some deaths occurring in the hospital setting can also be the subject of the judicial authorities' investigation as the management and fatal evolution of infection can be related to a medico-legal claim and a suit, and consequently, PMM analyses should be performed [27].

There are many syndromes capable of producing an SD. Among them, meningitis, meningoencephalitis, pneumonia, septic shock, gastroenteritis, abscesses, pyelonephritis and myocarditis. These can be caused by a wide variety of pathogens with indistinguishable clinical patterns and identical autopsy findings. Therefore, their diagnoses often need the combination of different molecular techniques or a large multiplex panel capable of detecting many pathogens. As in clinical microbiology, in forensic microbiology, sometimes despite performing a wide range of techniques, the aetiology of infection may remain unknown.

Although PMM can be affected by post-mortem translocation and sampling contamination, following sampling guidelines and strict aseptic measures during autopsy has shown its usefulness to minimise these effects [24]. In addition, the comprehensive combination of molecular and traditional methods – culture, antigenic tests, and serology – has made forensic microbiology an established tool in forensic pathology [28].

Recently, various molecular techniques have become an essential part of the routine work in forensic microbiology. The most frequently used until now have been the different variants of PCR, particularly the pathogen-targeted real-time PCR assays, aimed at detecting specific infectious agents [29], and, more recently, the commercial multiplex (syndromic) molecular panels [30].

In the last 10 years, within the clinical microbiology setting, Sanger sequencing of the 16S rRNA gene, which is ubiquitous across prokaryotes, has become a standard for classification and identification due to its universal distribution among bacteria [31]. The use of highly conserved primer binding sites to amplify one or more of the nine hypervariable regions (V1-V9) within the 16S rRNA gene has proved useful for the identification of species, as it can provide species-specific signature sequences and pan-bacterial PCR applied directly on samples. Although the analysis of this region by HTS has recently arrived at the microbiology lab, to date its use is mainly restricted to specific diagnostic and research applications. Current options involving HTS to identify bacteria are mostly related to the use of the 16S region alone or in combination with the 18S region and the ITS (internal transcribed spacer) 1–2 region for fungal detection. Different commercial kits offer such possibilities (IonTorrent, CD Genomics, among others). Whether the spread and popularisation of HTS in infectious disease diagnosis will displace or enhance gold standard methodologies is not yet known. Additionally, targeted HTS sequencing of the 16S-ITS-23S rRNA region has been proposed for the culture-independent identification of bacteria [32] using the MiSeq (Illumina) technology. This technique has shown its utility in clinical samples including blood cultures and urine samples among others. Although its reference database and complementary software are still under development, this method has the potential to increase the diagnostic yield to detect bacterial pathogenic species, in comparison with current techniques.

In forensic microbiology, it would also be desirable to get a pan-bacterial assay capable of detecting and correctly identifying bacteria responsible for fatal infections. However, the use of some broad-range PCR assays, such as the 16S PCR followed by Sanger sequencing (broad-range PCR), can be limited by post-mortem translocation and sampling contamination. In PMM, when the COD is due to infection, apart from the responsible pathogen, other contaminating micro-organisms present in the sample can also be isolated [33]. For this reason, when broad-range PCR is used to detect a pathogen in a post-mortem sample, an uninterpretable mixture of sequences is a frequent event; this situation can also occur under the scenario of a plausible lethal polymicrobial infection process such as peritonitis. Therefore, its use should be restricted to those cases with negative culture despite the suspicion of infection. Consequently, alternative strategies are needed in forensic microbiology to overcome these hindrances. As in clinical microbiology, HTS has been proposed to surmount them in forensic microbiology.

The application of HTS to a pan-bacterial system detection seems to be one of the most useful approaches to identify fatal bacterial infections. With the aim of detecting bacteria as responsible for COD, Cho et al. [34] have recently compared the resolution of the 16S region by Sanger sequencing (using the MicroSeq 500 16S rDNA microbial identification system – MSId) and NGS (with the MiSeq Illumina sequencing platform) in post-mortem specimens [34]. The MSId is tailored for the identification of bacteria isolated in cultures, and the MiSeq is a culture-independent metagenomic analysis system. The authors found the MiSeq to be more efficient in terms of time and costs. In addition, the larger library of the MiSeq and the simpler identification of difficult to culture bacteria allowed more accurate identification of bacterial species.

Above all, the ultimate aim of the application of meta-genomics in forensic microbiology is to provide a universal system for pathogen detection capable of identifying a diversity of micro-organisms. Different approaches have recently been designed for universal pathogen discovery. Among them, the work of Schlaberg and colleagues [35] is very promising, as they are aimed at universal pathogen detection in different settings. With this aim, they developed a fast, user-friendly, and interactive metagenomic sequence analysis tool (Taxonomer) able to classify sequencing reads of human, viral, bacterial, and fungal origin using probabilistic assignment algorithms with nucleotide and protein reference sequences [36]. The researchers validated shotgun metagenomic sequencing of RNA (RNA-seq) and compared it with pan-viral group (PVG) PCR for the detection of respiratory pathogens using specimens from children with community-acquired pneumonia and asymptomatic controls in whom viral or atypical bacterial pathogens had been detected in a previous study. The sequences obtained were analysed by Taxonomer. RNA-seq was shown to be unbiased, as demonstrated by the detection of divergent enteroviruses not identified by PVG PCR. Furthermore, Taxonomer enabled the identification of other pathogens different from viruses such as *Mycoplasma pneumoniae* and *Chlamydia trachomatis* [37].

In order to transfer these achievements to the forensic field, a pilot study in a mouse model of viral pneumonia was performed to assess the ability of RNA-seq to detect viral RNA and the stability of host mRNA transcription profiles during a 48-hour post-mortem interval [37, 38]. Their results demonstrated that RNA-seq may be used for direct pathogen detection and host response profiling. To take a further step, they have used RNA-seq to detect pathogens in post-mortem specimens from neonates with diagnosis of pneumonia and in infants without evidence of pulmonary disease. The whole process involved total RNA extraction from formalin-fixed paraffin embedded (FFPE) tissues, library preparation (with the KAPA Stranded RNA-Seq Kit with RiboErase, Roche), sequencing on a NextSeq 500 instrument (Illumina), sequencing analysis using Taxonomer [36] and alignment of confirmed results with curated reference sequences utilising Geneious (v8.1,Biomatters). The authors were able to detect pathogens in 6 of 13 (46.2%) post-mortem lung tissues from children with histopathologic diagnosis of pneumonia. These tissues had been stored for up to 15 years under routine conditions. Therefore, these results may enable comprehensive panmicrobial detection in archival samples useful for pathogen discovery in the context of SD. As an application of this technique, the authors detected gestational psittacosis as a cause of neonatal death in FFPE lung [39].

In conclusion, the above-mentioned studies show that, theoretically, implementation of NGS techniques and metagenomics can help find the pathogen considered responsible for CODs in previously unsolved cases and can thus contribute to a deeper comprehension of the post-mortem events.

12.5 Post-Mortem Interval Estimation by the Use of Thanatomicrobiome

A research area receiving increasing interest for forensic applications is the study of the microbiomes and their temporal changes in dead human bodies. Usually, they are grouped in the thanatomi crobiome, which corresponds to the microbiome of internal organs of cadavers, and the epinecrotic communities, the microbiome on surfaces of decaying remains [40].

Ecological succession of organisms in decomposing corpses follows some general patterns that have been used to estimate time since death under different circumstances. Probably, the best-known examples and common usages of this approach are found in forensic entomology [41, 42]. The composition and developmental stage of the fauna usually found in carcasses and corpses has been used to establish time since death for decades, with applications to calculate the minimum post-mortem interval (PMI) ranging from a few hours up to several months after death [43].

A similar approach, using the temporal changes in microbial communities associated with decomposing corpses, has been proposed [43–45]. Although the role of microbes on cadaver decomposition has been known for a long time [46], the advent of omics technologies has allowed the characterisation of complex microbiota and, as a result, several research groups have applied these methods to analyse the microbiome of decomposing corpses and its changes over time.

The initial analyses used swine carcasses as models for decomposing human cadavers [43, 44, 47]. Metcalf et al. used short (Illumina HiSeq) and long reads (Pacific Biosciences) of 16S rDNA and 18S rDNA to determine the taxonomy of bacteria and eukaryotes during 48 days after death of the experimental animals [45]. They found consistent shifts in the composition of bacterial and eukaryotic communities coincident with known stages of decomposition, thus suggesting that they could be used to estimate PMIs. For instance, in this experiment changes in the skin of the head were used to estimate PMI with a precision of 3.30 ± 2.52 days. Other studies analysed decomposition of 3 swine corpses in a temperate forest habitat, thus reflecting more "natural" conditions [43, 44]. They used pyrosequencing (454 Roche FLX) of PCR-amplified fragments targeting the V1-V3 region of the 16S rDNA and observed a significant decline in taxon richness with the time of decomposition. In addition, the specific composition of the bacterial community also correlated with the progression of decomposition, thus allowing the estimation of PMI.

Similar analyses have been extended to other mammals, including human corpses on different soil substrates [45]. The authors found that the bacterial and fungal communities, primarily driving decomposition were mainly derived from soil and that the main decomposers were ubiquitous and were of low abundance. Perhaps surprisingly, soil type (desert, grassland, or forest) was not a dominant factor in the development of the decomposing community. Nevertheless, the authors concluded that decomposer microbial communities could serve as forms of evidence for the time elapsed since death (PMI) and the location of this event.

These results clearly point to a future in which forensic evidence based on the analysis of microbiomes will be acceptable in courtrooms [48]. However, to our knowledge, no such cases have been accepted and reported yet.

12.6 Human Body Fluid Identification Using Microbial DNA

Determining the presence of biological traces and identifying their bodily origin is a critical task in forensic investigation, enabling the reconstruction of the events of a crime. For example, identifying semen or vaginal fluid can aid in determining the specific actions in the case of sexual assault. Following this identification of bodily origin, a stain can be further analysed with DNA markers to identify the individual from whom they originate [49, 50].

Often the first body fluid or tissue tests carried out at the crime scene are presumptive, helping to select samples that may then be subjected to confirmatory tests in the laboratory. However, both presumptive as well as confirmatory tests, which may be chemical, enzymatic, immunological, spectroscopic or microscopic, among others, may present limitations in terms of sensitivity and specificity. In order to overcome these, extensive research is being conducted for the application of human tissue-specific markers such as methylation, mRNA or microRNA markers [50–52].

Other than human markers, another promising avenue is that offered by microbial DNA. The human body harbours trillions of microbes that, in healthy states, form finely balanced ecosystems serving numerous functions such as in immunity and digestion [53]. Even fluids that were traditionally considered sterile, such as breast milk, have been found to contain rich microbial communities [54]. Interest in what constitutes a healthy microbiome, and in distinguishing healthy versus diseased states, has spurred numerous studies that have generated a wealth of metagenomic sequencing data, particularly those of the NIH-funded Human Microbiome Project [55, 56]. These sequencing studies have shown that microbial communities are tissue-specific, thus ushering in potential new markers for the identification of bodily source or habitat. Bacterial community profiling is typically carried out by sequencing variable regions of the 16S rRNA gene that, as mentioned earlier, is ubiqui-tous across the domains of bacteria and archaea. The read data obtained by sequencing are then clustered into operational taxonomic units (OTUs), and the OTU composition of samples is then compared. In their benchmark study, Costello et al. [57] examined the V2 region of the 16S rRNA gene of bacteria in the gut, oral cavity, nostril, hair, and various skin sites from 8 individuals. Their principal coordinates analysis (PCoA) of the communities at each site showed that samples grouped firstly by body site. Within a body site, samples are grouped according to individual. More recently, numerous other studies with larger sample sizes and targeting various body sites have confirmed this site-specificity of the microbiome [55–58]. The differentiation of microbial communities across body sites also appears to start early in the life of the neonate [59, 60].

The site-specificity of microbial communities renders them potentially valuable for forensic body fluid/tissue identification. Microbial markers offer several advantages compared to human DNA or RNA markers: for example, bacterial cells are numerous, out-numbering human cells at some body sites, and they are generally more robust than human cells. Thus, bacterial DNA may be present in high copy numbers, even when human DNA is minimal or no longer found [61].

To date, several studies have investigated the utility of bacterial 16S rRNA gene sequencing for body fluid/tissue identification. Most of these have focused on the selection of a limited set of markers for inclusion in multiplex assays or microarrays, with presence/absence patterns being used to establish the bodily origin of samples of vaginal, oral and faecal samples [62–65]. While these studies showed promising results, the selected bacterial markers were tested on small datasets, and occasionally produced false positives or false negatives. Thus, relying on very few or a limited set of bacterial markers alone may have important limitations. For example, utilising the presence/absence patterns of only very specific *Lactobacillus* taxa may be suboptimal for the identification of vaginal fluid, as discussed by Benschop et al. [64] in their study. Detailed investigation of vaginal microbiota has revealed the existence of at least six microbiome profiles or "community state types" (CSTs). While four of these are characterized by the dominance of one specific *Lactobacillus* subspecies (CST I, CST II, CST III and CST V), two have a relatively low abundance of *Lactobacillus* spp. and comprise diverse anaerobic bacteria [66, 67]. A non-targeted approach to detect a larger number of taxa with HTS may enable tapping into the diversity of microbial profiles across individuals, and therefore, more accurate body fluid/tissue prediction. Such an approach using 16S rRNA gene sequencing was recently used by Hanssen et al. [68] to detect saliva deposited on skin, with a correct classification of 94% of the samples examined.

Applying HTS to detect all bacterial markers possible at a given sequencing depth comes with certain challenges within the forensic setting, as we are not dealing with a restricted outcome in terms of the number of taxa to be analysed. Nonetheless, it provides a wealth of information, not only on the presence/absence of a larger number of taxa, but also quantitative values on the abundance of each taxon. This abundance information is valuable for the classification of samples and was also used by Hanssen et al. [68] in their study. The methods employed for classification based on microbiome HTS data have received some attention in recent years, with exciting work conducted with advanced computational approaches such as machine learning algorithms [69, 70]. However, most of these microbiome-based classification studies have been conducted using data from a single study. But, as more 16S rRNA sequence data becomes publicly available, we have the opportunity to pool together all this data generated by different laboratories in order to train the classifiers. It is unclear whether such data pooling will lead to improved accuracy, particularly given the differences in laboratory protocols across studies. Furthermore, for this classification to be useful in the forensic setting, the output should be given within a probabilistic framework, with a classification score and confidence in this score. A similar challenge was faced by molecular phylogeneticists in an investigation of a suspected case of intentional infection of hepatitis C. In their analyses,

González-Candelas and colleagues [8] used the genetic data from the viral strains of the potential victims to reconstruct phylogenetic trees, examining the likelihood of these under different hypotheses. They were thus able to obtain likelihood ratios that were used as evidence in the trial [8]. Similarly, the development of a probabilistic framework for the 16S rRNA sequence data is critical to enable microbiome-based classification as evidence in the forensic setting.

Despite the promising outlook so far, numerous questions remain to be answered before microbial community sequencing can be used in forensic laboratories. For example, while human microbiome studies have explored the stability of microbial communities across individuals and across time for a given body site [58, 71], the stability of the microbial signature for samples exposed outside the human habitat has not been explored in detail. Biological traces collected for forensic purposes may be found outside the human body, exposed to indoor or outdoor conditions, and on different substrates including textiles, a variety of surfaces, and soil among others. They may also be exposed for a period of time before a sample is collected by investigation teams. Once collected, the samples may also remain in storage for a given length of time before being handed over to the forensic geneticists. In such cases, it is unclear to what extent and for how long the microbial communities at a body site continue to show the characteristic composition of the body site of origin. That is, there may be changes in the taxonomic composition (either presence/absence or relative abundance of taxa), and it remains to be seen whether the changes still permit a reliable microbial signature for body tissue/fluid identification.

As progress is made in the investigation of this exciting new tool, we will need to dive deeper into the exploration of its validity and applicability in the forensic lab. It will be critical to establish a standardised laboratory protocol and bioinformatics workflow, selecting the specific software and algorithms to be applied. It would also be particularly interesting to explore the integration of microbial-based approaches with others, such as mRNA-based approaches, and the synergistic performance of these together.

12.7 The Human Microbiome to Identify Individuals

Microbial communities show a degree of specificity to the individual that makes them potentially useful for the forensic identification of individuals. However, there are several questions that need to be addressed in depth before we can bring these tools into the forensic arena. First, how unique is the microbial composition of an individual, can we really identify an individual only by examining microbial signatures? And how does individual identifiability differ across body sites? At what level of resolution, that is, at what taxonomic level, do we need to look at? Second, how stable are microbiomes within human body sites, or in other words, to what extent does the microbiome retain taxa useful for individual identification over time and space? Third, how stable is the microbial signature for individual identification when the microbial communities have been exposed outside the human body niche, and for how long?

The inter-individual variation in microbial composition at a given body site was highlighted in several human microbiome studies [57, 72, 73]. This variation was explored further, from a forensic perspective, by Fierer et al. [74] for skin microbial communities. The researchers wanted to know whether it was possible to characterise the microbial communities from touched surfaces and whether this characterisation could be used to match a surface to an individual. They sequenced the 16S rRNA genes from swabs of keys of three keyboards, and the fingertips of their users. Their results showed that, based on taxonomic composition, the keys and the fingertips of the keyboard owner could be matched. The microbial signature was still present even when the swabs were stored at room temperature for 2 weeks before extraction. In a different study, Meadow et al. [75] looked at the microbial composition in indoor air, comparing the "bacterial clouds" when the rooms had been occupied versus unoccupied, and comparing the clouds of different occupants. By sequencing regions of the 16S rRNA gene, the team showed that occupancy by an individual was reflected in the air, but not only that: the airborne microbial signature of the three subjects was statistically distinct, resulting in "personalised microbiome clouds". To find out how personal the microbiome really is, that is, to what extent an individual can be uniquely identified through his/her microbiome among hundreds of individuals, Franzosa et al. [76] used publically available 16S rRNA gene sequence data as well as whole meta-genome shotgun sequencing data from 242 individuals and various body sites. They constructed different types of personalised meta-genomic codes, based on taxon-level data (OTUs and species) or gene-level data (marker genes and kb-windows). Then the researchers tested the performance of these codes. Their analyses demonstrated that gene-level metagenomic codes, but not taxon-level metagenomic codes, can be used to identify an individual among hundreds, as a result of the differential strains unique to each person.

The presence of these individually identifying features of the microbiome is critical, but before we can apply these features, we need to assess their reliability across time: is there a core set of microbes that is stable and detectable at different time points? Sequencing studies examining the 16S rRNA variable regions of bacteria from different human body sites and through periods ranging from 3 months to 15 months indicate high temporal variation at the level of OTUs [58, 71]. Nonetheless, intra-individual variation remains smaller than inter-individual variation. Interestingly, the variation in microbial community richness and composition across time depends on the body site (for instance, gut communities were found to be more stable compared to those on the tongue) as well as on the individual [71]. In the investigation of Franzosa et al. [76], the researchers constructed personalised metagenomic codes from the microbial communities of body sites obtained at the first time point of sampling the site. The authors then investigated whether the taxa or strains from these codes were detectable in the same body site at the second sampling time, a month or more later. Their results illustrated, in agreement with previous studies, that the stability of the metagenomic code features differed across body sites. Using gene-level codes that tapped into strain variation, the researchers found correct matching for 86% of individuals when using codes from gut samples, while at other body sites this number was much lower (ca. 30% correct matching). In a recent study, Schmedes et al. [61] have explored a publicly available dataset comprising shotgun metagenomic sequencing reads from 17 skin sites of 12 subjects examined at three different time points [77]. The researchers identified bacterial markers common to each body site (across all individuals and time points) and focused on strain-level variation in order to develop a multiplex panel for forensic application. Their panel was shown to yield accurate matching (ranging from 92% to 100%) when tested on three body sites from eight individuals, thus holding promise for individual identification using skin samples. Furthermore, using this same panel, the researchers showed that they could identify the specific body site (foot, hand or manubrium) from which the skin microbiome originated (with up to 86% accuracy).

While several studies have addressed the stability of the microbiome, these have mainly focused on consistency when sampling from the individual. However, in forensic settings, it is important to also assess whether the microbial signature that can be used to identify an individual is reliable after it has been exposed to environmental conditions, outside the human host, for a given length of time. Few studies have explored this issue to date. Wilkins et al. [78] checked whether 16S rRNA gene-based microbial community composition data from household surfaces and air could be reliably matched to the skin of the occupants of the house, in the same season and different seasons. When the household and occupant samples were collected in the same season, the authors could accurately match 67% of subjects and household samples. However, accuracy decreased sharply when either the surface samples or the occupant samples were collected with a delay of several seasons, the trend being for lower accuracy when surface samples were collected after occupant skin samples. In the key-board study conducted by Fierer et al. [74], the investigators also checked the effects of storage at room temperature for the swabs collected from the skin of two individuals. This exposure was conducted for 3 days and for 14 days, before DNA extraction. Principal component analyses show that the owner and the keyboard samples continue to group even after such storage conditions. In both the Wilkins et al. [78] as well as the Fierer et al. [74] studies, microbial community data was based on 16S rRNA gene sequences. It would thus be interesting to explore how the microbial signature of samples changes over numerous time points when looking at strain-level variation obtained through shotgun sequencing and for larger sample sizes.

Microbial community composition can provide information not only on individual identity but also on ethnicity, geographical location, environment, lifestyle traits such as diet, and disease among other characteristics, all of which affect microbial community composition [56, 79–83]. For example, the use of cosmetics or topical antibiotics on skin will impact the diversity and composition of the bacteria on it [84].

Overall, microbiome studies exploring individual identifiability and phenotypic trait inferences show promising results, particularly when metagenomic data providing information on strain-level variation is used [61, 79, 85]. Further work needs to be done in settings where microbial DNA sequencing would be particularly valuable, that is, in settings when human DNA would be insufficient for traditional DNA profiling methods: for example, in the keyboard surface studies, we would expect human DNA to be present too. Human DNA STR typing remains the gold standard for accurate identity testing, and matching probabilities as well as allelic frequencies of populations have been extensively studied. Schmedes et al. [79] emphasised this point in their review, highlighting the uncertainties associated with methods used to type human DNA present in low copy numbers. Here, the possibility to obtain microbiome data if bacteria are present in higher quantities would open up an avenue to improve predictive values. However, few studies have explored such scenarios, or the statistical approaches that would need to be used when combining information from human DNA typing methods and microbial sequencing methods.

Finally, from a statistical perspective, and as with human body fluid/tissue identification, advances in predictive modelling and machine learning algorithms are anticipated to improve the prediction power of genomic tools, and further investigation in these areas will be of great interest and relevance to the forensic field.

12.8 Metagenomics and the Forensic Analysis of Soils

Soil is a promising tool in forensic science due to its ubiquity and transferability [86]. Soil can be a proof of evidence in the investigation of a crime when it occurs outdoors or is collected from different items related to it, such as shoes, tires or clothing [86, 87]. Provenance analysis of a particular soil specimen and comparisons of different soils samples related to a crime are the two main aims pursued in order to link the crime scene with the perpetrator. Although its use nowadays is not equally extended among the different forensic institutions, some of them consider it an established resource in criminal investigations [88]. Physical particles and the analysis of the inorganic soil component, which includes elemental/chemical analysis, mineralogy and other chemical methods applicable to both organic and inorganics components are frequently used as proofs of evidence in trials; additionally, investigation of biological traces such as botanical fragments, pollen or even diatoms could also be used [88]. The microbiological analysis of soils has also been used in particular cases as a complementary test in forensic investigations [89] and it is the subject of current research. Most of these studies have focused on the bacterial characterisation to discriminate soil samples, but others have done some research on fungi [90]. Some approaches have tried to identify the bacteria isolated on cultured soils specimens by Sanger sequencing of the 16S rRNA gene coupled with phylogenetic studies (Cordero JC, personal communication), but the main limitation being the biases derived from the lack of identification of non-viable bacteria, these techniques have been withdrawn and, consequently, the need to turn to independent-culture molecular techniques is now accepted.

According to Sensabaugh [91], a microbial profiling method for soils: (i) has to demonstrate its robustness and repeatability; (ii) must be able to differentiate two locations; and (iii) needs the use of objective statistical procedures to assess similarities and dissimilarities among specimens [91]. The following regions have been proposed to characterise the soil bacterial microbiome: ribosomal intergenic spacer analysis (RISA), terminal restriction fragment length polymorphism (TRFLP) of the rpoB gene, and the 16S rRNA gene. The 16S rRNA gene has been analysed using phylogenetic micro-arrays, TRFLP, and Next Generation Sequencing with Roche 454, Illumina MiSeq and Ion-Torrent PGM platforms. Habtom et al. [86] compared these methods to long-chain hydrocarbons (n-alkanes) and fatty alcohol profiling of the same soil samples. MiSeq, RISA and 16S TRFLP were the ones performing best, being able to discriminate between very similar soils [86]. The study designed by Jesmok et al. was aimed at determining the feasibility of distinguishing between bacterial profiles from differing habitat types, similar habitat types, and possible differences in sampling regarding time and space within the same habitat. They used different methods to measure dissimilarities: abundance charts, providing graphic quantification of which bacteria are present in a profile; two dissimilarity indices such Sørensen–Dice and Bray–Curtis, which were complementary; and non-metric multidimensional scaling (NMDS), allowing clustering bacterial profiles from a given location and distinguishing other profiles from one cluster. Statistical analysis was applied to NMDS clusters, yielding an objective measure of similarity or difference of the soil profiles. They stated that the conditions proposed for Sensabaugh et al. [91] regarding the adequacy of the method for forensic use were accomplished. As a conclusion of their work, they considered that, in order to achieve minimum forensic standards, more than one method of statistical analyses should be applied to the huge amount of data sets obtained in HTS experiments, coupling both objective and visual interpretation of the data [86, 87].

Fungal profiling by analysing the ITS I region by HTS can also be a useful strategy for soils discrimination due to its high discriminatory power and reproducibility using traces quantities of material. Young et al. [90] showed Ascomycota to be a robust fungal marker and that a scarce amount of soil mass did not affect identification, being able to obtain robust profiles with 150 mg soil [90]. In another study, this group also analysed eukaryotic microorganisms in soil by HTSg of the 18S rRNA in a mock crime scene scenario [92].

To facilitate further use of these strategies in the courtroom, standardisation of the HTS described techniques should encompass standardisation of other aspects of the procedure: (i) sampling methods, (ii) storage sampling conditions [92], and (iii) nucleic acid extraction methods. The analysis of soils in forensic science is still a challenge.

12.9 Omics and the Forensic Analysis of Drowning

The forensic diagnosis of drowning is one of the most challenging tasks for forensic pathologists. It combines a histopathological analysis, the use of blood strontium as an indicator as well as diatoms detection in closed organs of the victims of drowning [93, 94]. However, these techniques have limitations, and sometimes discriminating between a true drowning and submersion after death is not possible. Kakizaki and colleagues showed that detection of exogenous aquatic bacteria by 454-pyrosequencing in closed organs of drowning victims was of help to support the diagnosis of drowning when other techniques proved to be insufficient [95]. A recent study performed on rats has shown that the detection of aquatic microorganisms in the closed organs by HTS can be a marker for the diagnosis of death by drowning in combination with a higher expression level of pulmonary surfactant protein A in the lungs [96]. Three groups were included: drowning, post-mortem submersion, and control groups. The water, lungs, closed organs (kidney and liver), and cardiac blood in rats were assayed by HTS of the V3 and V4 regions of the 16S rRNA with a MiSeq platform. They also demonstrated that the microbiome of freshwater and seawater could be distinguished. Although scarce, these results are promising, and further assessing studies are needed.

12.10 Future Perspectives

From the above results and cases, it seems clear that Omics offer many new analytical strategies for their application in forensic microbiology. However, most of them are still under research and/or development and, although some studies are encouraging, dissemination of the results yielded by current studies and accessibility of the HTS methods to more forensic laboratories are preliminary conditions for its future application in the forensic arena. The availability of high-quality databases and the use of objective statistical tools in order to produce accurate interpretations of the results are also key factors for their forensic usage. Standardisation and validation are common concepts in human forensic genetics and other forensic specialities. Indeed, forensic laboratories follow strict quality criteria in terms of acceptability, validation and accreditation of the techniques in use, as well as in workflow, facilities design and chain of custody, all coupled with sampling guidelines for the analyses to be performed. European judicial expert laboratories are required to follow ISO-17025 standards regarding the operation of testing and calibration laboratories, as well as SWGDAM validation guidelines and assumption of the Daubert principle of validation, peer-review and acceptance by the relevant scientific community prior application to forensic casework [97, 98].

The future application of Omics techniques in the routine practice of the forensic experts will have to be encompassed by in-ternational standardisation and educational efforts that take these requirements into account. Some international entities and scientific societies related to forensics and microbiology are already aware of these needs. The ESCMID study group of Forensic and Post-mortem Microbiology (ESGFOR) is promoting educational and research activities in this field. Similarly, the European Network of Forensic Sciences Institutes (ENFSI) by means of its Animal Plant and Soil Traces (APST) working group is contributing to the exchange of knowledge about the application of Omics in forensic microbiology. Both groups are also joining efforts to disseminate investigation and promote further standardisation of methods. Forensic sciences will benefit from such initiatives, in order to incorporate the best-performing Omics techniques among the plethora of new methods into the casework of criminal cases.

Acknowledgments FGC was supported by project BFU2017-89594R from the MICIU (Spanish Government) and PROMETEO2016-122 from Generalitat Valenciana. AFR thanks the ESGFOR (ESCMID study group of Forensic and Post-mortem Microbiology) members for support and advice. We also thank Prof. Robert Schlaberg for giving us permission to cite his promising recent work about omics application in de-tecting infectious cause of death.

References

1. Budowle B, Schutzer SE, Einseln A, Kelley LC, Walsh AC, Smith JAL et al (2003) Public health. Building microbial forensics as a response to bioterrorism. Science 301(5641):1852–1853
2. Tucker JB, Koblentz GD (2009) The four faces of microbial forensics. Biosecur Bioterror 7(4):389–397
3. Sjödin A, Broman T, Melefors Ö, Andersson G, Rasmusson B, Knutsson R et al (2013) The need for high-quality whole-genome sequence databases in microbial forensics. Biosecur Bioterror 11(Suppl 1):S78–S86
4. Pilo P (2015) Improving exploitation of whole genome sequencing data for public health, Forensic Microbiology and Biosafety. EBioMed 2(11):1566–1567
5. Bertelli C, Greub G (2013) Rapid bacterial genome sequencing: methods and applications in clinical microbiology. Clin Microbiol Infect 19(9):803–813
6. Carriço JA, Sabat AJ, Friedrich AW, Ramirez M, ESCMID Study Group for Epidemiological Markers (ESGEM) (2013) Bioinformatics in bacterial molecular epidemiology and public health: databases, tools and the next-generation sequencing revolution. Euro Surveill 18(4):20382
7. Abecasis AB, Geretti AM, Albert J, Power L, Weait M, Vandamme A-M (2011) Science in court: the myth of HIV fingerprinting. Lancet Infect Dis 11(2):78–79
8. González-Candelas F, Bracho MA, Wróbel B, Moya A (2013) Molecular evolution in court: analysis of a large hepatitis C virus outbreak from an evolving source. BMC Biol 11:76
9. Domingo E, Martin V, Perales C, Grande-Pérez A, García-Arriaza J, Arias A (2006) Viruses as quasispecies: biological implications. Curr Top Microbiol Immunol 299:51–82
10. Yang Y, Xie B, Yan J (2014) Application of next-generation sequencing technology in forensic science. Genomics Proteomics Bioinformatics 12(5):190–197
11. González-Candelas F. (2010) Molecular phylogenetic analyses in court trials. In: Encyclopedia of life sciences. John Wiley & Sons, Ltd, New York
12. Campo DS, Xia G-L, Dimitrova Z, Lin Y, Forbi JC, Ganova-Raeva L et al (2016) Accurate genetic detection of hepatitis C virus transmissions in outbreak settings. J Infect Dis 213(6):957–965
13. Gonçalves Rossi LM, Rossi LMG, Escobar-Gutierrez A, Rahal P (2016) Multiregion deep sequencing of hepatitis C virus: an improved approach for genetic relatedness studies. Infect Genet Evol 38:138–145
14. Caraballo Cortes K, Rosińska M, Janiak M, Stępień M, Zagordi O, Perlejewski K et al (2018) Next-generation sequencing analysis of a cluster of hepatitis C virus infections in a haematology and oncology center. PLoS One 13(3):e0194816
15. Montoya V, Olmstead A, Tang P, Cook D, Janjua N, Grebely J et al (2016) Deep sequencing increases hepatitis C virus phylogenetic cluster detection compared to Sanger sequencing. Infect Genet Evol 43:329–337

16. Dudas G, Carvalho LM, Bedford T, Tatem AJ, Baele G, Faria NR et al (2017) Virus genomes reveal factors that spread and sustained the Ebola epidemic. Nature 544(7650):309–315
17. Faria NR, Quick J, Claro IM, Thézé J, de Jesus JG, Giovanetti M et al (2017) Establishment and cryptic transmission of Zika virus in Brazil and the Americas. Nature 546(7658):406–410
18. Zuckerkandl E, Pauling L (1965) Molecules as documents of evolutionary history. J Theor Biol 8(2):357–366
19. Drummond AJ, Ho SYW, Phillips MJ, Rambaut A (2006) Relaxed phylogenetics and dating with confidence. PLoS Biol 4(5):e88
20. Suchard MA, Lemey P, Baele G, Ayres DL, Drummond AJ, Rambaut A (2018) Bayesian phylogenetic and phylodynamic data integration using BEAST 1.10. Virus Evol 4(1):vey016
21. Drummond AJ, Bouckaert RR (2015) Bayesian evolutionary analysis with BEAST. Cambridge University Press. 260 p
22. Sagulenko P, Puller V, Neher RA (2018) TreeTime: maximum-likelihood phylodynamic analysis. Virus Evol [Internet] 4(1). Available from: https://doi.org/10.1093/ve/vex042
23. Rambaut A, Lam TT, Max Carvalho L, Pybus OG (2016) Exploring the temporal structure of heterochronous sequences using TempEst (formerly Path-O-Gen). Virus Evol 2(1):vew007
24. Fernández-Rodríguez A, Cohen MC, Lucena J, Van de Voorde W, Angelini A, Ziyade N et al (2015) How to optimise the yield of forensic and clinical post-mortem microbiology with an adequate sampling: a proposal for standardisation. Eur J Clin Microbiol Infect Dis 34(5):1045–1057
25. Prtak L, Al-Adnani M, Fenton P, Kudesia G, Cohen MC (2010) Contribution of bacteriology and virology in sudden unexpected death in infancy. Arch Dis Child 95(5):371–376
26. Andréoletti L, Lévêque N, Boulagnon C, Brasselet C, Fornes P (2009) Viral causes of human myocarditis. Arch Cardiovasc Dis 102(6–7):559–568
27. Burton JL, Saegeman V, Arribi A, Rello J, Andreoletti A, Cohen M, Fernandez-Rodriguez A (submitted) Post-mortem microbiology sampling following death in hospital: An ESGFOR task force consensus statement. In: On behalf of ESGFOR Joint working group of the European Society of Clinical Microbiology and In-fectious Diseases (ESCMID). J Clin Pathol
28. Saegeman V, Cohen MC, Alberola J, Ziyade N, Farina C (2017) ESCMID study Group for Forensic and Postmortem Microbiology, et al. how is post-mortem microbiology appraised by pathologists? Results from a practice survey conducted by ESGFOR. Eur J Clin Microbiol Infect Dis 36(8):1381–1385
29. Fernández-Rodríguez A, Alcalá B, Alvarez-Lafuente R (2008) Real-time polymerase chain reaction detection of Neisseria meningitidis in formalin-fixed tissues from sudden deaths. Diagn Microbiol Infect Dis 60(4):339–346
30. Daş T, Sargan A, Yağmur G, Yildirim M, Topal CS, Gürler AS et al (2016) Viral pneumonias in forensic autopsies: evaluation and classification of histopathologic changes with microbiologic correlation. Am J Forensic Med Pathol 37(4):255–263
31. Patel JB (2001) 16S rRNA gene sequencing for bacterial pathogen identification in the clinical laboratory. Mol Diagn 6(4):313–321
32. Sabat AJ, van Zanten E, Akkerboom V, Wisselink G, van Slochteren K, de Boer RF et al (2017) Targeted next-generation sequencing of the 16S-23S rRNA region for culture-independent bacterial identification – increased discrimination of closely related species. Sci Rep 7(1):3434
33. Morris JA, Harrison LM, Partridge SM (2007) Practical and theoretical aspects of postmortem bacteriology. Curr Diagn Pathol 13(1):65–74
34. Cho Y, Lee MH, Kim HS, Park M, Kim M-H, Kwon H, Eom K, Kim J-B, Lee KL, Lee YH, Lee DS. Sequencing methods for application of forensic microbiology. 64. Poster abstracts TOPIC 02 Non-Human, Microbiome; 8 August-2 September 2017; 27th Congress of the International Society for Forensic Genetics (ISFG)
35. Schlaberg R, Queen K, Simmon K, Tardif K, Stockmann C, Flygare S et al (2017) Viral pathogen detection by metagenomics and pan-viral group polymerase chain reaction in children with pneumonia lacking identifiable Etiology. J Infect Dis 215(9):1407–1415

36. Flygare S, Simmon K, Miller C, Qiao Y, Kennedy B, Di Sera T et al (2016) Taxonomer: an interactive metagenomics analysis portal for universal pathogen detection and host mRNA expression profiling. Genome Biol 17(1):111
37. Paul L, Comstock J, Edes K, Schlaberg R. Hypothesis-free pathogen detection by next-generation RNA sequencing from archived, postmortem lung tissues of neonates with fatal pneumonia. Session: innovations in molecular pathogen detection. 28th ECCMID; 21–24 April 2018; Madrid, Spain
38. Paul L, Edes K, Salama M, Fujinami R, Schlaberg R. Post-mortem stability of viral RNA and host mRNA expression profiles in an animal model of viral pneumonia. Session: measles and more respiratory viruses. 28th ECCMID; 21–24 April 2018; Madrid, Spain
39. Paul L, Comstock J, Edes K, Schlaberg R (2018) Gestational Psittacosis resulting in neonatal death identified by next-generation RNA sequencing of Postmortem, formalin-fixed lung tissue. Open Forum Infect Dis. In press
40. Javan GT, Finley SJ, Abidin Z, Mulle JG (2016) The Thanatomicrobiome: a missing piece of the microbial puzzle of death. Front Microbiol 7:225
41. Byrd JH, Castner JL (2009) Forensic entomology: the utility of arthropods in Legal Investigations. 2nd edn. CRC Press, 705 p
42. Amendt J, Richards CS, Campobasso CP, Zehner R, Hall MJR (2011) Forensic entomology: applications and limitations. Forensic Sci Med Pathol 7(4):379–392
43. Campobasso CP, Pietro Campobasso C, Di Vella G, Introna F (2001) Factors affecting decomposition and Diptera colonization. Forensic Sci Int 120(1–2):18–27
44. Pechal JL, Crippen TL, Eric Benbow M, Tarone AM, Dowd S, Tomberlin JK (2013) The potential use of bacterial community succession in forensics as described by high throughput meta-genomic sequencing. Int J Legal Med 128(1):193–205
45. Metcalf JL, Xu ZZ, Weiss S, Lax S, Van Treuren W, Hyde ER et al (2016) Microbial community assembly and metabolic function during mammalian corpse decomposition. Science 351(6269):158–162
46. Parkinson RA, Dias K-R, Horswell J, Greenwood P, Banning N, Tibbett M, et al (n.d.) (2009) Microbial Community Analysis of Human Decomposition on Soil. In: Ritz K., Dawson L., Miller D. (eds) Criminal and Environmental Soil Forensics. Springer, Dordrecht. https://doi.org/10.1007/978-1-4020-9204-6_24
47. Metcalf JL, Wegener Parfrey L, Gonzalez A, Lauber CL, Knights D, Ackermann G et al (2013) A microbial clock provides an accurate estimate of the postmortem interval in a mouse model system. elife 2:e01104
48. Metcalf JL, Xu ZZ, Bouslimani A, Dorrestein P, Carter DO, Knight R (2017) Microbiome tools for forensic science. Trends Biotechnol 35(9):814–823
49. Gefrides L, Welch K (2010) Forensic biology: serology and DNA. In: A. Mozayani and C. Noziglia (eds.), The Forensic Laboratory Handbook Procedures and Practice, pp 15–50; DOI 10.1007/978-1-60761-872-0_2, © Springer Science+Business Media, LLC 2011
50. Virkler K, Lednev IK (2009) Analysis of body fluids for forensic purposes: from laboratory testing to non-destructive rapid confirmatory identification at a crime scene. Forensic Sci Int 188(1–3):1–17
51. Kayser M, de Knijff P (2011) Improving human forensics through advances in genetics, genomics and molecular biology. Nat Rev Genet 12(3):179–192
52. Sijen T (2015) Molecular approaches for forensic cell type identification: on mRNA, miRNA, DNA methylation and microbial markers. Forensic Sci Int Genet 18:21–32
53. Sender R, Fuchs S, Milo R (2016) Revised estimates for the number of human and bacteria cells in the body. PLoS Biol 14(8):e1002533
54. Hunt KM, Foster JA, Forney LJ, Schütte UME, Beck DL, Abdo Z et al (2011) Characterization of the diversity and temporal stability of bacterial communities in human milk. PLoS One 6(6):e21313
55. Lloyd-Price J, Mahurkar A, Rahnavard G, Crabtree J, Orvis J, Hall AB et al (2017) Strains, functions and dynamics in the expanded human microbiome project. Nature 550(7674):61–66

56. The Human Microbiome Project Consortium (2012) Structure, function and diversity of the healthy human microbiome. Nature 486(7402):207–214
57. Costello EK, Lauber CL, Hamady M, Fierer N, Gordon JI, Knight R (2009) Bacterial community variation in human body habitats across space and time. Science 326(5960):1694–1697
58. Caporaso JG, Lauber CL, Costello EK, Berg-Lyons D, Gonzalez A, Stombaugh J et al (2011) Moving pictures of the human microbiome. Genome Biol 12(5):R50
59. Chu DM, Ma J, Prince AL, Antony KM, Seferovic MD, Aagaard KM (2017) Maturation of the infant microbiome community structure and function across multiple body sites and in relation to mode of delivery. Nat Med 23(3):314–326
60. Costello EK, Carlisle EM, Bik EM, Morowitz MJ, Relman DA (2013) Microbiome assembly across multiple body sites in low-birthweight infants. MBio 4(6):e00782–e00713
61. Schmedes SE, Woerner AE, Budowle B (2017) Forensic human identification using skin microbiomes. Appl Environ Microbiol [Internet]. Available from: https://doi.org/10.1128/AEM.01672-17
62. Giampaoli S, Berti A, Valeriani F, Gianfranceschi G, Piccolella A, Buggiotti L et al (2012) Molecular identification of vaginal fluid by microbial signature. Forensic Sci Int Genet 6(5):559–564
63. Choi A, Shin K-J, Yang WI, Lee HY (2014) Body fluid identification by integrated analysis of DNA methylation and body fluid-specific microbial DNA. Int J Legal Med 128(1):33–41
64. Benschop CCG, Quaak FCA, Boon ME, Sijen T, Kuiper I (2012) Vaginal microbial flora analysis by next generation sequencing and microarrays; can microbes indicate vaginal origin in a forensic context? Int J Legal Med 126(2):303–310
65. Quaak FCA, de Graaf M-LM, Weterings R, Kuiper I (2017) Microbial population analysis improves the evidential value of faecal traces in forensic investigations. Int J Legal Med 131(1):45–51
66. Gajer P, Brotman RM, Bai G, Sakamoto J, Schütte UME, Zhong X et al (2012) Temporal dynamics of the human vaginal microbiota. Sci Transl Med 4(132):132ra52
67. Ravel J, Gajer P, Abdo Z, Schneider GM, Koenig SSK, McCulle SL et al (2011) Vaginal microbiome of reproductive-age women. Proc Natl Acad Sci U S A 108(Suppl 1):4680–4687
68. Hanssen EN, Avershina E, Rudi K, Gill P, Snipen L (2017) Body fluid prediction from microbial patterns for forensic application. Forensic Sci Int Genet 30:10–17
69. Statnikov A, Henaff M, Narendra V, Konganti K, Li Z, Yang L et al (2013) A comprehensive evaluation of multicategory classification methods for microbiomic data. Microbiome 1(1):11
70. Knights D, Costello EK, Knight R (2011) Supervised classification of human microbiota. FEMS Microbiol Rev 35(2):343–359
71. Flores GE, Caporaso JG, Henley JB, Rideout JR, Domogala D, Chase J et al (2014) Temporal variability is a personalized feature of the human microbiome. Genome Biol 15(12):531
72. Turnbaugh PJ, Ley RE, Hamady M, Fraser-Liggett CM, Knight R, Gordon JI (2007) The human microbiome project. Nature 449(7164):804–810
73. Fierer N, Hamady M, Lauber CL, Knight R (2008) The influence of sex, handedness, and washing on the diversity of hand surface bacteria. Proc Natl Acad Sci U S A 105(46):17994–17999
74. Fierer N, Lauber CL, Zhou N, McDonald D, Costello EK, Knight R (2010) Forensic identification using skin bacterial communities. Proc Natl Acad Sci U S A 107(14):6477–6481
75. Meadow JF, Altrichter AE, Bateman AC, Stenson J, Brown GZ, Green JL et al (2015) Humans differ in their personal microbial cloud. PeerJ 3:e1258
76. Franzosa EA, Huang K, Meadow JF, Gevers D, Lemon KP, Bohannan BJM et al (2015) Identifying personal microbiomes using metagenomic codes. Proc Natl Acad Sci U S A 112(22):E2930–E2938
77. Oh J, Byrd AL, Park M, Kong HH, Segre JA, NISC Comparative Sequencing Program (2016) Temporal stability of the human skin microbiome. Cell 165(4):854–866
78. Wilkins D, Leung MHY, Lee PKH (2017) Microbiota fingerprints lose individually identifying features over time. Microbiome 5(1):1

79. Schmedes SE, Sajantila A, Budowle B (2016) Expansion of microbial forensics. J Clin Microbiol 54(8):1964–1974
80. Schnorr SL, Candela M, Rampelli S, Centanni M, Consolandi C, Basaglia G et al (2014) Gut microbiome of the Hadza hunter-gatherers. Nat Commun 5:3654
81. David LA, Maurice CF, Carmody RN, Gootenberg DB, Button JE, Wolfe BE et al (2014) Diet rapidly and reproducibly alters the human gut microbiome. Nature 505(7484):559–563
82. Lozupone CA, Stombaugh JI, Gordon JI, Jansson JK, Knight R (2012) Diversity, stability and resilience of the human gut microbiota. Nature 489(7415):220–230
83. Shaw L, Ribeiro ALR, Levine AP, Pontikos N, Balloux F, Segal AW et al (2017) The human salivary microbiome is shaped by shared environment rather than genetics: evidence from a large family of closely related individuals. MBio 8(5):e01237–e01217
84. Ross AA, Doxey AC, Neufeld JD (2017) The Skin Microbiome of Co-habiting Couples. mSystems [Internet] 2(4). Available from: https://doi.org/10.1128/mSystems.00043-17
85. Clarke TH, Gomez A, Singh H, Nelson KE, Brinkac LM (2017) Integrating the microbiome as a resource in the forensics toolkit. Forensic Sci Int Genet 30:141–147
86. Habtom H, Demanèche S, Dawson L, Azulay C, Matan O, Robe P et al (2017) Soil characterisation by bacterial community analysis for forensic applications: a quantitative comparison of environmental technologies. Forensic Sci Int Genet 26:21–29
87. Jesmok EM, Hopkins JM, Foran DR (2016) Next-generation se-quencing of the bacterial 16S rRNA Gene for forensic soil comparison: a feasibility study. J Forensic Sci 61(3):607–617
88. Dawson LA, Hillier S (2010) Measurement of soil characteristics for forensic applications. Surf Interface Anal 42(5):363–377
89. Concheri G, Bertoldi D, Polone E, Otto S, Larcher R, Squartini A (2011) Chemical elemental distribution and soil DNA fingerprints provide the critical evidence in murder case investigation. PLoS One 6(6):e20222
90. Young JM, Weyrich LS, Cooper A (2016) High-throughput Sequenc-ing of trace quantities of soil provides reproducible and dis-criminative fungal DNA profiles. J Forensic Sci 61(2):478–484
91. Sensabaugh GF (2009) Microbial Community Profiling for the Characterisation of Soil Evidence: Forensic Considerations. In: Ritz K., Dawson L., Miller D. (eds) Criminal and Environmental Soil Forensics. Springer, Dordrecht. https://doi.org/10.1007/978-1-4020-9204-6_4
92. Young JM, Weyrich LS, Breen J, Macdonald LM, Cooper A (2015) Predicting the origin of soil evidence: high throughput eukaryote sequencing and MIR spectroscopy applied to a crime scene scenario. Forensic Sci Int 251:22–31
93. Azparren JE, Fernandez-Rodriguez A, Vallejo G (2003) Diagnosing death by drowning in fresh water using blood strontium as an indicator. Forensic Sci Int 137(1):55–59
94. Azparren JE, Perucha E, Martínez P, Muñoz R, Vallejo G (2007) Factors affecting strontium absorption in drownings. Forensic Sci Int 168(2–3):138–142
95. Kakizaki E, Ogura Y, Kozawa S, Nishida S, Uchiyama T, Hayashi T et al (2012) Detection of diverse aquatic microbes in blood and organs of drowning victims: first metagenomic approach using high-throughput 454-pyrosequencing. Forensic Sci Int 220(1–3):135–146
96. Lee S-Y, Woo S-K, Lee S-M, Ha E-J, Lim K-H, Choi K-H et al (2017) Microbiota composition and pulmonary surfactant protein expression as markers of death by drowning. J Forensic Sci 62(4):1080–1088
97. Murch RS, Bahr EL (2011) Validation of microbial forensics in Scientific, legal, and policy contexts. In: Budowle B, Schutzer SE, Breeze RG, Keim PS, Morse SA. Microbial Forensics. pp 649–663. Academic Press
98. Kuiper I (2016) Microbial forensics: next-generation sequencing as catalyst. EMBO Rep 17(8):1085–1087

The manufacturer's authorised representative in the EU is Springer Nature Customer Service Centre GmbH, Europaplatz 3, 69115 Heidelberg, Germany. If you have any concerns regarding our products, please contact ProductSafety@springernature.com

Printed and bound by CPI Group (UK) Ltd, Croydon, CR0 4YY

30/09/2026

02225294-0006